PRACTICAL DESIGN OF
FLIGHT CONTROL SYSTEMS
FOR
LAUNCH VEHICLES AND MISSILES

PRACTICAL DESIGN OF FLIGHT CONTROL SYSTEMS FOR LAUNCH VEHICLES AND MISSILES

NV KADAM

Emeritus Scientist
Director, Weapon Systems

ALLIED PUBLISHERS PRIVATE LIMITED

New Delhi • Mumbai • Kolkata • Chennai • Bangalore • Hyderabad

ALLIED PUBLISHERS PRIVATE LIMITED

1/13-14 Asaf Ali Road, **New Delhi**–110002
Ph.: 011-23239001 • E-mail: delhi.books@alliedpublishers.com

17 Chittaranjan Avenue, **Kolkata**–700072
Ph.: 033-22129618 • E-mail: cal.books@alliedpublishers.com

15 J.N. Heredia Marg, Ballard Estate, **Mumbai**–400001
Ph.: 022-42126969 • E-mail: mumbai.books@alliedpublishers.com
No. 25/10, Commander-in-Chief Road, Ethiraj Lane (Next to Post Office)
Egmore, **Chennai**–600008
Ph.: 044-28223938 • E-mail: chennai.books@alliedpublishers.com

Hebbar Sreevaishnava Sabha, Sudarshan Complex-2
No. 24/1, 2nd Floor, Seshadri Road, **Bangalore**–560009
Ph.: 080-22262081 • E-Mail: bngl.books@alliedpublishers.com

Sri Jayalakshmi Nilayam, No. 3-4-510, 3rd Floor (Opp. State Bank of India)
Barkatpura, Telangana, **Hyderabad**–500027
Ph.: 040-27551811, 2755 1812 • E-mail: hyd.books@alliedpublishers.com

Website: www.alliedpublishers.com

© 2009, Allied Publishers Pvt. Ltd.

Second Thoroughly Revised Edition, 2019

ISBN: 978-93-87997-81-3

Preface

Indian satellite launch vehicle project was initiated during early 1970s. Though NASA reports were available, no good book was available at that time which discussed in a comprehensive manner the various aspects of design of the flight control systems for launch vehicles or missiles. There were no experienced scientists in India to guide in the design of control systems. I had to struggle to develop formulation and design methodology and fix the design parameters till the first good book by AL Greensite came in the market. I had to learn many aspects in a hard way and through flight trials. All this work has been already documented as departmental reports and sometimes as Conference papers. With the lapse of several years, many of these reports will now be untraceable and even if traceable, may not be made available for scientists not belonging to the department since those will contain data about the vehicles. These circumstances will lead to loss of experience gained over the years in a hard way and the new generation scientists have to learn their way once again though learning process will be fast enough since many of the design features are already in use and many senior scientists are now available for providing the required guidance.

With this book I have made a humble effort to consolidate in single place what I have learnt during my active professional period. Many references to the departmental reports have been cited in the book not with a view that the readers may refer to them but mainly as an acknowledgment to the original contributor. Many of the concepts, ideas and results for which I have been personally responsible, I have attempted to bring in this work in a cohesive manner and without giving the actual data so that the concerned authorities will have less difficulty in permitting publication of this work. The readers will have no need to refer to original reports for understanding the concept and theoretical background.

The presently available books on flight control systems mainly discuss the material given in Chapter IV and partly in Chapter II of this book. No other book is found to discuss the major part of the material given in Chapter III, V and VI. The Chapter VI discusses the validation aspects and flight trial experiences only briefly and there is enough scope for expanding it, particularly the hardware-in-loop simulation part. I sincerely feel that scientists and professionals working in this field will be definitely benefited by this work.

I sincerely thank Shri Prahlada, Director, DRDL for recommending my appointment as Emeritus Scientist and extending full support which enabled me to undertake this project without making any announcement and Dr. VK Saraswat, Director, RCI for encouraging me to undertake this task.

I thank Mrs. V Sridevi for translating the manuscript into a computerized document, part of which has been quite tedious due to many mathematical expressions. I also thank my scientist colleagues Shri V Srinivasa Rao and Shri M Manickavasagam for helping me in generating many results and figures, Shri M Raghvendra Rao & Amba Das, Shri AK Kaushik & Shri Saji Skaria for supplying sketches.

I thank Shri M. Natarajan, SA to RM for permitting publication of this book and his encouraging remarks.

HYDERABAD
2009

NV Kadam
Email Id: nvkadam@rediffmail.com
Mobile: 9849352452

Dedicated to
My wife Mrs. Revati
and Sons
Sameer and Kartik

Contents

Chapter 1: Preliminaries and Mission Considerations

Chapter 2: Generalized Equations of Motion

Chapter 3: Control Systems Design–1: Configuration and Sizing

Chapter 5: Control Systems Design–3: Analysis and Design of ON-OFF Reaction Control Systems

Chapter 6: Design Validation and Flight Trial Experiences

Nomenclature

a_x, a_y, a_z	Components of acceleration along x, y, z axes
C_D	Aerodynamic drag coefficient
$C_{N\alpha}, C_{Y\beta}$	Aerodynamic normal force coefficients in pitch and yaw planes
C_P	Aerodynamic centre of pressure
C_B^A	Transformation matrix from A frame to B frame
d_z	Dead zone for reaction control system
$\bar{F}, F_c$	Force, control force
$\bar{g}$	Gravity vector
$\bar{H}$	Angular momentum
I	Moment of inertia matrix
$\hat{I}$	Instantaneous moment of inertia matrix for a deflected vehicle
K_s, K_R, K_a	Forward gain, rate feed back gains and acceleration feedback gain
L_R	Distance of nozzle (or engine) CG from gimballing point
L_G	L_G location of nozzle throat (engine gimbal point) from nose tip
L_C	Distance between vehicle CG and control force location
l	Length measured from nose tip to a station on the vehicle
M	Mass
M_R	Mass of gimbaled engine
M_{sj}	Slosh mass for j^{th} pendulum
M_α	Angular acceleration due to aerodynamic torque per unit angle of attack
m_i	Generalised mass for i^{th} bending mode
n_1	Rise slope of acceleration due to reaction control force
n_2	Falling slope of acceleration due to reaction control force
p, q, r	Body angular rates about x, y and z axes
Q	Aerodynamic pressure (N/m^2)

$Q \equiv [q_0 q_1 q_2 q_3]$	Quaternion
S	Reference area for aerodynamic coefficients
R	Radial distance from centre of earth
R_e	Radius of earth
T	Thrust and also used for torque at times
T_E	Thrust due to one engine
t	time
T_p	Period of limit cycle
$u\ v\ w$	Components of velocity along body axes
$\overline{X}$	State vector
$\alpha,\ \alpha_w$	Angle of attack, angle of attack due to wind
β	Side slip angle
γ	Flight path angle
$\delta,\ \delta_{pk},\ \delta_{yk}$	Engine deflection angle, Kth engine deflection in pitch plane and yaw plane
ω	Angular frequency (r/s)
ω_n	Actuator natural frequency
ω_c	Control frequency
ζ_n	Damping ratio for actuator
ξ	Generalised deflection of mass element from nominal position due to bending, sloshing and/or engine gimballing
$\underline{\Phi}^T \equiv (\phi_x,\ \phi_y,\ \phi_z)$	Bending mode shape vector
$\overline{\Psi}^T \equiv (\Psi_x,\ \Psi_y,\ \Psi_z)$	Torsional mode shape vector
λ	Deflection angle for slosh pendulum
$\left.\begin{array}{l}\lambda_p \\ \lambda_y\end{array}\right\}$	Deflection angle for slosh pendulum in pitch and yaw planes
μ_c	Angular acceleration per unit control deflection
μ_α	Angular acceleration per unit angle of attack
μ_d	Angular acceleration due to disturbance torque
$\theta,\ \theta_c$	Pitch attitude angle, attitude control command in pitch plane
ψ	Yaw attitude angle
$\overline{\rho}$	Instantaneous position vector of a vehicle mass element from vehicle CG.

CHAPTER 1

Preliminaries and Mission Considerations

1.1 INTRODUCTION

The flight vehicles under consideration for this work are mainly launch vehicles and missiles. However, some of the considerations may be equally applicable for the automatic control of the aircrafts as well.

All science students learn about the projectiles in the early years of college. Using the kinematic equations of motion, it is very easy to calculate the trajectory and the range of the projectile for a given initial velocity. However, the task will appear formidable if one is asked to design a system to put a small round or a pebble (say 1 cm in diameter) in a hole of 5 cm dia. at a distance of 25 m. Inspite of our knowledge that the kinematic equations are quite accurate, one will develop cold feet about succeeding in the first attempt.

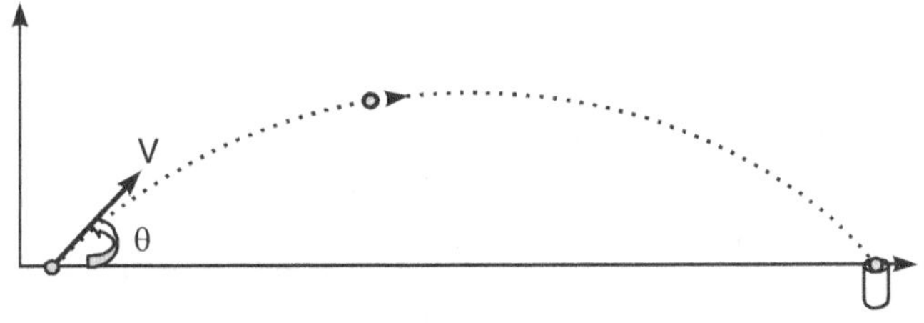

Fig. 1.1

Several doubts will arise. The distance, exit velocity, launch angle or the acceleration due to gravity may not be very accurate. Similarly, the atmospheric drag, effect of wind etc. also may not be accurately known. Exactly same thing will happen on a larger scale when a missile is to be designed for delivering a payload or a warhead thousands of kilometers away with a small circular error probability. Feedback control and guidance systems are designed to achieve this task instead of depending on unrealizable accuracies of various parameters to meet the objectives. The system depends heavily on accurate sensors and actuators. For a simple case, the sensors or navigation system gives the current position of the missile or rocket. The guidance computer finds out the error between the actual position and the desired instantaneous position of the vehicle for realizing the final objective and generates demands on control system to correct the position. The control sensors sense the current missile orientation and control computer generates demands on actuation system to generate the required forces and moments acting on the missile so that the missile position is corrected as per the guidance requirement. The actuation system uses another feedback control system to sense the present position of the control surfaces or gimbaled nozzles and generate currents to drive them to the required position to generate control forces as required by the control system. Fig. 1.2a shows various errors pictorially. Fig. 1.2b shows the general sketch of the three feedback control loops. Though the final mission objective of the launch vehicle or missile depends on the accuracy of the guidance loop, its performance depends on the accuracy and the speed of response of the autopilot loop which in turn depends on the accuracy and speed of response of the actuator loop. Hence, while designing, say, an autopilot loop the designer needs to meet the requirements of guidance loop and also lay down the specifications on the actuator loop.

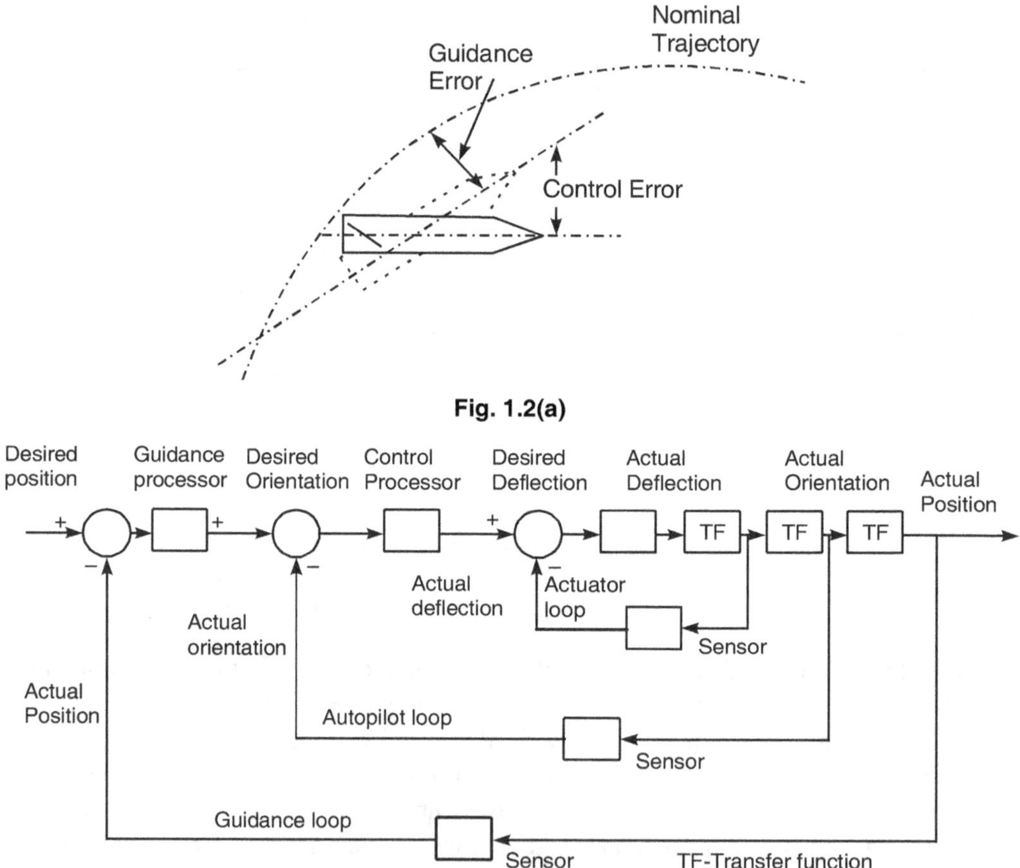

Fig. 1.2(a)

Fig. 1.2(b) Guidance, Control and Actuator Loops

The response of a feedback control system is decided by the poles and zeros of closed loop transfer function. When a large number of poles and zeros are present, it is difficult to quickly assess the performance of the system. Hence, for first cut design one generally attempts to find an equivalent second or third order system for which the system response characteristics can be quickly assessed. This helps in fixing the various design parameters. In general, the poles near to imaginary axis in the complex s-plane, dominate the overall transient response. If some of the poles have real part 5 to 6 times the real part of dominant poles, the response due to them dies down quickly and they may be neglected to start with. The same consideration is used while designing the above three loops. Thus, the guidance loop dominant poles will be decided by the time available for errors to settle down and the autopilot dominant poles needs to be kept 5 to 6 times away from the imaginary axis. Similarly, to enable autopilot design as a second or third order system, actuator poles need to be 5 to 6 times farther from imaginary axis in comparison to autopilot poles. This, however, is associated with cost, size, weight and power requirements and the separation of 5 to 6 times will be in most practical cases brought down to realizable levels making various compromises.

1.2 FLIGHT VEHICLES

The typical flight vehicles which include launch vehicles and missiles are shown in Fig. 1.3.

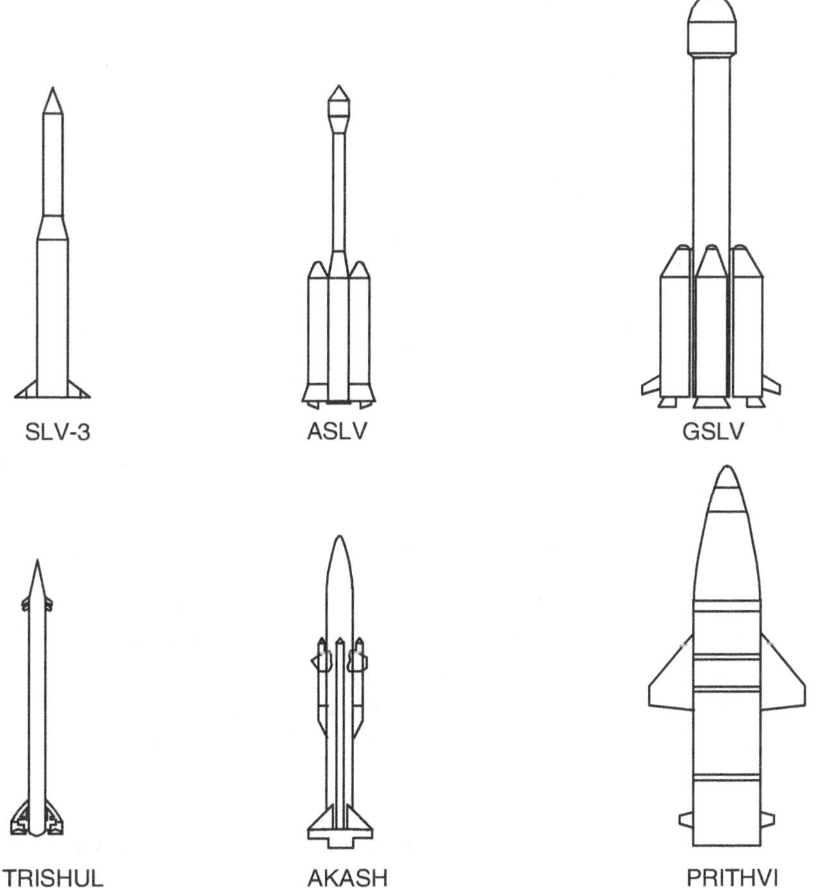

Fig. 1.3

The launch vehicle configuration starting with a simple configuration becomes more and more complex due to desire of increasing the propulsion energy by strapping on number of booster stages on the main core vehicle. The aerodynamics accordingly becomes more complex. However, most of the launch vehicles follow a simple trajectory which is computed well in advance. The vehicles are designed to maximize the payload for the given take off weight. Naturally, the inert weight or structure factor is reduced to the maximum extent. The trajectory is designed to have minimum aerodynamic loads during atmospheric phase. The trajectories however will have the dispersions due to several imperfections such as variation in thrust-time curve, aerodynamic coefficients, autopilot errors etc. These are corrected by the guidance system once the vehicle is out of atmosphere. There is sufficient time for correcting the trajectory. Hence, demand on speed of response on autopilot is not very high. However, the same thing is not true for missiles, specially so for tactical missiles where the missile is expected to intercept the target within the available time and with required accuracy. Thus, the demands on speed of guidance, autopilot and actuator are high in order to meet the mission objectives.

1.3   CONTROL EFFECTORS

Various methods are used in both the launch vehicles as well as missiles to generate control force required for controlling the vehicle's orientation or lateral acceleration.

1.3.1   Aerodynamic Control

In this, the surfaces (fins) on the vehicle are rotated in order to generate aerodynamic forces and moment on the vehicle. Some of the schemes are shown in Fig.1.4.

The control force is given by $F_c = C_{N\delta}\delta QS$ with appropriate sign conventions for δ and F_c. The moment about CG is given by $Mc = C_{N\delta}\delta QSl_c$ where l_c is moment arm.

The direction of the control force is determined by its orientation w.r.t. the air velocity. The turning moment acting on the vehicle can then be easily known from the knowledge of centre of gravity and the point of application of control force.

This scheme can be used only in atmospheric phase of flight and is ineffective out of atmosphere.

The actual aerodynamic force depends on the net aerodynamic angle of attack of the control surface which includes control deflection and vehicle body angle of attack. However, this effect can be incorporated in simulation and only δ dependence can be used for autopilot linear design. Further, the relation is usable upto a specified maximum level of δ beyond which aerodynamic stall takes place, meaning, control force starts decreasing for any further increase in δ.

The critical specifications for this type of control is δ_{max}, $\dot{\delta}_{max}$, bandwidth and the hinge moment against which the actuator is expected to work.

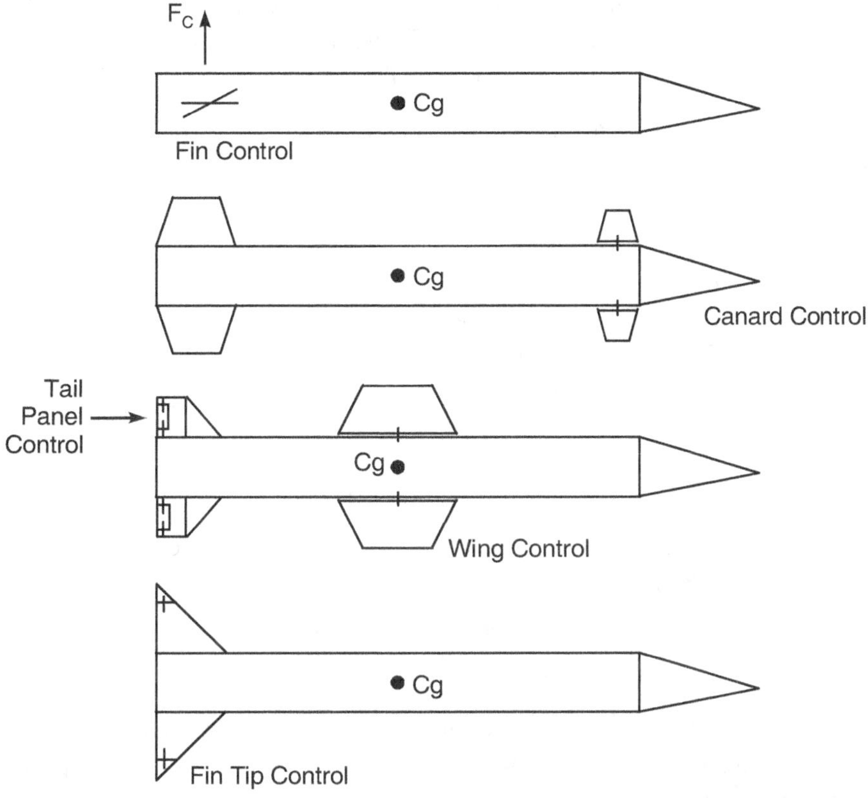

Fig. 1.4

1.3.2 Thrust Vector Control

In this scheme, the main thrust direction is changed from the nominal position. Thus, a component perpendicular to the vehicle is created for controlling the vehicle orientation. Following are some of the schemes and are shown schematically in Fig. 1.5 to Fig. 1.10.

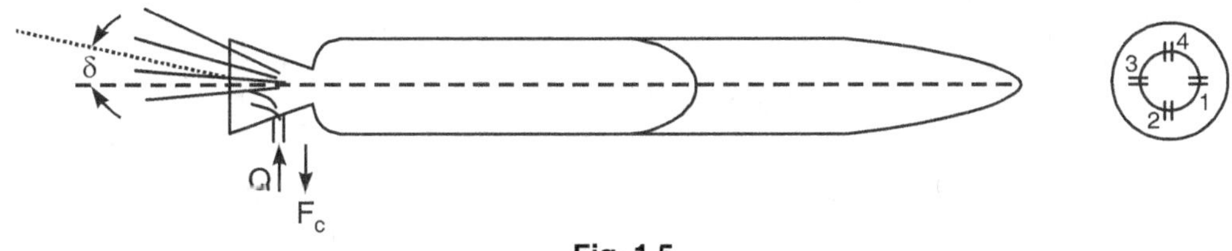

Fig. 1.5

1.3.2.1 Secondary Injection Thrust Vector Control System

This system is used in case of solid propellant stages. In this, a reactive fluid such as stronsium perchlorate is injected in the main nozzle gases at high pressure. Due to expansion of the gases at high temperature, the main flow of gases gets deflected, thus, creating a side force on the vehicle. The characteristics of side force (F_c) w.r.t. the injectant flow rate (Q) must be determined in static tests for further use in control calculations.

The fluid rate is controlled by a pintle valve. Thus, the control force can be related to the pintle valve stroke.

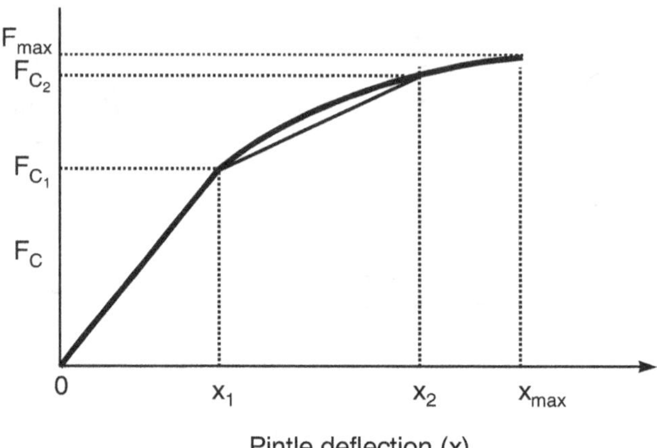

Fig. 1.6. Side force (F_c) vs Pintle deflection

Fig. 1.6 gives a typical curve for side force as a function of pintle valve deflection. The side force may be varying as function of main thrust level. For control design purpose, the band can be approximated by a mean curve.

Fig. 1.6 gives a typical mean curve with three slopes. The autopilot design, as a standard practice, is done for small perturbations. Therefore, the first slope of the SITVC characteristics can be used for design purpose. However, the entire curve (all the three slopes curve) should be incorporated in the trajectory simulation program when extreme disturbance cases also need to be simulated. Thus

$$F_{s_I} = K_{D_s} x \qquad \text{for} \qquad x < x_1$$

$$= F_{c_1} + K_{D_{s1}}(x - x_1) \qquad \text{for} \qquad x_1 \leq x \leq x_2$$

$$= F_{c_2} + K_{D_{s2}}(x - x_2) \qquad \text{for} \qquad x_2 \leq x \leq x_{max} \qquad \text{... (1.1)}$$

The pintle valve can move only in one direction. Hence, for generating negative control force, the input command is given to the diametrically opposite port. Thus, if signal to generate positive force is given to Port 1, the signal for generating negative force needs to be given to Port 3. Roll control force cannot be generated by this system.

This system needs adequate storage of the injectant fuel which is indicated by control impulse. Thus, main specifications for this type of system are x_{max}, x_{max}, bandwidth, maximum control force and control impulse.

The stronsium perchlorate is a highly corrosive fluid. Hence, the fluid is loaded just before the launch. Hence, this scheme is not preferred for missiles, where preparation time should be as short as possible.

1.3.2.2 Flexible Nozzle Thrust Vector Control

This scheme is also used for solid propellant stages. The schematic of the system is shown in Fig. 1.7.

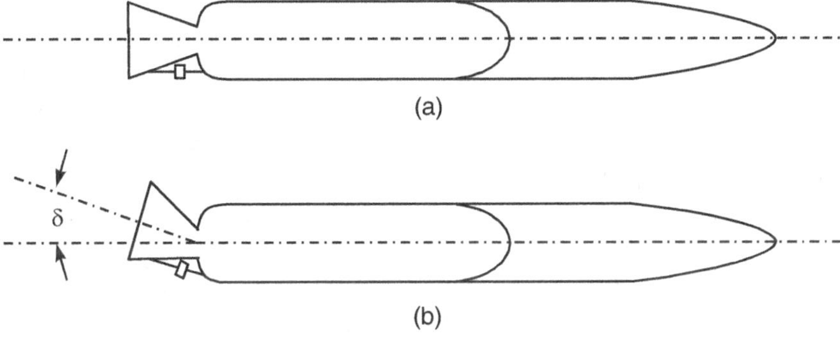

(a)

(b)

Fig. 1.7

The nozzle joint portion inside the motor chamber is sealed using special rubber boots which allows movement of the nozzle. The nozzle is deflected by actuator for generating the required force.

No roll control force can be generated by this scheme.

Further, when the motor chamber pressure rises on ignition of the motor, the nozzle end moves a little outward. If there is no input command to actuator, its stroke length is fixed. In this situation shifting of nozzle end under pressure introduces a tilt in the nozzle with reference to vehicle axis. To avoid this tilt, a mirror image sensor *i.e.*, a potentiometer or LVDT is put exactly on opposite side of nozzle. The potentiometer senses the shift in the nozzle end. This signal is given as an additional input to the actuator in addition to the command coming from the autopilot. Hence, even when there is no autopilot command, if the mirror image senses any shift of the nozzle end, the same signal goes to actuator and it changes the actuator stroke by an equal amount thus eliminating any tilt in the nozzle axis in relation to missile axis.

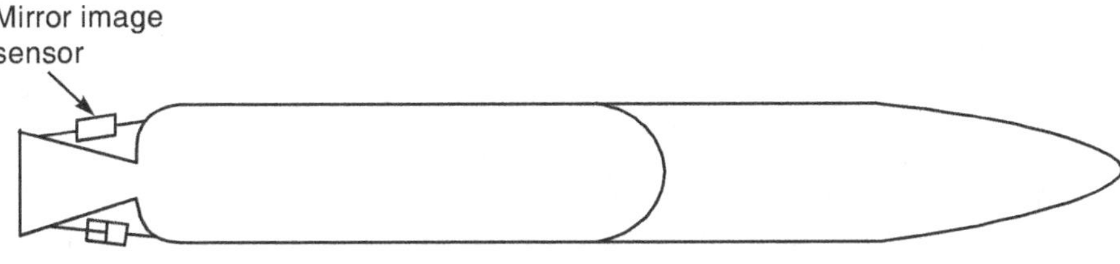

Fig. 1.8

The critical specifications for this system are:

$$\delta_{max}, \ \dot{\delta}_{max}, \ \text{and bandwidth.}$$

The nozzles have significant mass and when the actuator is attempting to move the nozzle, equal and opposite torque (which is significant in this case) acts on the main vehicle. This is generally termed as tail-wags-dog effect. This needs to be properly accounted for in the design of autopilot otherwise it could lead to unstable oscillations. Further, the support structure for mounting actuator needs adequate stiffness to avoid significant deflections under actuator force and possible structural resonances.

1.3.2.3 Jet Vane Control

This system is also used in case of solid propellant stages. In this scheme, small vanes made of special materials are placed inside the main flow of exhaust gases. The exhaust gas stream is deflected by rotating the diametrically opposite vanes simultaneously. The vane material must withstand the high temperature and the force of exhaust gases. This system is generally used for a short time during initial take off phase and till the vehicle acquires a reasonable velocity at which aerodynamic controls can take over the control of the vehicle. The jet vanes are subsequently disconnected and dropped which otherwise can create a disturbing torque due to unequal erosion etc. It may be possible to generate a small roll moment in this case by rotating the jet vanes in opposite direction or unsymmetrical deflection.

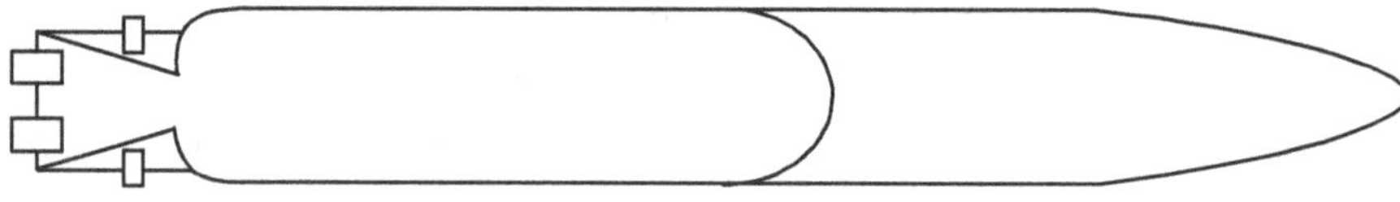

Fig. 1.9

1.3.2.4 Gimballed Engine Thrust Vector Control

This is used in case of liquid propellant stages. The engine(s) are mounted on gimbals which are mounted on a thrust frame which transfers engine thrust to the missile. If there is only one engine, only pitch and yaw control can be accomplished by deflecting the engine in two planes.

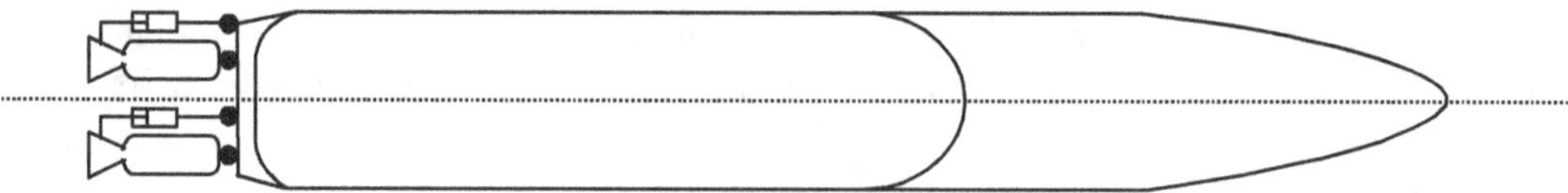

Fig. 1.10

When two engines are available the roll control can also be accomplished by giving differential deflection out of engine plane for generating roll control torque. The engine mass being significant, the tail-wags-dog effect needs to be considered in all the pitch, yaw and roll control systems. The main specifications for this system are also δ_{max}, δ_{max} and the bandwidth.

1.3.3 Power Sources

The control effectors discussed above use either the electric, electro hydraulic or pneumatic power source.

Fig. 1.11 gives a line diagram of pneumatic actuation system.

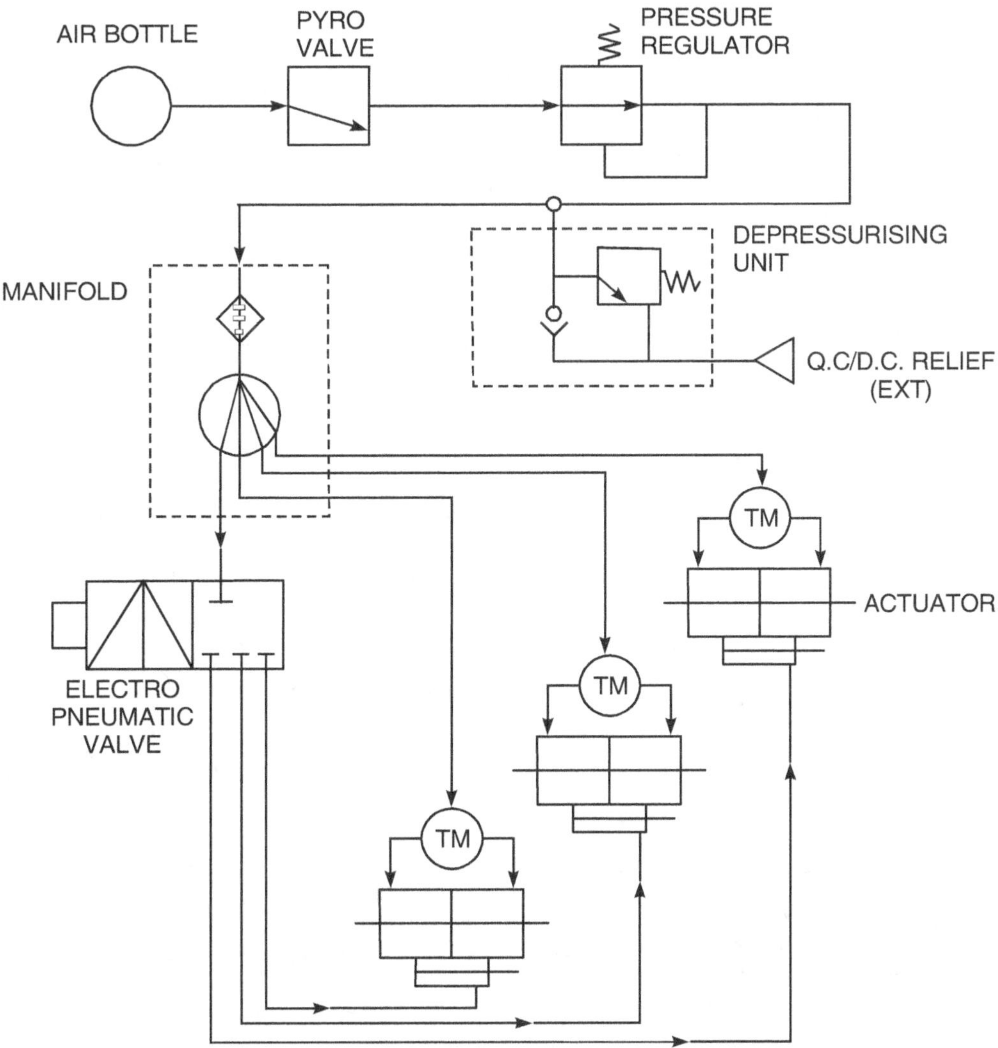

Fig. 1.11 Schematic Pneumatic Actuation system

Fig. 1.12 gives a line diagram for electro hydraulic actuation system.

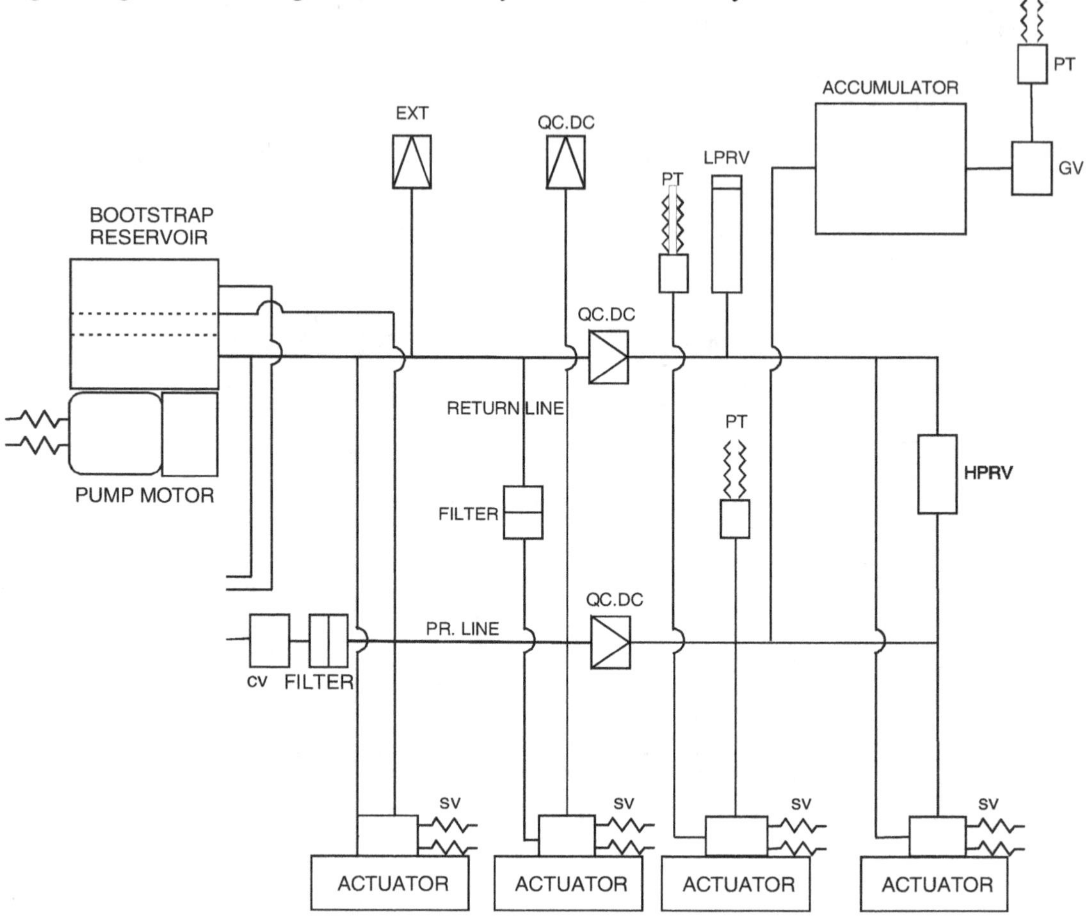

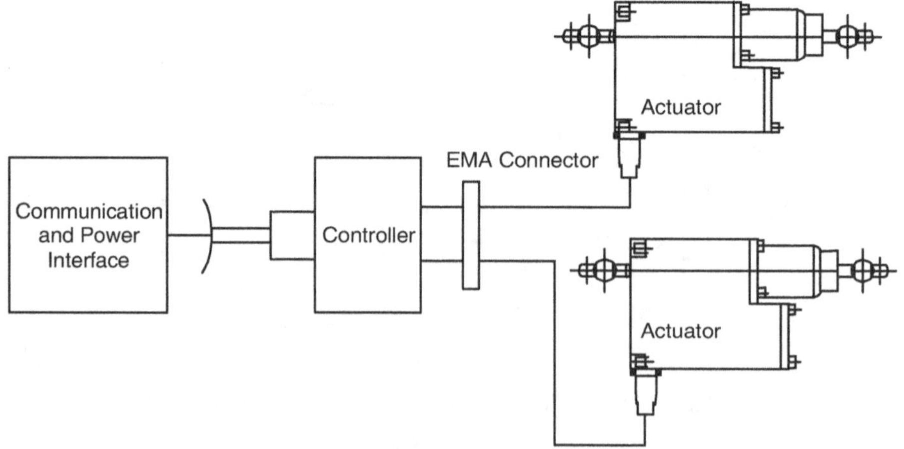

Fig. 1.12 Schematic–Electro hydraulic Actuation system

Fig. 1.13 gives a line diagram of electro mechanical actuation system.

Fig. 1.13 Schematic of Electromechanical Actuation system

1.3.4 Reaction Control

The aerodynamic control can be used only when there is sufficient aerodynamic pressure. The thrust vector control can be used only when the main stage is burning. However, there are regions of the trajectory where aerodynamic pressure is negligible or there is a coasting phase provided for various reasons (which will be discussed subsequently). In such cases, auxiliary or small reaction control motors are used to generate the control forces. Fig. 1.14 shows the illustrative mountings of these reaction motors.

Fig. 1.14

These reaction control motors are ON-OFF type. The fuel flow to these motors is controlled by solenoid valves which are normally closed and open whenever a voltage pulse is provided to the valves. A typical response of control force to the input voltage is shown in Fig.1.15.

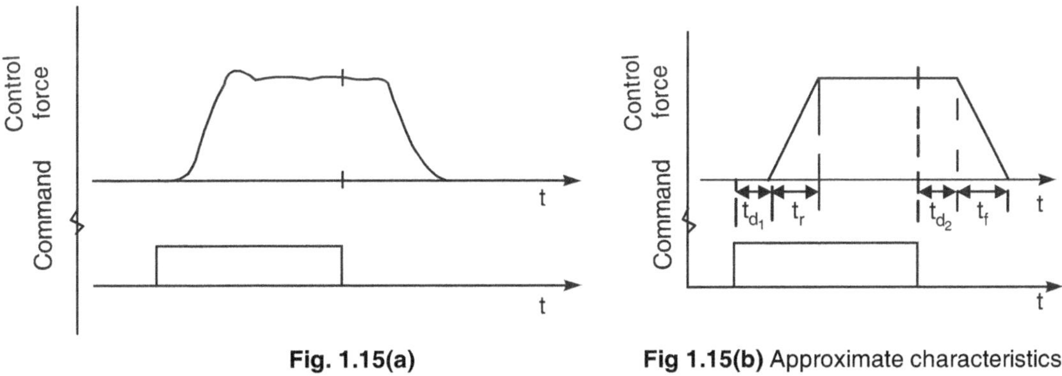

Fig. 1.15(a)   **Fig 1.15(b)** Approximate characteristics

Fig. 1.15(b) shows the approximated response characteristics and is defined by four parameters in this work.

t_{d1} – On delay

t_r – Rise time

t_{d2} – cut off delay

t_f – fall time

The on delay may include the valve delay, time required for the fuel to flow into the motor and ignition delay etc.

The reaction control is of two types. Cold gas reaction control and hot gas reaction control system.

In cold gas reaction control system, the gas at high pressure is made to expand in the motor chamber thereby generating the reaction force.

Hot gas reaction control systems use normal liquid propellants. Here also, there are mono-propellant systems and bi-propellant systems. In case of mono-propellant system, the propellant gets ignited in the presence of a catalyst. In case of bi-propellant system, fuel and oxidizer are brought together in the motor chamber and the ignition takes place in the presence of oxidizer. Since there are two separate flow lines in this case, there will be separate valves, regulators, storage tanks, pyro valves for opening the flow etc. and thus the hardware weight increases. However, one can generate high levels of forces. In case of cold gas, the force levels are small, say 5 to 10 kg and in case of mono-propellant systems, the force level may be upto 25 kg or so. In case the required control force is higher, one has to necessarily go for bi-propellant reaction control system.

The schematic diagrams for different types of reaction control systems are given in Figs. 1.16, 1.17 and 1.18. For control system designer, the important specifications for reaction control system are Force level, Total impulse, Thrust time characteristics *i.e.*, t_{d1}, t_{d2}, t_r and t_f.

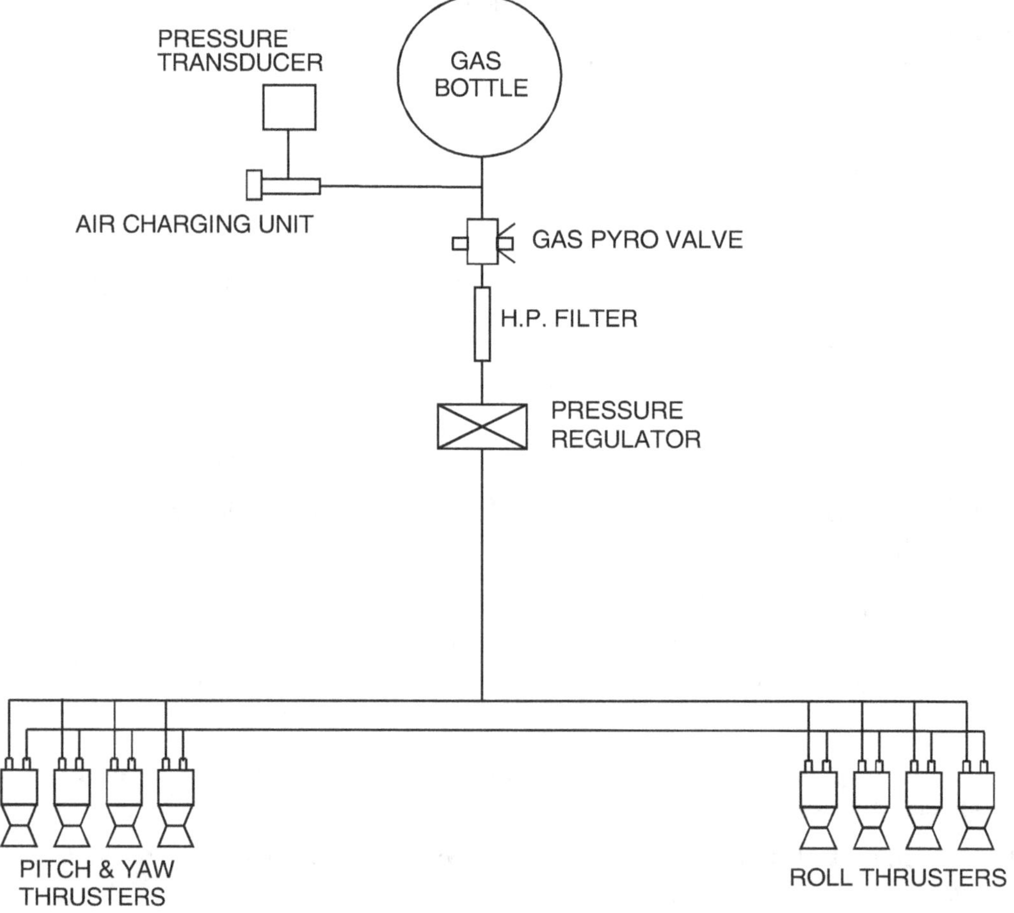

Fig. 1.16 Cold Gas Reaction Control System

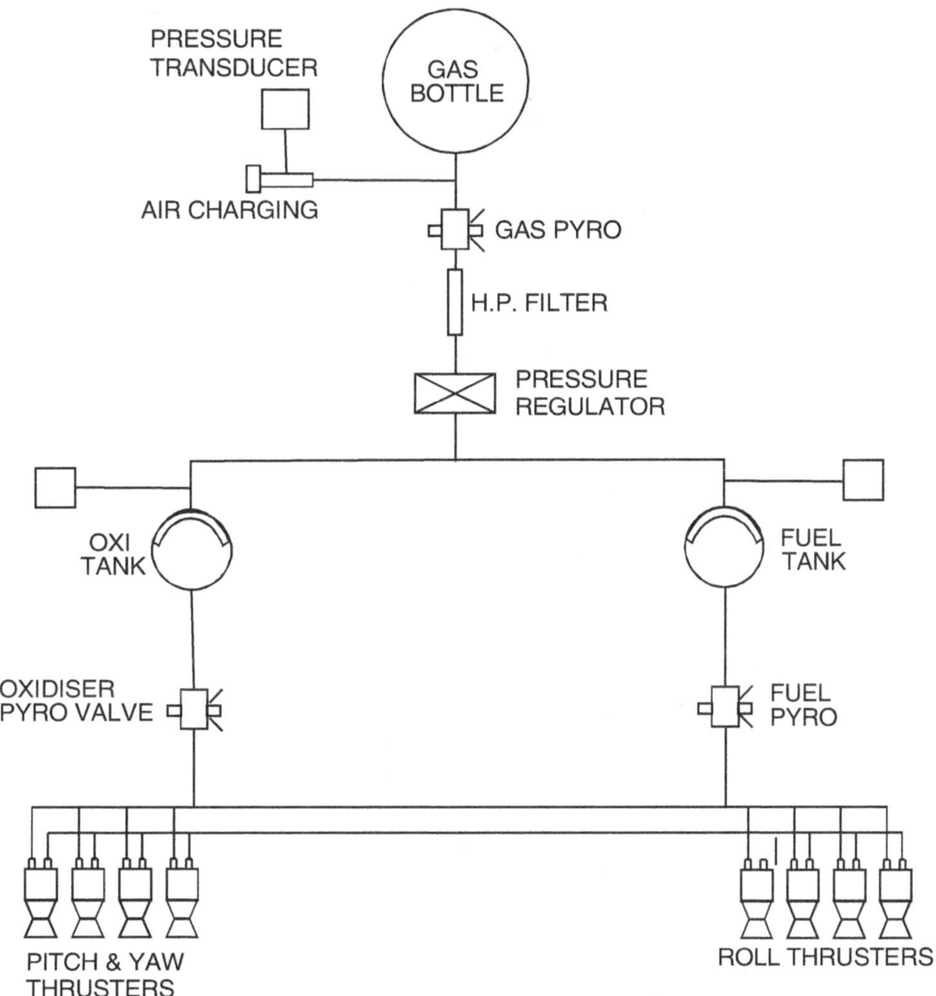

Fig. 1.17 Bi-Propellant Reaction Control System

One of the important characteristics of the reaction control system is the "shortest pulse width" which can be generated by the system. This will depend on delay times and rise and fall times. If these values are large, the system may not be able to respond to narrow voltage pulses. Even if there is some response, the force may not rise to a full level. The maximum specific impulse of the propellants will not be realized in such cases and much of the fuel will be spent inefficiently. The system which can respond to very narrow pulses is a better system and is required wherever high accuracy control system is required. In subsequent chapters, it will be shown that the larger on delay and rise times result in larger errors and larger cut off delay and fall time result in higher value of angular rates and hence higher total impulse. The reaction control system, therefore, sometimes is characterized by the value of smallest impulse bit it can deliver efficiently.

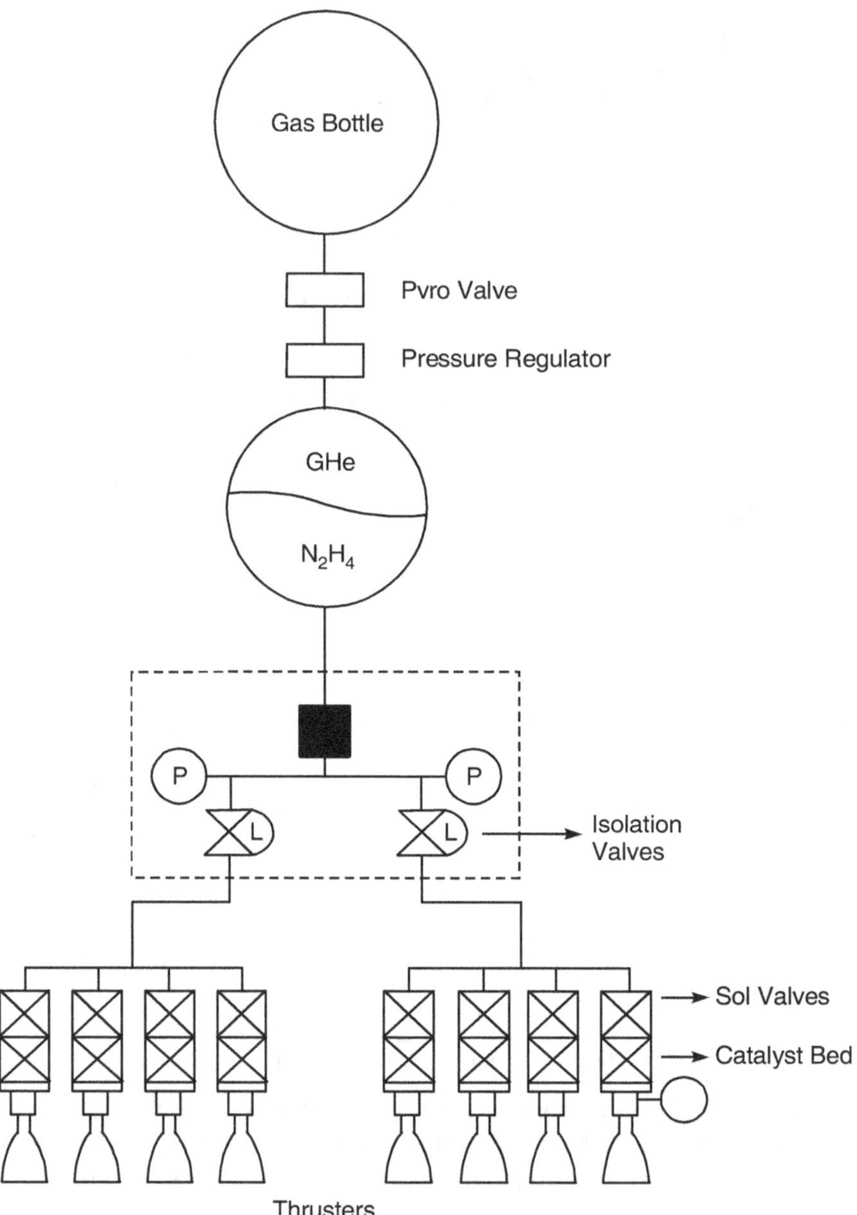

Fig. 1.18 Mono propellant RCS

1.4 EXTERNAL FORCES AND MOMENTS

The state of the vehicle at any instant is determined by the history of all the external forces and moments acting on the vehicle. The forces and moments being vector quantities, it is essential to define the vehicle axes system so that the appropriate signs are given to external forces and moments.

Fig.1.19 shows the vehicle axes system which is generally followed in the literature though one may define a system in a different way for some special considerations.

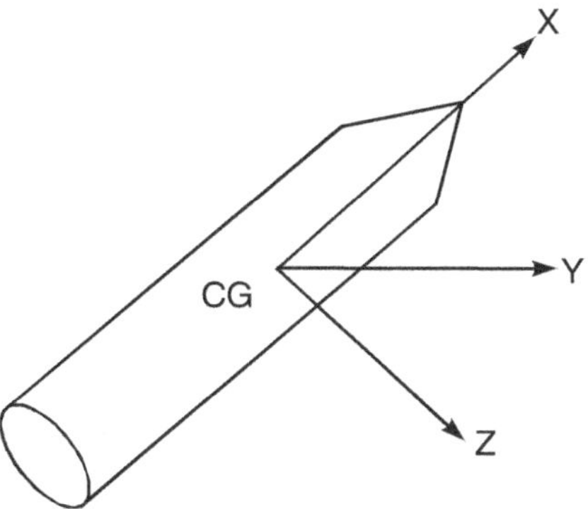

Fig. 1.19

Here, the origin is taken at the instantaneous centre of gravity.

X   in forward direction

Y   towards right while looking in the direction of X

Z   Downward and completing the right hand triad

There are several sources which lead to the disturbance forces and moments acting on the vehicle and resulting into a change in its inertial state. All the external forces and moments can be classified as due to

1. Propulsion

2. Aerodynamic

3. Gravitational

4. Control

Other forces acting on the vehicle which may be termed as inertial forces are due to sloshing of liquid propellant and gimballed engine inertia. However, as shown in subsequent chapter, those are the result of realignment of internal masses of the vehicle and their accelerations and not considered here as external forces and moments. The computation of various forces and moments is given in this development for the most general case which can then be simplified to give the equations for the particular case on hand by giving values to the various parameters. However, it is essential to first describe the methodology for reducing the forces and moments from a general case to the simple body axes frame for writing the final equations of motion.

1.4.1 Euler Angles and Pio Diagrams

It is known that the two right handed axes systems can be aligned with each other by giving three successive rotations about one of the axes system about its intermediate positions after each rotation with a condition that no two successive rotations are given about the same axis. Refs. 2 and 3 give 12 such possible sequences (Table 1.1) for aligning one frame with the other and methodology to find a preferred one sequence when another sequence is known.

TABLE 1.1: 12 Possible Euler Sequences

1-2-1	1-2-3
1-3-1	1-3-2
2-1-2	2-1-3
2-3-1	2-3-2
3-1-2	3-1-3
3-2-1	3-2-3

Let $i_1\ j_1\ k_1$ and $i_2\ j_2\ k_2$ be two systems where the second system is obtained by a single rotation $\quad$ about k_1 axis. The coordinates of point P in two frames are given by (x_1, y_1, z_1) and (x_2, y_2, z_2).

$$x_2 = x_1 \cos\psi + y_1 \sin\psi$$

$$y_2 = -x_1 \sin\psi + y_1 \cos\psi$$

$$z_2 = z_1$$

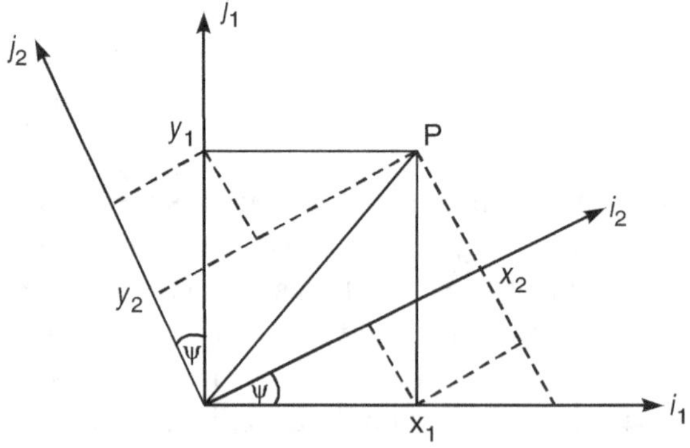

Fig. 1.20

$$\begin{bmatrix} x_2 \\ y_2 \\ z_2 \end{bmatrix} = \begin{bmatrix} \cos\psi & \sin\psi & 0 \\ -\sin\psi & \cos\psi & 0 \\ 0 & 0 & 1 \end{bmatrix} \begin{bmatrix} x_1 \\ y_1 \\ z_1 \end{bmatrix} \qquad \dots (1.2)$$

Similarly, if θ is the rotation angle about y_2 and ϕ is the rotation angle about x_3, obtained after θ rotation, the final coordinates x_4, y_4, z_4 are given by:

$$\begin{bmatrix} x_4 \\ y_4 \\ z_4 \end{bmatrix} = \begin{bmatrix} 1 & 0 & 0 \\ 0 & \cos\phi & \sin\phi \\ 0 & -\sin\phi & \cos\phi \end{bmatrix} \begin{bmatrix} \cos\theta & 0 & -\sin\theta \\ 0 & 1 & 0 \\ \sin\theta & 0 & \cos\theta \end{bmatrix} \begin{bmatrix} \cos\psi & \sin\psi & 0 \\ -\sin\psi & \cos\psi & 0 \\ 0 & 0 & 1 \end{bmatrix} \begin{bmatrix} x \\ y \\ z \end{bmatrix}$$

$$= \begin{bmatrix} c\theta c\psi & c\theta s\psi & -s\theta \\ -c\phi s\psi + s\phi s\theta\, c\psi & c\phi c\psi + s\phi s\theta s\psi & s\phi c\theta \\ s\phi s\psi + c\phi s\theta c\psi & -s\phi c\psi + c\phi s\theta s\psi & c\phi c\theta \end{bmatrix} \begin{bmatrix} x_1 \\ y_1 \\ z_1 \end{bmatrix} \qquad \dots (1.3)$$

where $\qquad c = \cos\,(\ )$

and $\qquad s = \sin\,(\ )$

R.L. Pio gives a pictorial representation of this transformation as given in Fig. 1.21.

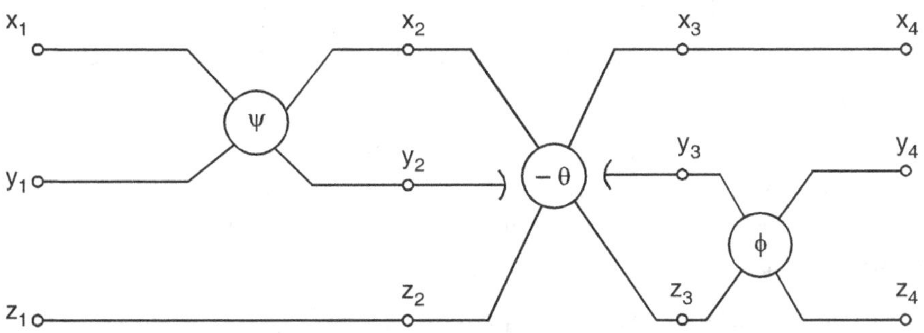

Fig. 1.21

This representation helps one to write down the transformation matrix between $X_4Y_4Z_4$ or any other intermediate system in terms of $X_1Y_1Z_1$ by simply inspection of the diagram instead of going through the multiplication of all the 3 matrices. This is particularly useful when one is interested to know only direction cosines of one of the axes or only a particular element of the transformation matrix and saves trouble of multiplication of 3 matrices.

The rules for finding the pictorial representation and for finding the matrix elements are given in the box.

For illustration, transformation matrix evaluation is shown below:

$$x_4 = x_1 \cos \psi \cos \theta + y_1 \sin \psi \cos(-\theta) + z_1 \sin(-\theta)$$

$$y_4 = x_1(-\sin \psi \cos \phi + \cos \psi \sin \theta \sin \phi) + y_1(\cos \psi \cos \phi + \sin \psi \sin \theta \sin \phi) + z_1(\cos \theta \sin \phi)$$

$$z_4 = x_1(\cos \psi \sin \theta \, \cos \phi + \sin \psi \sin \phi) + y_1(\sin \psi \sin \theta \cos \phi - \cos \psi \sin \phi)$$

$$+z_1(\cos \theta \cos \phi) \qquad ... (1.4)$$

RULES FOR PIO REPRESENTATION AND TRANSFORMATION

1. In Fig. 1.2.1, the three dots in a vertical array $(x_1, y_1, z_1$ or x_2, y_2, z_2, etc.) show an orthogonal set of unit vectors at any stage of rotation.

2. The unit vector (or axis) about which the rotation is given is shown by the continuous line (*e.g.* z_1 to z_2 or x_3 to x_4). The axes which are rotated through an angle are shown passing through an angle (*e.g.* ϕ) given in the circle ($y_2 - y_3$ continuity is shown as in Fig. 1.21).

3. The rotation sequence is represented as, say, 312 meaning first rotation about 3rd axis, 2nd rotation about first axis and 3rd rotation about 2nd axis.

4. The rotation about the first and third axis is shown through a positive angle and the rotation about second or middle axis is shown through a negative angle (*e.g.* – in Fig. 1.21).

5. Any axis at the output (*e.g.* x_4) is given in terms of starting coordinate axes (or intermediate coordinate axes, if required) by tracing all possible paths from each starting axis. The direction cosine between the two axes will have as many additive terms as there are paths between them.

   For example : Path between x_1 and x_4 is $x_1 - x_2 - x_3 - x_4$

   Path between y_1 and x_4 is $y_1 - x_2 - x_3 - x_4$

   Path between z_1 and x_4 is $z_1 - z_2 - x_3 - x_4$

6. While traversing the path, multiply by cosine of the angle of rotation if the path comes back to the same side of axis (*e.g.* $x_1 - x_2$ or $x_2 - x_3$) and multiply by sine of the angle if the path goes to other side (*e.g.* $y_1 - x_2$ or $z_1 - x_3$ or $x_2 - z_3$ etc.) If the movement is from lower side to higher side *i.e.* positive slope, multiplication is with positive sine of the angle and if the slope is downwards multiplication is by negative of the sine of the angle.

1.4.2 Propulsion Force and Moment

PSLV Configuration is considered here as a general case.

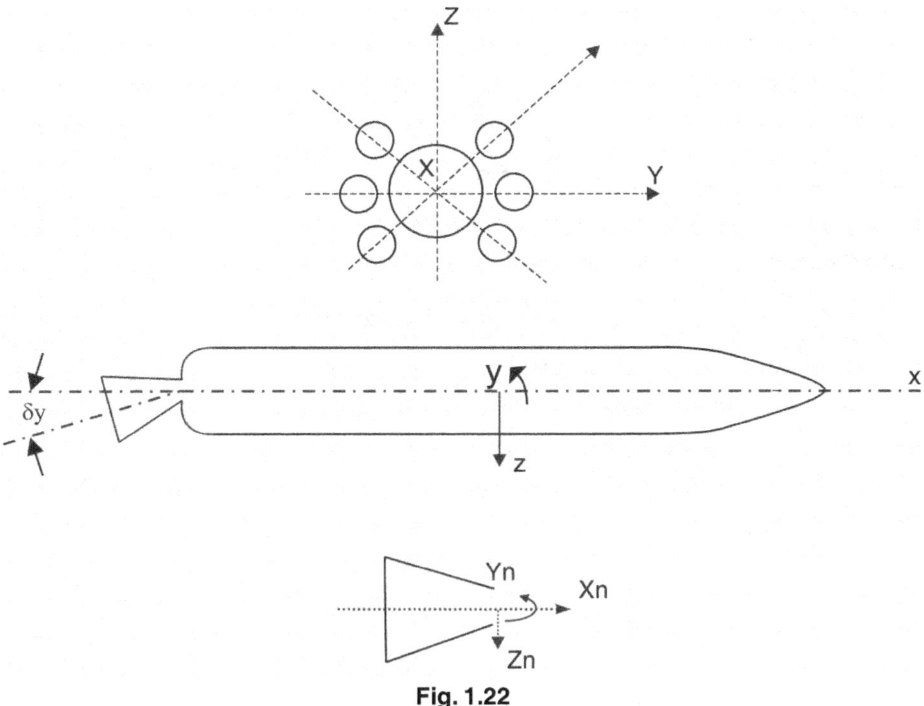

Fig. 1.22

We define following coordinate systems:

NOZZLE COORDINATE SYSTEM

The propulsion thrust is assumed to be acting along X_n. When it is said that thrust misalignment is present, it is the nozzle alignment with reference to the main booster or the corresponding strap on motor. Thrust misalignment is described by two angles δ_x and δ_y about the x and y axis of the motor as described under strap on axes system. As a general case, these are given by δ_{xi} and δ_{yi} for ith booster. The point of application of the thrust is assumed to be at nozzle throat where the nozzle axis intersects the booster longitudinal axis. The vector $\bar{R}_{ni}$ with reference to main vehicle nose tip gives the location of the nozzle throat.

STRAPON AXES SYSTEM

Let the origin of the strapon axes system be defined at the nozzle throat point which is assumed to be on the longitudinal axis of symmetry of the strapon booster. OX axis points towards the nose of strapons. OY and OZ axes are shown by the marked points on the strapons.

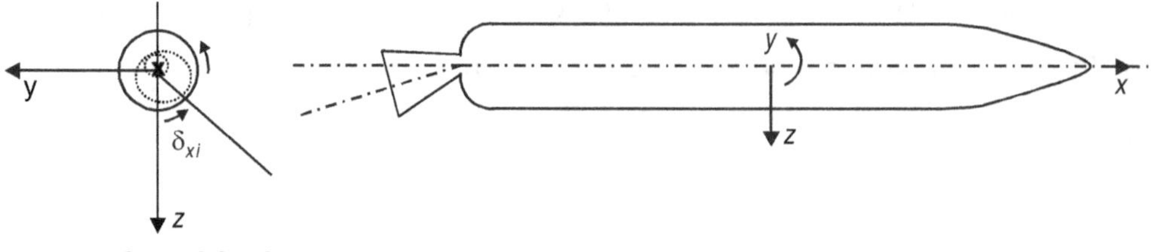

Fig. 1.23

If there are no misalignments and only the pure cant angle is present, the OZ axis is assumed to lie in the plane of OX and nozzle axis.

When the misalignments are present, those are defined by δ_x rotation about x axis and δ_y about the shifted y axis after δ_x rotation. The angle δ_y can be considered as a resultant angle of (cant angle + misalignment angle) in the same plane.

For getting the thrust components in strapon, one needs to transform from nozzle axis to strapon axis by $(-\delta_y)$ about y_n and $(-\delta_x)$ about changed x_n which is now parallel to strapon axis. The transformation matrix is given by

$$C\!\underset{S}{\overset{N}{\downarrow}} = \begin{bmatrix} 1 & 0 & 0 \\ 0 & \cos(-\delta_x) & \sin(-\delta_x) \\ 0 & -\sin(-\delta_x) & \cos(-\delta_x) \end{bmatrix} \begin{bmatrix} \cos(-\delta_y) & 0 & -\sin(-\delta_y) \\ 0 & 1 & 0 \\ \sin(-\delta_y) & 0 & \cos(-\delta_y) \end{bmatrix}$$

$$= \begin{bmatrix} \cos\delta_y & 0 & \sin\delta_y \\ \sin\delta_y \sin\delta_x & \cos\delta_x & -\sin\delta_x \cos\delta_y \\ -\sin\delta_y \cos\delta_x & \sin\delta_x & \cos\delta_x \cos\delta_y \end{bmatrix} \qquad \text{... (1.6)}$$

The thrust components in ith strapon axis are then given by

$$\begin{bmatrix} Tsx_i \\ Tsy_i \\ Tsz_i \end{bmatrix} = C\!\underset{S}{\overset{N}{\downarrow}} \begin{bmatrix} T_i \\ 0 \\ 0 \end{bmatrix} = \begin{bmatrix} T_i \cos\delta_{yi} \\ T_i \sin\delta_{yi} \sin\delta_{xi} \\ -T_i \sin\delta_{yi} \cos\delta_{xi} \end{bmatrix} \qquad \text{... (1.7)}$$

The most general position of the strapon with reference to the central booster is now given as in Fig. 1.24.

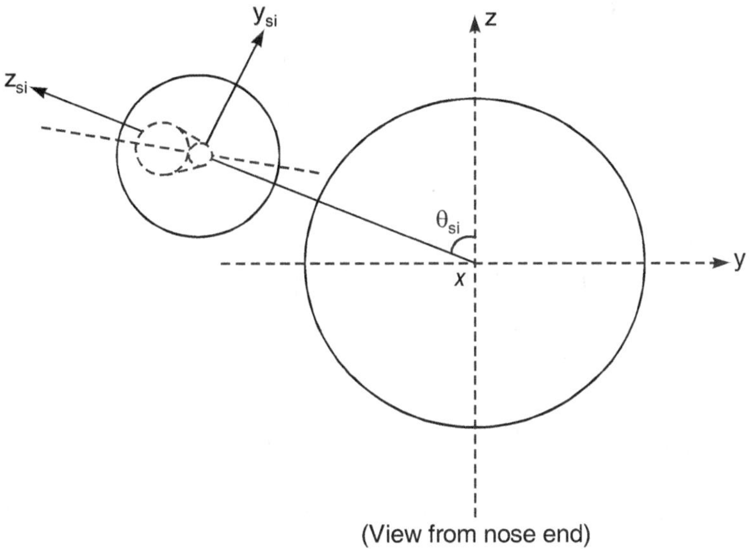

Fig. 1.24

The location of the *i*th strapon is given by θ_{si}.

For $\theta_{si} = 0$, the strapon axes system is parallel with the central booster stage. In ideal condition, the Z_{si} axis passes through the plane formed by origin of strapon axes system and the x axis of central booster. In general case, the strapon may have alignment error η_{xi}, η_{yi}, η_{zi} about the strapon axes system. Since these are small angles, it will be possible to make small angle approximations while transforming. Further, for small angles, the sequence of rotation can also be ignored without affecting results significantly.

Thus neglecting the products of two and three small angles, the transformation between ideal and actual strapon axes systems is given by

$$C\begin{smallmatrix}S_iI\\ \downarrow\\ S_iA\end{smallmatrix} = \begin{bmatrix} 1 & \eta_z & -\eta_y \\ -\eta_z & 1 & \eta_x \\ \eta_y & -\eta_x & 1 \end{bmatrix} \quad ; \quad \begin{matrix} \text{I– Ideal} \\ \text{A– Actual} \end{matrix} \qquad \dots (1.8)$$

and

$$C\begin{smallmatrix}S_iA\\ \downarrow\\ S_iI\end{smallmatrix} = \begin{bmatrix} 1 & -\eta_z & \eta_y \\ \eta_z & 1 & -\eta_x \\ -\eta_y & \eta_x & 1 \end{bmatrix} = \left[C\begin{smallmatrix}S_iI\\ \downarrow\\ S_iA\end{smallmatrix} \right]^T \qquad \dots (1.9)$$

The thrust components from strapon axes to central booster axes system can now be obtained by rotating through $(-\theta_{si})$ about strapon x_{si} axis. Thus,

$$C\!\!\underset{B}{\overset{S_iA}{\downarrow}} = \begin{bmatrix} 1 & 0 & 0 \\ 0 & \cos(-\theta_{si}) & \sin(-\theta_{si}) \\ 0 & -\sin(-\theta_{si}) & \cos(-\theta_{si}) \end{bmatrix} \begin{bmatrix} 1 & -n_z & n_y \\ n_z & 1 & -n_x \\ -n_y & n_x & 1 \end{bmatrix}^i = C\!\!\underset{B}{\overset{S I_i}{\downarrow}} C\!\!\underset{S_iI}{\overset{S A}{\downarrow}} \qquad \text{...(1.10)}$$

Then the components of propulsion forces on the vehicle can be written as

$$\overline{F}_{Ti} = \begin{bmatrix} T_X \\ T_Y \\ T_Z \end{bmatrix} = \sum_{i=0}^{n} \left[C\!\!\underset{B}{\overset{S_iI}{\downarrow}} \right]\left[C\!\!\underset{S_iI}{\overset{S_iA}{\downarrow}} \right]\left[C\!\!\underset{S_iA}{\overset{N_i}{\downarrow}} \right]\begin{bmatrix} T_i \\ 0 \\ 0 \end{bmatrix} \qquad \text{...(1.11)}$$

where $i = 0$ indicates the central booster and $C\!\!\underset{S_0B}{\overset{S_0I}{\downarrow}}$ and $C\!\!\underset{S_0I}{\overset{S_0A}{\downarrow}}$ are identity matrices. Let $\overline{R}_i$ denote the vector in vehicle body axes frame giving the location of nozzle throat of ith strap-on. Then the torque acting on the vehicle is given by

$$\overline{M} = \overline{R}_i \times \overline{F}_{Ti} \qquad \text{...(1.12)}$$

The expressions for force and moments are written assuming the vehicle as a rigid body. When the vehicle flexibility is considered as given in subsequent chapter, the $\overline{R}_i$ and $\overline{F}_{Ti}$ will get additional components due to flexing of the vehicle and those will be discussed at appropriate place.

1.4.3   Aerodynamic Force and Moment

When the vehicle is flying through the sensible atmosphere, it experiences aerodynamic forces and moments. These forces are axial drag, side force and normal force along x, y and z axes respectively and rolling, pitching and yawing moments.

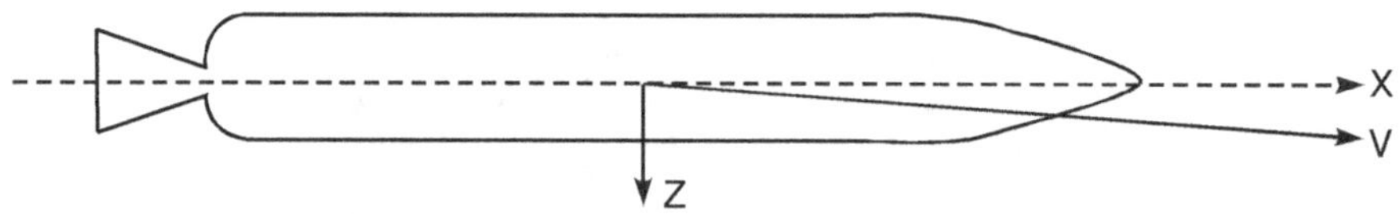

Fig. 1.25

Fig.1.25 shows a simple configuration for defining the aerodynamic forces and moments. The angle of attack (α) and side slip angle (β) are defined by:

$$\alpha = \tan^{-1}\frac{V_z}{V_x} \qquad \text{...(1.13)}$$

$$\beta = \tan^{-1}\frac{V_y}{V_x} \qquad \qquad \dots (1.14)$$

The aerodynamic forces are then given by:

$$F_x = -\frac{1}{2}\rho v^2 C_D S = -QSC_D$$

$$F_y = -C_{N_\beta}\,\beta QS + C_{N_r}r\frac{d}{2v}QS + C_{N_\beta}\,\beta\frac{d}{2v}QS$$

$$F_z = -C_{N_\alpha}\,\alpha QS + C_{Nq}q\frac{d}{2v}QS + C_{N\alpha}\,\alpha\frac{d}{2v}QS \qquad \dots (1.15)$$

Let the resultant aerodynamic normal force act at a point L_{cpp} in pitch plane and side force act at point L_{cpy} in yaw plane and let the origin be considered at the centre of gravity L_{CG} of the vehicle w.r.t. nose tip. Then the roll, pitch and yaw moments are given as follows:

$$M_x = C_lQSd + C_{lp}p\frac{d}{2v}QSd$$

$$M_y = -C_{N_\alpha}\,\alpha(L_{cpp} - L_{cg})QS + C_{mq}q\frac{d}{2v}.QSd + C_{m\alpha}\alpha\frac{d}{2v}QSd$$

$$M_z = C_{y\beta}\beta(L_{cpy} - L_{cg})QS + C_{mr}r\frac{d}{2v}QSd + C_{m\beta}\beta\frac{d}{2v}.QSd \qquad \dots (1.16)$$

The second and third terms in F_y, F_z, M_y and M_z provide the damping terms and in most of the cases those can be neglected in the design stage of control systems where the damping is mainly provided by sensing the vehicle angular rates and using as feedback signals in the control law.

In the above equations, the aerodynamic forces are considered as resultant forces. The actual aerodynamic pressure over the entire vehicle will not be uniform and needs to be separately calculated. The aerodynamic force distribution along the length of the vehicle will be required while modeling the flexible vehicle dynamics and will be considered in the next chapter.

The above equations provide for the normal configurations. There can be disturbance forces and moments due to various misalignments, projections on the vehicle body due to antennas, or signal or power cable tunnels provided along the length of the body. The various such disturbances need to be estimated and catered for while sizing the control power plants.

ATMOSPHERIC WINDS

One of the important sources of external disturbance is atmospheric winds. The winds need to be simulated to check the adequacy of control system for all launch vehicles and missiles.

The wind data is collected over several years and various statistical patterns are obtained. For example,

(*i*) Average winds for different months

(*ii*) Percentile winds for various months or quarters. (A 95 percentile wind means, 95 percent of the time wind will not exceed the percentile wind)

(*iii*) Synthetic wind for different key heights.

Similarly, wind shear and wind gust values for various altitudes are collected over the years.

A wind profile is specified by the wind velocity and the azimuth of the wind direction as a function of altitude.

A synthetic wind for a key height is generated by taking the average wind velocity at all other altitudes and percentile wind at the key altitude. A specified percentile wind shear value is used to connect the key altitude wind with the average wind above and below the key altitude.

Fig.1.26 gives a typical wind profile used in simulation.

For calculating the disturbance due to wind, first obtain the components of wind velocity and total relative velocity of the vehicle with reference to air. The disturbance forces and moments are then calculated as net aerodynamic forces and moments by calculating α and β using the resultant velocity.

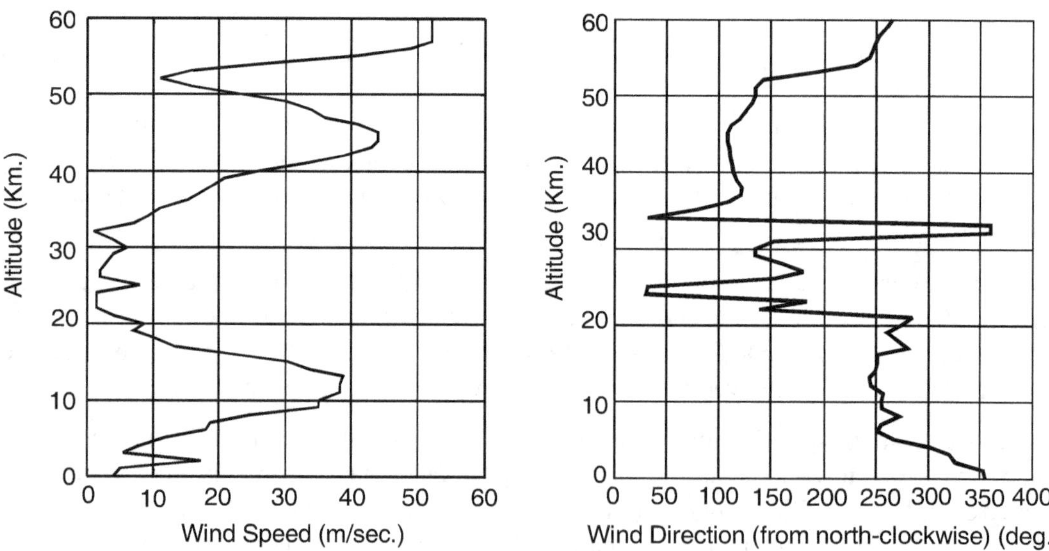

Fig. 1.26 (a) Typical wind profile

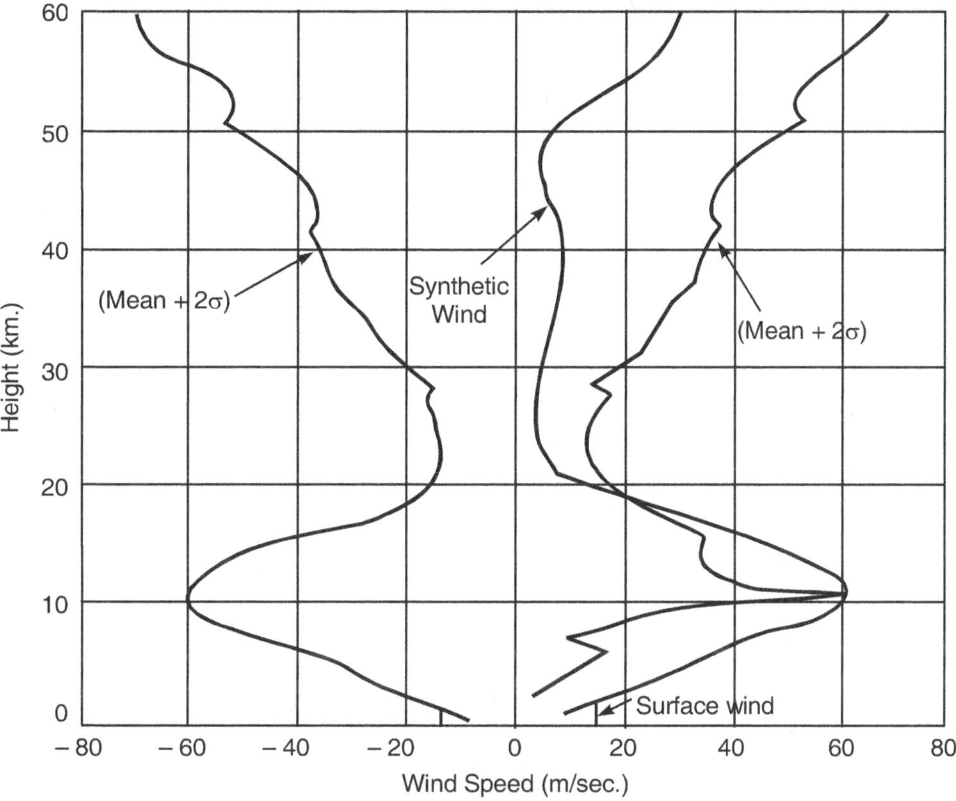

Fig. 1.26(b) Typical Synthetic Wind Profile

1.4.4 Gravity Force and Moment

The gravity force acts along the local vertical. Let the vehicle orientation be given by the Euler angles , and ϕ with reference to local vertical frame of reference. Then, the components of gravity force in body axes frame are given by equation 1.17 (using transformation given by equation 1.4).

$$\begin{bmatrix} F_{gx} \\ F_{gy} \\ F_{gz} \end{bmatrix} = -mg \begin{bmatrix} \cos\theta\cos\psi \\ -\cos\phi\sin\psi + \sin\phi\sin\theta\cos\psi \\ \sin\phi\sin\psi + \cos\phi\sin\theta\cos\psi \end{bmatrix} \qquad \dots (1.17)$$

If the origin of the coordinate frame is taken at the instantaneous centre of gravity, the moment arm is zero and hence there is no moment due to gravity about the centre of gravity of the vehicle.

Eq 1.17 gives gravity force components for a spherical earth model in which case the gravity vector passes though centre of earth. For a more general case *i.e.,* oblate earth the gravity acceleration will have two components—one along the local vertical and the other in local horizontal plane in North South direction.

Fig. 1.27 shows the oblate earth (Ref. 20) and acceleration due to gravity is given by

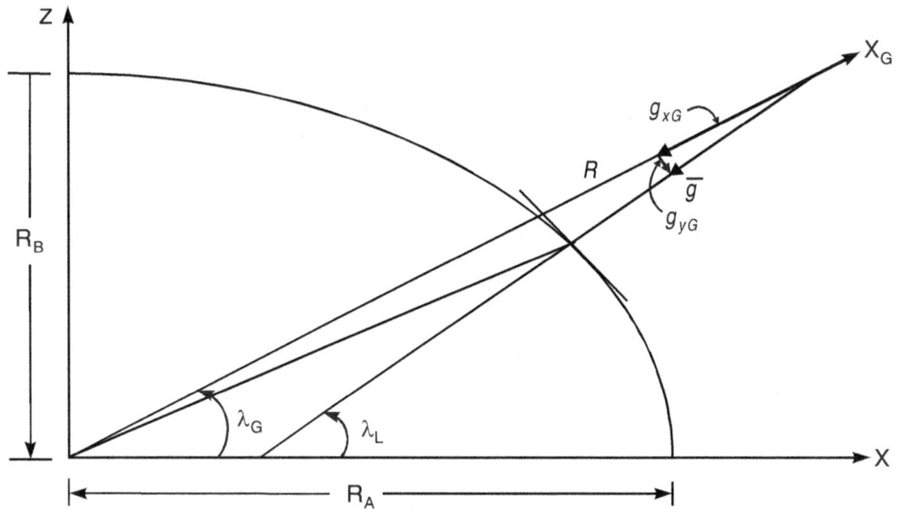

Fig. 1.27

$$g_{xG} = -\frac{\mu}{R^2}\left[1 + \frac{3J_2 P_2}{2}\left(\frac{R_e}{R}\right)^2 + 2J_3 P_3\left(\frac{R_e}{R}\right)^3 + \frac{5J_4 P_4}{8}\left(\frac{R_e}{R}\right)^4\right]$$

$$g_{yG} = \frac{\mu}{R^2}\left[3J_2 P_5\left(\frac{R_e}{R}\right)^2 + \frac{3J_3 P_6}{2}\left(\frac{R_e}{R}\right)^3 + \frac{5J_4 P_7}{2}\left(\frac{R_e}{R}\right)^4\right]$$

$$g_{zG} = 0 \quad (z_g \text{ is towards east})$$

where
$$\mu = 3.986005 * 10^{14} \text{ m}^3/\text{sec}^2$$

$$R_e = 6.378135 * 10^6 \text{ m}$$

$$J_2 = 1.08263 * 10^{-3}$$

$$J_3 = 2.532153 * 10^{-3}$$

$$J_4 = 1.61098 * 10^{-6}$$

$$P_2 = 1 - 3\sin^2 \lambda_g$$

$$P_3 = 3\sin \lambda_g - 5\sin \lambda^3{}_g$$

$$P_4 = -3 + 30\sin^2 \lambda_g - 35\sin^4 \lambda_g$$

$$P_5 = \sin \lambda_g \cos \lambda_g$$

$$P_6 = \cos \lambda_g (5\sin^2 \lambda_g - 1)$$

$$P_7 = \sin \lambda_g \cos \lambda_g (7\sin^2 \lambda_g - 3) \qquad \ldots (1.18)$$

1.4.5 Control Force and Moment

The control forces may be calculated depending on the type of control force generation scheme *i.e.*, whether it is aerodynamic control, SITVC, or thrust vector control system. The control moments will depend upon the location of the control force with reference to centre of gravity of the vehicle. In general, the control force location need not be only on the X-axis of the vehicle but it could be by SITVC or flexible nozzle or gimbaled engines provided on strapon stages.

Let F_{cyi} and F_{czi} be the control force provided by each control effector and its location be given by $\overline{R}_{ci}$ with reference to centre of gravity. Then, the net control forces and moments are given by

$$F_{cx} = 0$$

$$F_{cy} = \Sigma F_{cyi}$$

$$F_{cz} = \Sigma Fc_{zi}$$

$$\overline{M}_c = \Sigma \overline{R}_{ci} \times \overline{F}_{ci} \qquad \ldots (1.19)$$

For simple case of a single booster, $i = 1$ and no summation will be required.

1.5 TRAJECTORY/MISSION PLANNING

1.5.1 Launch Vehicles

1.5.1.1 Optimal Staging

Launch vehicle trajectory is designed to place a given payload in the required orbit. There is always a demand to increase the payload capability of the launch vehicle. Inversely, there is a demand to reduce the take off weight of the launch vehicle for a given payload and orbit. The total weight of the vehicle consists of payload weight, propellant weight and structure weight. As it is impractical to continuously shed the structure weight as the propellant gets consumed, the practical way of increasing the ratio of (payload/take off weight) is to split the total vehicle in number of stages and separate the empty stage after the propellant in that stage is consumed. Theoretically, the ratio (payload/take off weight) would increase with increase in number of stages. However, in practice additional weights have to be provided to incorporate separation systems, interstages, retro rockets for providing separation distance etc. Hence, in practice most of launch vehicles restrict to 3 to 4 stages only and there is a tendency to reduce the number of stages as far as possible. Ref. 7, 8 give the methodology for optimizing the sizes of various stages.

1.5.1.2 *Pitch Program Optimization*

After freezing the vehicle configuration, the trajectory is shaped to give the maximum payload. The simplified or 3 degrees of freedom equations, as given below are used for optimizing the trajectory.

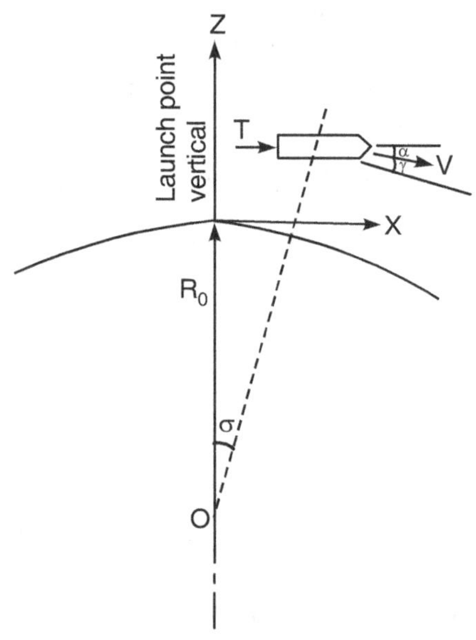

Fig. 1.28

$$v = \frac{(T-D)\cos\alpha}{m(t)} - g\sin\gamma \qquad\qquad \dots (1.20)$$

$$\gamma = \frac{(T-D)\sin\alpha + C_{N_\alpha}\alpha QS}{vm(t)} - \frac{g}{v}\cos\gamma \qquad\qquad \dots (1.21)$$

$$\alpha = \theta + \sigma - \gamma \qquad \ldots (1.22)$$

$$g = g_o \left(\frac{R}{R_o + h}\right)^2 \; ; \; h = \sqrt{x^2 + z^2} - R_0 \qquad \ldots (1.23)$$

$$\sigma = \sin^{-1}\left(\frac{x}{\sqrt{x^2 + z^2}}\right) \qquad \ldots (1.24)$$

$$Q = \frac{1}{2}\rho v^2, \rho \text{ is available as a function of } h \text{ } i.e. \text{ altitude}$$

$$\text{from standard atmosphere model} \qquad \ldots (1.25)$$

$$x = v\cos(\theta - \alpha) \qquad \ldots (1.26)$$

$$z = v\sin(\theta - \alpha) \qquad \ldots (1.27)$$

with initial conditions as $x_0 = 0$, $z_0 = R_0$, $v_0 = 0$, $\gamma = 90°$, 90 for vertical launch. The component of velocity due to earth's rotation can be added to x based on direction of launch.

Most of the launch vehicles have very low initial acceleration. As can be seen from $\dot{\gamma}$ equation, the velocity appears in denominator. Hence, $\dot{\gamma}$ will be high due to gravity acceleration at low initial velocity if the velocity is inclined with reference to horizontal and there will be a rapid loss in the vehicle capability. Hence, the launch vehicles (or vehicles with low thrust/weight ratio) are launched vertically. The time, for which vehicle travels vertically may be optimized but most of the time it is decided by the distance travelled such that the lowest part of the vehicle is well above the launch tower or holding structure to avoid any interferences taking into account the thrust variations on lower side.

The pitch program or as a function of time is a control variable for optimizing the trajectory and optimization is carried out with following constraints:

1. Vertical travel distance to clear the launch tower.

2. Initial pitch down rate not less than a specified value from range safety considerations (vehicle to clear the range as early as possible).

3. Gravity turn or near zero trajectory during high aerodynamic pressure region to minimize aerodynamic loads on the vehicle or minimization of Q factor.

4. Stage separation altitude to ensure low dynamic pressure at stage separation to help safe separation and minimize separation disturbances and requirement on retro rocket thrust levels.

5. Aerodynamic heating

6. Safe region for the impact of separated stages on earth.

7. Adequate coast period to minimize the dynamic pressure at stage separation.

Ref. 9, 10 give the computer program for obtaining optimized pitch program.

1.5.1.3 Disturbances in Stage Separation Region—Separation Disturbances

There is a general realization among the designers after facing initial difficulties that it is desirable to size the stages in such a way that the launch vehicle (or a long range missile) is almost out of sensible atmosphere or the dynamic pressure is very low at the time of separation. This helps in ensuring:

(*i*) Minimum disturbances at the time of separation.

(*ii*) Quick capture of the vehicle attitude for the remaining vehicle after separation.

However, this condition may not be possible sometimes when it is desired to enhance the payload capability of the existing launch vehicle by strapping on additional available boosters stages on the main vehicle to increase the propulsive energy and the separation of strapon stages may have to be done in high dynamic pressure region as in case of ASLV.

One must exercise extra care in such cases to ensure that adequate control force is available to counter the separation disturbances, wind disturbances and thrust misalignments.[11, 12] Further, when multiple strapons are present, one must consider disturbances due to unequal thrust of diametrically opposite located strapons and the case wherein one of the diametrically opposite located strapon booster burns out earlier than the other and large unequal thrust may create a large turning moment for a short time.

To minimize turning moments due to such large unequal thrust, the strapon nozzles are provided with a cant angle.[13, 14] The cant angle is designed in such a way that the thrust line passes very nearly through the centre of gravity of the total vehicle at strap on burn out. This reduces the moment arm thus reducing the turning moment on the vehicle. Since the cant angle is associated with loss in forward thrust, the magnitude of cant angle will be decided by a trade off between loss in forward thrust and maximum turning moments due to unequal thrust.

It is important to ensure that the vehicle is under full control when strapons are separated in high Q region and 'no control zone' is not permitted.

1.5.1.4 Disturbances In High Dynamic Pressure Region

The trajectory is designed to have near zero angle of attack during high dynamic pressure region. The nominal design is generally done assuming there is no wind. In actual practice, vehicle will experience angle of attack due to following factors:

(*i*) Wind

(*ii*) Thrust variation (highest and lowest thrust vs. time profiles)

(*iii*) Control errors (error in following the pitch program *i.e.*, $(\theta_c - \theta)$ and $(\psi_c - \psi)$

The maximum $Q\alpha$ needs to be estimated incorporating the above factors in trajectory simulation program. The vehicle needs to be designed to take up these aerodynamic loads. However, higher loads on the vehicle needs stronger vehicle structure which leads to poorer structure factor and consequent payload loss. If this lateral aerodynamic loads are resulting in unacceptable increase in structure factor, various schemes as discussed below may be adopted to minimize the aerodynamic loads.

1.5.1.5 Different Schemes for Reducing Aerodynamic Loads

(*i*) **Wind Biasing The Pitch Program**

The additional aerodynamic loads on the vehicle can be minimized by optimizing the pitch program incorporating the desired wind model in the problem formulation. Thus, a separate pitch program can be designed for different winds and selection of the pitch program can be made based on the expected launch window and corresponding winds. Wind biasing can also be used to minimise the trajectory dispersion to minimize guidance requirement.

(*ii*) **Load Relief Control**[15, 16, 17, 18]

In this scheme, vehicle lateral acceleration or angle of attack and side slip angles are sensed and fed back to modify the control law. Less weightage is given to vehicle attitude and more weightage is given to angle of attack or lateral acceleration which gives a measure of aerodynamic loads on the vehicle. The control system, thus, attempts to align the vehicle attitude so as to reduce the aerodynamic load.

Since the vehicle orientation is slightly changed from the required optimal orientation, vehicle will have more trajectory dispersion. This dispersion needs to be corrected by the closed loop guidance scheme, once the high dynamic pressure region is passed.

(*iii*) θ **vs. Height Pitch Program**[19]

One important parameter which causes trajectory dispersion in vertical plane (apart from wind) is vehicle thrust.

Fig. 1.29(a) shows a typical θ vs. t program and Fig. 1.29(b) shows a height vs. time profile for the pitch program given in Fig.1.29. Fig 1.30 gives θ vs. height obtained using the same data.

The vehicle attitude (θ) in this scheme is obtained as a function of altitude H by simulating the trajectory for a nominal thrust. The θ vs. *H* pitch program is then used for generating commands to pitch control system. Thus, θ is interpolated for the current height from this pitch program and is used as θ_c.

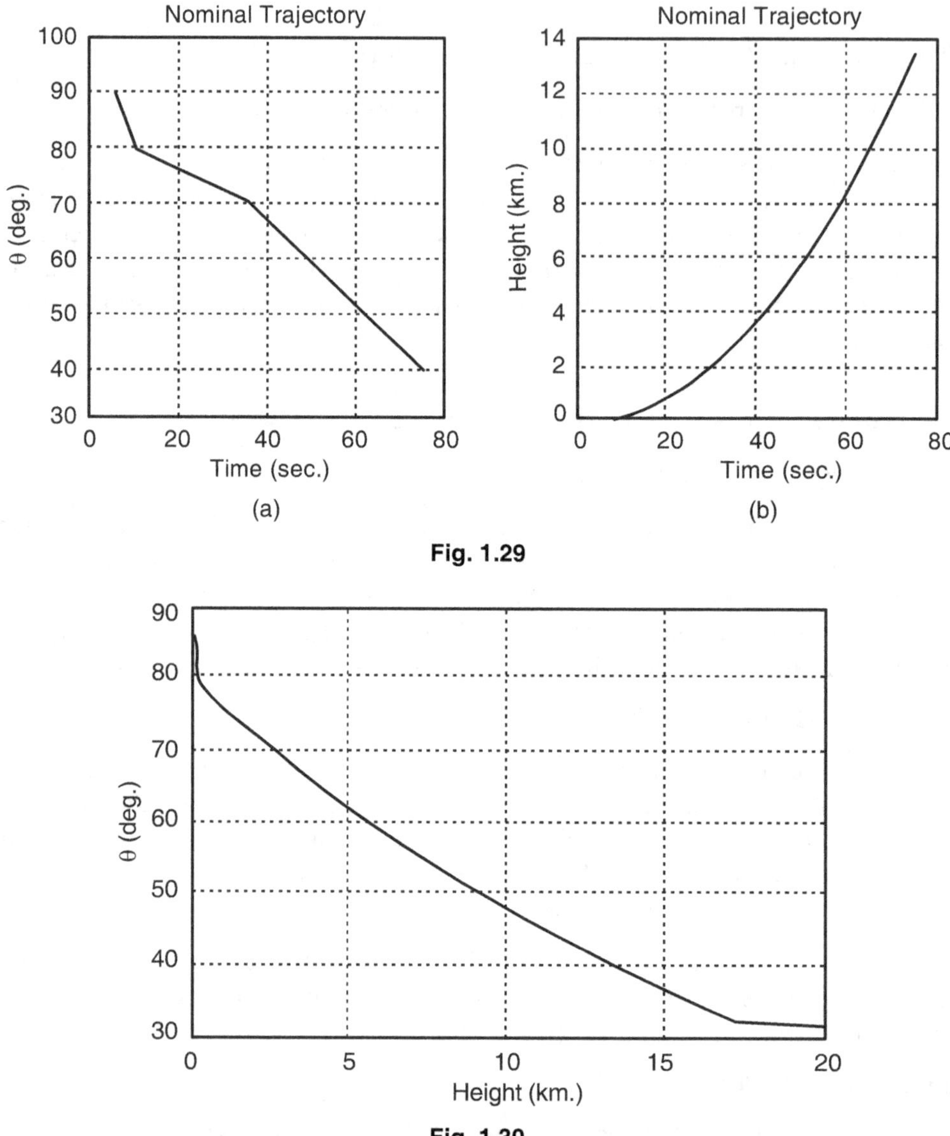

Fig. 1.29

Fig. 1.30

It may be observed that θ is decreasing with altitude. Thus, if there is a tendency for the vehicle to be at a higher altitude than the nominal altitude at any instant, lower θ will be commanded requiring vehicle to come towards the nominal trajectory.

It has been observed that (Ref. 19) a dispersion of about 1 km in the trajectory when θ vs. t pitch program is used has been brought down to nearly 100 to 200 m when θ vs H pitch program is used. This also reduces the vehicle angle of attack and thus the factor Q causing aerodynamic load. Fig. 1.31 gives trajectory dispersion for a typical surface to surface missile for θ versus time and θ versus height pitch program when the thrust is perturbed.

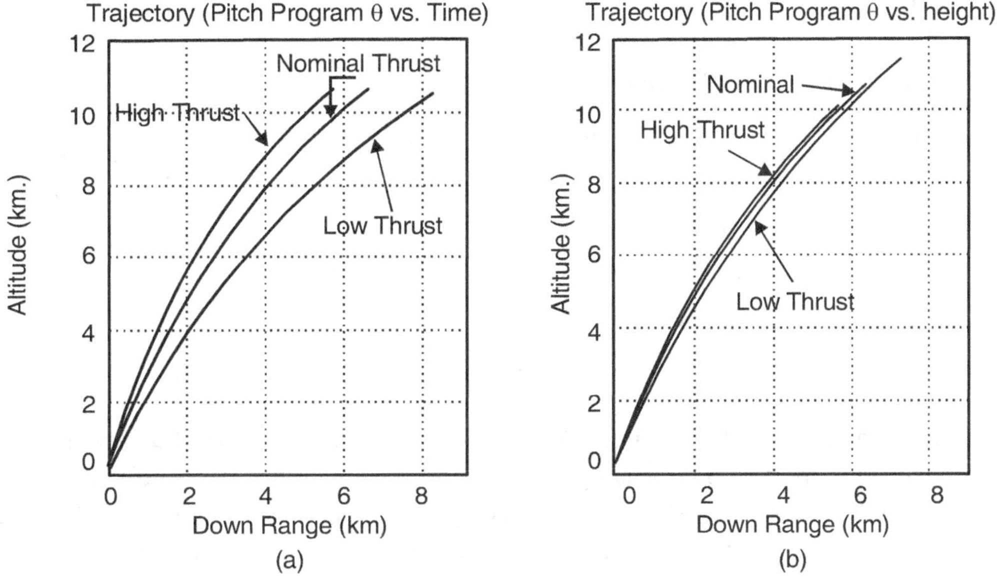

Fig. 1.31 Trajectory Dispersion for   vs time and   vs. height Pitch Program

(*iv*) **Improving Control Errors**

The attitude errors $(\theta_c - \theta)$ and $(\psi_c - \psi)$ are a function of thrust misalignments and external disturbances. Generally, these errors result in only minor variations in trajectory (Fig. 1.32) and can be improved by providing proportional + integral control without much difficulty.

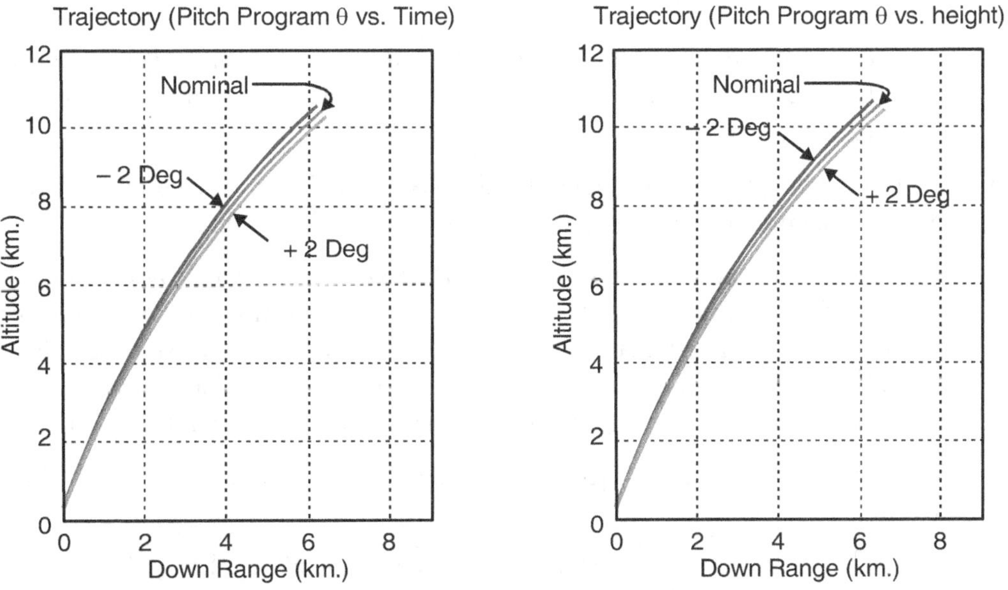

Fig. 1.32 Sensitivity of Trajectory with Attitude Error

1.5.2 Missiles

1.5.2.1 Different Types of Missiles

The missiles have a variety of mission profiles. Depending on their missions, they are classified as:

- Surface to surface missiles
- Surface to air missiles
- Air to Surface missiles
- Air to Air missiles

The classification can be further extended to account for missiles launched against ships, missiles launched from ships or underwater platforms (submarines). Though much of the actual control loop design is similar for various types of vehicles, the vehicle configuration has to take into account special needs of different categories of vehicles.

1.5.2.2 Surface to Surface Missiles

Medium and long range surface to surface missiles have more or less similar trajectory design considerations as in case of launch vehicles. However, the missile trajectories will have the following special considerations:

(*i*) Missiles need to be available for launch during the entire year. Hence, the design must cater for any winds during the year.

(*ii*) Missile may be aimed at any target within the missile range capability. Hence, one must be able to generate the trajectory quickly when the launch point and target point coordinates are given.

(*iii*) Missile could be fixed on the launcher and its axes pointing in a certain direction. The target at which missile is being aimed at, may be in a different direction.

Hence, the missile control systems must have a provision of roll manoeuvre for vertically launched missiles to align the missile axes in the desired direction. Further, the software must have a provision such that the missile pitches down in the desired plane only even if, the roll manoeuvre is not completed during the vertical rise phase and axes system not properly aligned before pitching down starts to avoid loss in missile capability.

(*iv*) Control system must give satisfactory performance for any chosen trajectory within the missile range capability.

(*v*) Longer range missile terminal stages experience high temperatures during re-entry. Missile structure must be designed for the same.

(*vi*) Appropriate control must be provided for re-entry region which would work satisfactorily at high temperature and high dynamic pressure conditions.

(*vii*) The missile needs to deliver the payload at target with a very good accuracy. Therefore, closed loop guidance is essential.

A. Closed Loop Guidance out of Atmosphere

For medium and long ranges, multi-stage missile will be used to have a reasonable size of the missile. In such cases, the first stage uses only a fixed pitch program, in order to avoid missile manoeuvres in high dynamic pressure and closed loop guidance is used when missile goes out of sensible atmosphere (Fig. 1.33a).

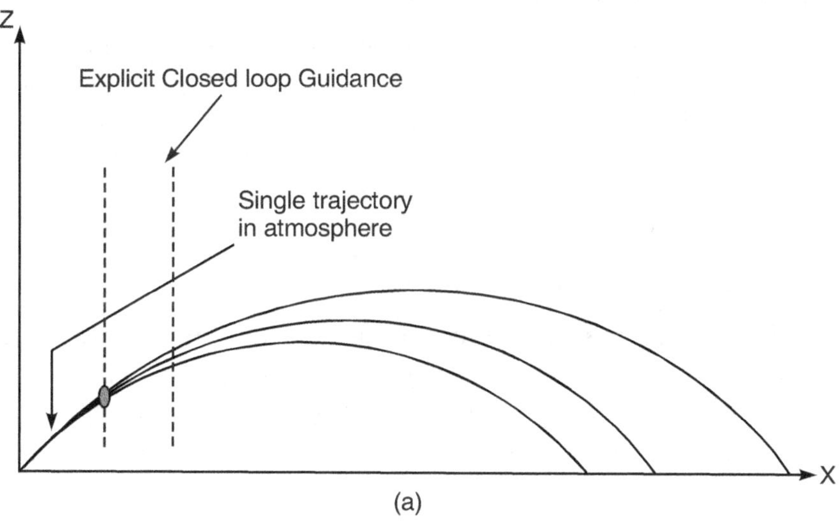

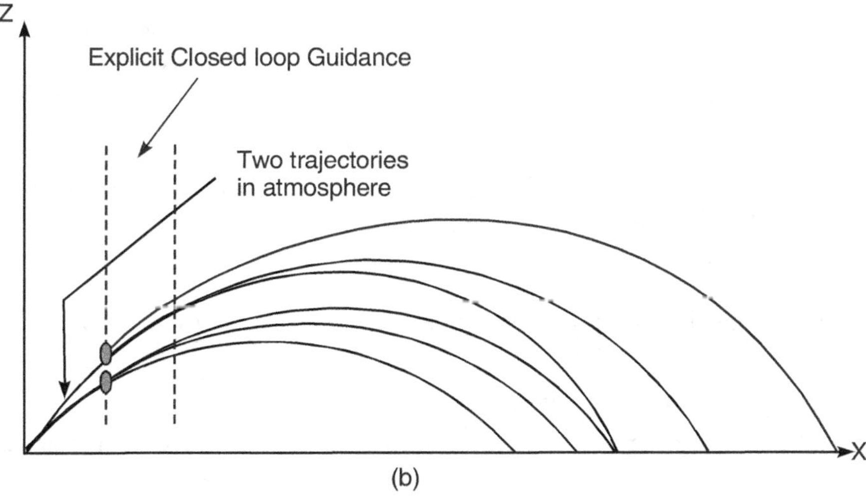

Fig. 1.33

One may use two or three different fixed pitch programs each for a given range bracket (Fig. 1.33b) so that the closed loop guidance when initiated in higher stages does not require very high trajectory corrections and corresponding high missile body rates.

A velocity to be gained guidance scheme with a thrust cut off provision is used when there is liquid propellant stage (Ref. 21). However, there is a general tendency to avoid liquid propellant stages for missiles due to longer propellant filling and preparation time for launch.

Solid propellant stages generally do not have thrust termination facility though there are references of such provisions in some missiles. Hence, energy management guidance schemes (Ref. 22) are devised for missiles with solid propellant stages and a small module with liquid propulsion is provided as velocity trimming package to correct the final velocity of the missile.

The energy management guidance assumes that the propulsion stage will be giving highest bound of performance. If it actually gives that performance, velocity trimming package gives only a minimal essential additional velocity (just to assist separation of the burnt out stage). The size of the velocity trimming package is decided by the velocity which will be required to be provided if the main propulsion system gives lowest bound performance so that the mission will be still met without compromise.

B. Closed Loop Guidance within Atmosphere

For medium and short range missiles, the missile will be having only a single stage. Either a shaped trajectory with a nominal following guidance scheme (Fig.1.34) (implicit guidance) or a ballistic trajectory (Fig.1.35) with a velocity to be gained (VG guidance) with thrust cut off or energy management guidance with velocity trimming feature is used.

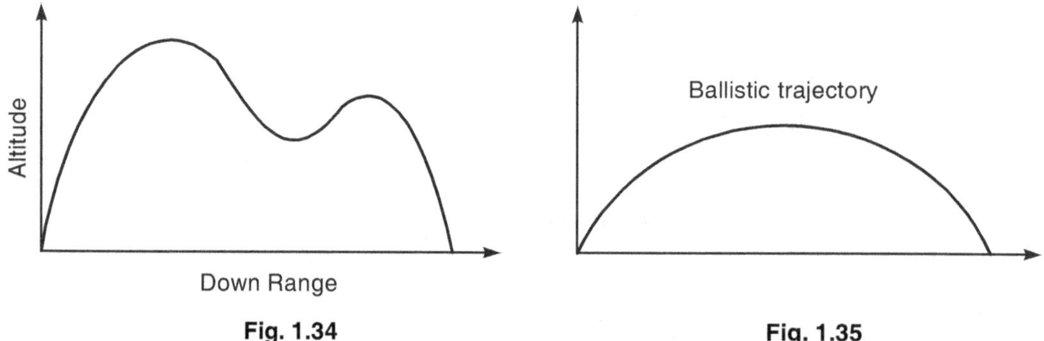

Fig. 1.34   **Fig. 1.35**

The nominal trajectories for different ranges are designed with following constraints. (Ref. 23).

1. Terminal velocity magnitude and flight path angle
2. Lowest dynamic pressure not to be below a specified value (to avoid providing additional RCS control)
3. Highest angle of attack not to exceed a specified value

4. Trim condition control deflection not to exceed a specified value

5. Maximum demanded lateral acceleration or (lateral load) not to exceed a specified value.

For any given target range the actual trajectory is interpolated from the immediately lower range and higher range trajectory data sets.

The control and guidance design needs to be validated for the interpolated trajectory from the following considerations.

1. Satisfactory stability margins throughout the flight

2. Angle of attack and control deflections are within the available range and no saturation occurs

3. Guidance errors are satisfactory.

The missiles which are designed to follow ballistic trajectory are generally designed for low lateral accelerations. Hence, the missile pitch program is designed such a way that it follows nearly zero   trajectory (gravity turn trajectory) in high dynamic pressure region and the maximum angle of attack region occurs at much lower dynamic pressure (say less than 25% of peak dynamic pressure).

Such trajectories are designed for the lowest, highest and mid range of the missile capability.

Fig. 1.36 shows the typical pitch program as a function of time and corresponding   profiles.

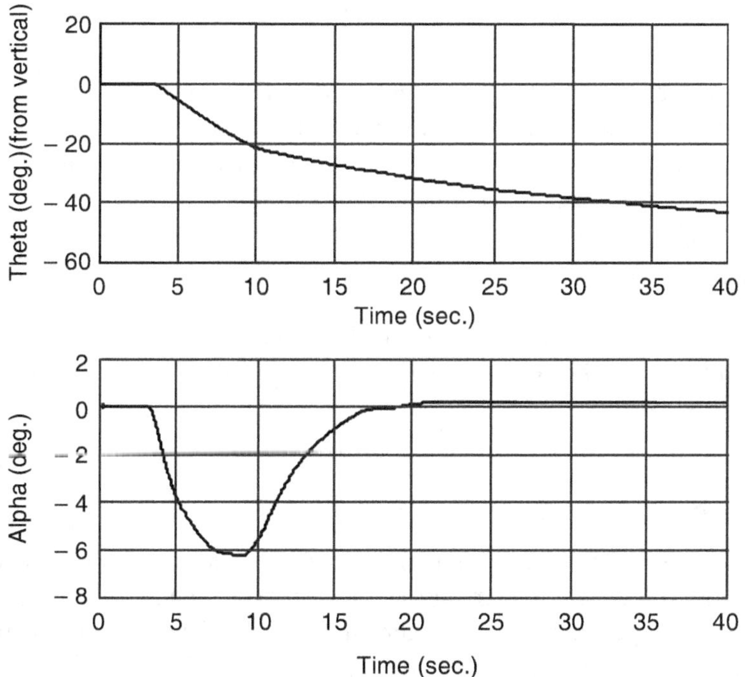

Fig. 1.36

Let the pitch program be represented as a polynomial given by:

For range $R = R_L$ (lowest range)

$$\theta_L = a_{Lo} + a_{L1}t + a_{L2}t^2 + a_{L3}t^3 \qquad\qquad t \le t_1$$

$$= b_{Lo} + b_{L1}t + b_{L2}t^2 + b_{L3}t^3 \qquad\qquad t > t_1 \qquad\qquad \dots (1.28)$$

For Range $R = R_H$

$$\theta_H = a_{Ho} + a_{H1}t + a_{H2}t^2 + a_{H3}t^3 \qquad\qquad t \le t_1$$

$$= b_{Ho} + b_{H1}t + b_{H2}t^2 + b_{H3}t^3 \qquad\qquad t > t_1 \qquad\qquad \dots (1.29)$$

The pitch program for any range R within R_L and R_H is given by interpolating the coefficients a and b as:

$$a_{R_i} = a_{L_i} + \left(\frac{a_{H_i} - a_{L_i}}{R_H - R_L} \right)(R - R_L) \qquad\qquad \dots (1.30)$$

$$b_{R_i} = b_{L_i} + \left(\frac{b_{H_i} - b_{L_i}}{R_H - R_L} \right)(R - R_L) \qquad\qquad \dots (1.31)$$

Simulation studies show that (Ref. 24) the pitch program obtained by interpolating coefficients as above give fairly accurate range of the interpolated trajectory and near zero   during high dynamic pressure region as desired. This reduces the load on closed loop guidance significantly and the closed loop guidance then has to take care of only the dispersions due to propulsion variations and wind variations.

C. Terminal Errors

The terminal or impact errors are usually called Circular Error Probability (CEP) and is an important parameter for the medium and long range missiles. Following are the main contributors to this error:

1. Guidance errors

2. Navigation errors.

In case of explicit guidance for a ballistic missile, the terminal errors will be determined by the accuracy of the vehicle parameters at the thrust cut off such as height, velocity and fight path angle. The actual parameter values immediately after the thrust cut off may not be the same as required by the guidance algorithm. The errors will be propagated into the final errors from the intended impact point.

In case of implicit guidance scheme, the guidance errors are determined by the stiffness of the guidance loop, autopilot performance and the vehicle lateral acceleration capability of the missile.

The navigation errors depend on the accuracy of the sensors used in the inertial navigation system. These errors are not known to the guidance system and hence cannot be corrected by guidance loop. The inertial navigation errors are basically cubic function of time and hence , the errors becomes large as the flight time increases.

These errors can be corrected only by updating the navigation data using external information such as Global Positioning System (GPS) or determining the vehicle position based on the known features on land and updating the navigation data accordingly.

1.5.2.3 Surface to Air and Air to Air Missiles

These missiles are primarily meant for intercepting high speed targets such as aircrafts and missiles. Since one would like to intercept the incoming targets as far away as possible, these missiles have very little time to take off (reaction time) from the instant of detecting the target and traveling as much distance as possible for intercepting it. Hence, these missiles need quick reaction time and high acceleration and coast velocities. The reaction time, the acceleration and velocity levels of the missile and the desired maximum range for the intercept determine the target detection range capability of the radar for different speeds of incoming target. The present day fighters have upto 8 to 9 g normal acceleration capability and they can take sudden manoeuvres upto this level when they detect the attacking missiles. The missile must be able to kill such targets and hence requires two to three times the target manoeuvrability to ensure low miss distances.

The missiles mainly use radio command guidance and seeker based homing guidance. In command guidance, the radar tracks both the target and the missile. The line of sight angle and angular rates are computed from the tracked data and the information is used to generate lateral acceleration (latax) command using the proportional navigation (PN) or augmented PN guidance laws.

In seeker based homing missiles, the line of sight rate is obtained from the seeker rate gyros and the lateral acceleration (latax) command is generated using line of sight rate ($\dot{}$) as

$$a_z = N'V_c\lambda_p \qquad\qquad \dots (1.32)$$

where V_c is the closing velocity between missile and target.

The total time available for the missile to correct the initial heading errors, put the missile on the collision course and intercept the target is determined by the maximum detection range capability of the seeker and the closing speed V_c. Ref. 25 shows that the normalized guidance time to correct initial heading error is 7 to 8 where the normalized time means the ratio of actual time and the guidance time constant. Thus, for a seeker having a locking-on range of 15 km and closing speed of 1500 m/s. guidance has 10 secs to correct heading error and minimize the miss distance. This determines the maximum allowable guidance time constant. The autopilot time constant needs 5 to 6 times shorter than the guidance time constant.

Thus, the seeker lock-on range capability and the closing speed of the missile finally decide the guidance and autopilot speed of response for the interceptor missiles.

The missile trajectory is decided by the target altitude and speed and also the time of target detection and hence can be any possible trajectory within the volume of space usually called the 'kill zone' of the missile. Hence, the control system design needs to have gain adaptation feature so that the control system has adequate stability margins within the entire kill zone of the missile.

The single shot kill probability of the missile is determined by the guidance accuracy or miss distance and the warhead lethal radius (Ref 26). Faster autopilot response enables faster guidance loop and thus better miss distance even in the presence of target manoeuvres.

1.6   CLOSING REMARKS

This chapter gives brief description of the various control systems used both in launch vehicles and missiles. It describes various types of forces and moments acting on the missile and transformation scheme for transforming the forces and torques from one coordinate frame to another coordinate frame.

The chapter then discusses various mission aspects applicable for launch vehicles and medium to long range missiles and also for surface to air and other types of missiles. These aspects are discussed in sufficient details so that the requirement on the control system design are properly understood. The discussion clearly brings out that the control systems design for launch vehicle is mainly driven by the considerations of stability in the presence of structural flexibility and propellant sloshing and minimization of structural loads whereas the control systems design for missiles is mainly driven by the stability and speed of response.

A brief outline of the remaining chapters of this book is as follows:

Chapter II discusses the generalized model for structural flexibility, propellant sloshing and engine inertia effects. It also demonstrates how various models can be derived from the generalized model without much additional effort.

Chapter III discusses general aspects of control system design. These include resolution of errors, quaternion and Euler angle approach for control law formulation, different control loop configurations, various aspects related to the sizing of control systems and various considerations in planning appropriate control systems for various segments of vehicle trajectory.

Chapter IV first discusses the objectives and specifications for the flight control systems and then discusses the methodology for the control loop design for both attitude control and latax control system. It also addresses the issues related to vehicle flexibility, propellant sloshing and engine inertia effects, control gains adaptation features so that the design is applicable for any trajectory within the vehicle's capability. It also discusses the digital control systems and its advantages in making the system adaptive in nature.

Chapter V discusses the on-off reaction control system in its various phases such as within the atmosphere, out of atmosphere phase, in the presence of a constant disturbance and when the disturbance is absent and illustrates the design procedure by giving one example.

Chapter VI gives briefly the validation aspects of the flight control system design and discusses some of the flight trial experiences.

To the authors knowledge, there is no book on the flight control systems design which elaborates topics discussed in Chapter III and V. Further, the material in Chapter II will bring uniformity in modeling and the simplified model for slosh and engine gimballing will make this model a 'common sense' model rather than a mathematical exercise.

Many of the books on flight control systems mainly concentrate on the material discussed only in Chapter IV and that too in a limited manner. The author strongly believes that this book will give a first comprehensive picture of practical aspects of flight control system design for launch vehicles and missiles.

REFERENCES

1. **A.L. Greensite:** Control Theory Vol. II—Analysis and Design of Space Vehicle Flight Control Systems. Spartan Books, 1970.

2. **R.L. Pio:** Symbolic representation of coordinate transformation, IEEE Trans. On Aerospace and Navigational Electronics Vol. ANE – 11, pp. 128—134 June 1964.

3. **R.L. Pio:** Euler angle transformation, IEEE Trans on automatic control Vol. II No.4, Oct 1966.

4. **B.A. Appleby, T.E. Reed:** Dynamic stability of space vehicles Vol. VIII – Atmospheric disturbances that affect flight control systems, NASA CR 942, 1967.

5. **K.L. Handoo, N. Somarajan:** Analysis of Madras Upper Atmosphere wind, data and construction of synthetic wind profiles for SLV-3 (first report), STR(D)-VA-003/77-78, VSSC, Trivandrum, 1977.

6. **Atul Nautiyal:** Analysis of winds over Balasore and construction of synthetic wind profiles DRDL.6100.1011.000, March 1993.

7. **J.F. White:** Flight performance handbook for "Powered flight operations", John Wiloy & Sons, New York, 1963.

8 **Atul Nautiyal:** Computer Program for optimum vehicle sizing and sensitivity coefficients, DRDL.6100.1023.000 March 1995.

9. **V. Adimurthy:** Launch Vehicle Trajectory Optimization, Acta Astronautica. Vol. 15, 1987 pp. 845–850.

10. **S. Vathsal, K.A.P. Menon, R. Swaminathan:** Minimax approach to trajectory optimization to

multistage launch vehicles, IEEE trans. on Aerospace and Electronic Systems. Vol. AES – 13, No. 2, Mar 1977 pp. 179–187.

11. **N.V. Kadam:** Report of the Committee on control system analysis and specifications for ASLV, VSSC-ASLV-TN-25-81, Oct. 1981.

12. **A.D. Murthy, N.V. Kadam, E.M. George:** Studies on fin requirement for ASLV, VSSC/CGD/TM/01/81 Apr. 1981

13. **N.V. Kada, A.D. Murthy:** A note on the nozzle cant angle for ASLV, VSSC/CGD/TR/01/81 Feb. 1981.

14. **N.V. Kadam:** Effect of differential cant angle for ASLV, VSSC/CGD/TN/01/82, Jan. 1982.

15. **Robert Harris:** Analysis and design of space vehicle flight control systems, Vol. XIV : Load relief-NASA CR-833.

16. **R.F. Hoelkar** Theory of artificial stabilization of missile and space vehicles with exposition of four control principles, NASA TN D – 555, June 1961.

17. **M.H. Rheinfurth:** The alleviation of aerodynamic loads on rigid vehicles, NASA - TM X 53397

18. **D.L. ST. John:** Simplified criteria for minimum load and drift control of large booster vehicles, NASA CR - 74461.

19. **D.K. Avasthi, N.V. Kadam:** 6 DOF Simulation study of Prithvi TVC phase control system, DRDL.3300.1004.513, 31 Oct. 1985.

20. **Frank J. Reagan, Satya M. Anandakrishnan:** Dynamics of Atmospheric Re-entry, AIAA Education Series, 1992.

21. **A. Nautiyal, S.A. Warrier:** An explicit guidance scheme for oblate earth gravitational field, DRDL.6100.1002.000, July 1987.

22. **Tessy Thomas:** Guidance scheme for solid propelled vehicle during atmospheric phase, Defence Science Journal—Special Issue on guidance and control of missiles, Vol. 55, No. 3, Jul. 2005, pp. 253–264.

23. **N. Prabhakar:** RDTOP Program for trajectory optimisation.

24. **M. Manickavasagam:** Pitch program design for configuration A1, DRDL.6100.1034.014, Sep. 2001.

25. **Paul Zarchan:** Tactical and strategic missile guidance, Vol. 199 – Progress in Astronautics and Aeronautics AIAA Inc. Virginia- 20191-4344.

26. **J.J. Jerger:** Systems Preliminary Design, Grayson Merrill Series on Principles of Guided Missile Design, 1960.

CHAPTER 2

Generalized Equations of Motion

2.1 INTRODUCTION

Performance of a launch vehicle or a missile is evaluated using six degrees of freedom (6DOF) trajectory simulation program which incorporates all nonlinearities of control systems and external disturbances such as winds, wind gusts, misalignments etc. The 6DOF program also incorporates various levels of modeling complexities such as structural flexibility, propellant sloshing and gimballed engine dynamics.

Number of authors (Ref. 1, 2) have discussed the methods of determining the natural vibration frequencies and mode shapes for free-free beams with applications to launch vehicles. Application of finite element techniques for determination of mode shapes is also now a common practice among the professionals. Large number of authors have also discussed the control structure interaction problem in launch vehicles (Ref. 1, 2, 3, 4, 5, 6). A.L. Greensite[2] gives a detailed treatment of various aspects of modeling the structural flexibility, propellant sloshing and gimballed engine dynamics and their effect on the stability of control system for launch vehicles. Greensite's model is a linearised model using single plane flexibility and is useful for short period analysis of control system. The model is valid only when the structural mode shapes and frequencies are computed for the reduced vehicle which means the remaining vehicle after gimballed engine and slosh model masses are removed (though Ref. 2 does not explicitly mention so) from it and the motion is considered about the centre of gravity of the reduced vehicle.

Lester and Collins[4] have given a planar trajectory simulation model incorporating structural flexibility, propellant sloshing and gimballed engine inertia effects in a single plane. This is a nonlinear model and assumes that the structural modes are computed for the vehicle after removing only the gimballed engines but, propellant and slosh mode masses are assumed frozen and the motion is considered about the centre of gravity of total vehicle. K.L. Handoo[5] has developed a model for 6DOF trajectory simulation incorporating flexibility and slosh which also assumes that the modes are computed for reduced vehicle. All these models used Lagrangian approach for deriving the equations.

The dynamic equations for all the above models are highly coupled due to the facts that the mode shapes and frequencies are not computed for the total vehicle and the motion is considered about the *CG* of reduced vehicle. For example, translational mode is affected by rotational mode and elastic modes and one elastic mode is affected by all other modes as well as rigid body translational and rotational modes. Even though one may argue that the structural coupling due to engine inertia may be small, the theoretical appeal of modal orthogonality is lost. Further, the mode shape data in actual practice may be obtained for vehicle including engine (and not for reduced vehicle) but is being incorrectly (theoretically) used for the above models.

The mathematical equations given in the above formulations are highly complex and physical significance of various terms is not clear. Due to this, it is very difficult or time consuming to assess the validity of signs of various terms or eliminate the inadvertent error made while writing the equations. Further, one needs to rederive the equations if the configuration is changed or if an additional feature is to be incorporated in the model. For example, one needs a complete rederivation if one intends to incorporate propellant sloshing effect in the already available model incorporating structural flexibility.

It will be very desirable if one is able to write down the applicable equations without any difficulty by only inspecting the various terms and by interpreting them either as gimballed engine, flexibility, propellant sloshing or all of them occurring simultaneously. The present work proposes to do exactly the same. (Ref. 6,7)

Another anamoly observed by the author in many 6DOF programs is the incorrect use of the principle of conservation of momentum, *i.e.*,

$$\frac{d}{dt}(I\omega) = I\,\dot{\omega} + \dot{I}\,\omega = \text{External Torque}$$

In almost all programs came across by this authors, the $\dot{I}$ is calculated as the rate of change of moment of inertia due to burning of the propellant (and consequent change of mass). This is incorrect use of the principle of conservation of momentum since it is applicable for a closed system. However, the simulation results will not be much affected by its use since the term can very well be ignored also. The mathematical equations derived in this chapter demonstrate the correct computation of $\dot{I}$ which is due to changing the shape of the body without change of mass.

The present model is thus, a generalized model with following important and appealing features:

(*i*) The generalized model has the same form as the normal 6DOF equations of motion with some additional terms.

(*ii*) The translational, rotational and flexible structural modes are completely decoupled. The coupling occurs only through external forces and moments.

(*iii*) The same generalized equations of motion can be used to write down without much extra effort either the flexible mode equation or slosh mode equation or gimballed engine dynamic equations.

(*iv*) The translational equations can be used to write down planar vibration equations and the rotational equations can be used to write down torsional mode equations.

Thus, all these modeling features can be written down in the same setting.

(*v*) The model can be used very easily either for a single booster or multiple strapon stages. It can be very easily used to incorporate propellant sloshing and gimballed engines on strapon stages as well.

(*vi*) The model gives a very simple, lucid, easy to remember and a common sense model for writing down propellant sloshing and gimballed engine dynamics model without any compromise on accuracy. It is claimed that there is no better simplified model available in the literature than the one given here. (Ref. 6,7)

(*vii*) The model can be easily used to derive Greensite's model which uses modeshapes for reduced vehicle and which gives coupling between various modes. No great effort is required for deriving coupled mode model from the non-coupled mode model, thus, demonstrating the elegance of this generalized model.

(*viii*) The physical significance of various terms in the model is obvious. Hence, validation of the model and detection of inadvertent errors is very easy.

2.2 DERIVATION OF THE DYNAMIC EQUATIONS

The derivation uses the 'tracking' (or some times referred as 'shadow') coordinate system. The notations are as follows:

N – Origin of inertial coordinate system

O – Origin of tracking coordinate system which moves with the flight vehicle. It is the point on the body in an unflexed condition

P – Any point on the body whose position vector is given by

$$\bar{\rho} = \bar{r} + \bar{\bar{\xi}} = (x + \xi_x)i + (y + \xi_y)j + (z + \xi_z)k \qquad \ldots (2.1)$$

$$\hat{x}i + \hat{y}j + \hat{z}k$$

where $\bar{r}$ gives the position vector of the elemental mass in an unflexed condition and $\bar{\bar{\xi}}$ gives the instantaneous shift due to structural flexibility (Fig. 2.1) (or other effects as discussed subsequently).

$$\bar{\mu} = ui + vj + wk \qquad \text{is the linear velocity of 'o'}$$

$$\bar{\omega} = pi + qj + rk \qquad \text{is the angular velocity of the body.}$$

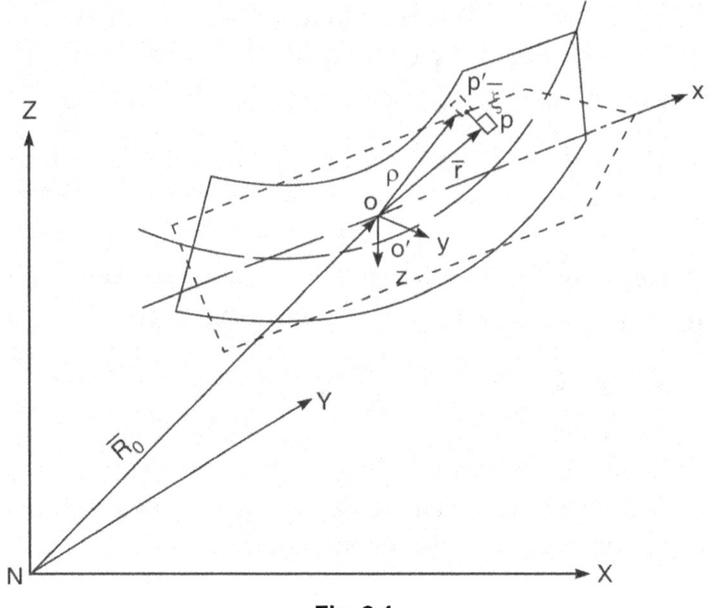

Fig. 2.1

The velocity of the elemental mass p is given by

$$\overline{v} = \overline{\mu} + \frac{d\overline{\rho}}{dt} = \overline{\mu} + \frac{\partial \overline{\rho}}{\partial t} + \overline{\omega} \times \overline{\rho} \qquad \qquad \dots (2.2)$$

The acceleration of the elemental mass is given by

$$\overline{a} = \frac{d\overline{v}}{dt}$$

$$\overline{a} = \frac{\partial \overline{\mu}}{\partial t} + \overline{\omega} \times \overline{\mu} + \frac{\partial^2 \overline{\rho}}{\partial t^2} + 2\overline{\omega} \times \frac{\partial \overline{\rho}}{\partial t} + \frac{\partial \overline{\omega}}{\partial t} \times \overline{\rho} + \overline{\omega} \times (\overline{\omega} \times \overline{\rho}) \qquad \dots (2.3)$$

The vector $\overline{r} = xi + yj + zk$ gives the position of the elemental mass in an unflexed condition.

It does not vary with time. Hence, the time derivatives of x, y and z are zero. (If origin is taken at *CG* and one accounts for variation of *CG* due to burning of propellant, appropriate derivatives for these variables may be taken)

$$\frac{\partial \overline{\rho}}{\partial t} = \hat{x}i + \hat{y}j + \hat{z}k = \xi_x i + \xi_y j + \xi_z k \qquad \text{if } x = y = z = 0 \qquad \dots (2.4)$$

$$\frac{\partial^2 \overline{\rho}}{\partial t^2} = \hat{x}i + \hat{y}j + \hat{z}j = \xi_x i + \xi_y j + \xi_z k \qquad \text{if } x = y = z = 0 \qquad \dots (2.5)$$

$$\bar{a} = \frac{\partial \bar{\mu}}{\partial t} + \bar{\omega} \times \bar{\mu} + \frac{\partial^2 \bar{\xi}}{\partial t^2} + 2\bar{\omega} \times \frac{\partial \bar{\xi}}{\partial t} + \frac{\partial \bar{\omega}}{\partial t} \times (\bar{r} + \bar{\xi}) + \bar{\omega} \times \left[\bar{\omega} \times (\bar{r} + \bar{\xi}) \right] \qquad ...(2.6)$$

if $x = y = z = x = y = z = 0$

OR

$$\begin{bmatrix} a_x \\ a_y \\ a_z \end{bmatrix} = \begin{bmatrix} u + qw - rv \\ v + ru - pw \\ w + pv - qu \end{bmatrix} + \begin{bmatrix} -(q^2 + r^2) & -r + pq & q + rp \\ r + pq & -(r^2 + p^2) & -p + rq \\ -q + pr & p + qr & -(p^2 + q^2) \end{bmatrix} \begin{bmatrix} \hat{x} \\ \hat{y} \\ \hat{z} \end{bmatrix}$$

$$+ 2 \begin{bmatrix} 0 & -r & q \\ r & 0 & -p \\ -q & p & 0 \end{bmatrix} \begin{bmatrix} \hat{x} \\ \hat{y} \\ \hat{z} \end{bmatrix} + \begin{bmatrix} \hat{x} \\ \hat{y} \\ \hat{z} \end{bmatrix} \qquad ...(2.7a)$$

If $x = y = z = 0 = x = y = z$

$$\begin{bmatrix} a_x \\ a_y \\ a_z \end{bmatrix} = \begin{bmatrix} u + qw - rv \\ v + ru - pw \\ w + pv - qu \end{bmatrix} + \begin{bmatrix} -(q^2 + r^2) & -r + pq & q + rp \\ r + pq & -(r^2 + p^2) & -p + rq \\ -q + pr & p + qr & -(p^2 + q^2) \end{bmatrix} \begin{bmatrix} {}_x x + \xi \\ {}_y y + \xi \\ {}_z z + \xi \end{bmatrix}$$

$$+ 2 \begin{bmatrix} 0 & -r & q \\ r & 0 & -p \\ -q & p & 0 \end{bmatrix} \begin{bmatrix} \xi x \\ \xi y \\ \xi z \end{bmatrix} + \begin{bmatrix} \xi x \\ \xi y \\ \xi z \end{bmatrix} \qquad ... (2.7b)$$

Multiplying by an elemental mass through out and writing $\bar{f}$ $\bar{a}dm$, eqn. 2.7b can be written as

$$\begin{bmatrix} f_x \\ f_y \\ f_z \end{bmatrix} = \begin{bmatrix} u + qw - rv \\ v + ru - pw \\ w + pv - qu \end{bmatrix} dm + \begin{bmatrix} -(q^2 + r^2) & -r + pq & q + rp \\ r + pq & -(r^2 + p^2) & -p + rq \\ -q + pr & p + qr & -(p^2 + q^2) \end{bmatrix} \begin{bmatrix} (x + \xi_x) \, dm \\ (y + \xi) \, dm \\ (z + \xi) \, dm \end{bmatrix}$$

... equation continued on next page

$$+\,2\begin{bmatrix} 0 & -r & q \\ r & 0 & -p \\ -q & p & 0 \end{bmatrix}\begin{bmatrix} \xi_x\,dm \\ \xi_y\,dm \\ \xi_z\,dm \end{bmatrix}+\begin{bmatrix} \xi_x\,dm \\ \xi_y\,dm \\ \xi_z\,dm \end{bmatrix}\qquad \ldots (2.8)$$

The angular momentum about the origin of inertial system can be written as

$$\bar{H}_0 = \int(\bar{R}_0 + \bar{\rho})\times(\bar{R}_0 + \bar{\rho})\,dm$$

The torque about the origin of inertial system is given by

$$\frac{d\bar{H}_0}{dt} = \int \bar{R}_0 \times\left(\bar{R}_0 + \bar{\rho}\right)dm + \int\bar{\rho}\times\bar{R}_0\,dm + \int\bar{\rho}\times\bar{\rho}\,dm$$

$$= \int\bar{R}_0 \times \bar{f}_1 + M\bar{\rho}_{cg}\times\bar{R}_0 + \int\bar{\rho}\times\bar{\rho}\,dm \qquad \ldots (2.9)$$

The first term on the right hand side gives the total moment of all forces acting on the body about the origin of inertial frame. The second and third terms give torque about origin of the body axes frame. Thus, the torque acting on the body is:

$$\bar{T} = M\bar{\rho}_{cg}\times\bar{R}_0 + \int(\bar{\rho}\times\bar{\rho})\,dm \qquad \ldots (2.10)$$

Writing
$$\frac{d\bar{\rho}}{dt} = \frac{\partial\bar{\rho}}{\partial t} + \bar{\omega}\times\bar{\rho}$$

$$\frac{d^2\bar{\rho}}{dt^2} = \frac{\partial^2\bar{\rho}}{\partial t^2} + 2\bar{\omega}\times\frac{d\bar{\rho}}{dt} + \frac{\partial\bar{\omega}}{\partial t}\times\bar{\rho} + \bar{\omega}\times(\bar{\omega}\times\bar{\rho}) \qquad \ldots (2.11)$$

and using equation 2.4 and 2.5

$$\frac{\partial^2\bar{\rho}}{\partial t^2} = \frac{\partial^2\bar{\xi}}{\partial t^2} - 2\frac{\partial\bar{\xi}}{\partial t}\times\bar{\omega} - (\bar{r}+\bar{\xi})\times\frac{\partial\bar{\omega}}{\partial t} + \left[\bar{\omega}\ (\bar{r}+\bar{\xi})\right]\bar{\omega} - (\bar{\omega}\cdot\bar{\omega})(\bar{r}+\bar{\xi}) \quad \ldots (2.12)$$

Using Equation (2.12) and noting that

$$\bar{R}_o = \begin{bmatrix} u + qw - rv \\ v + ru - pw \\ w + pv - qu \end{bmatrix} = \begin{bmatrix} A_x \\ A_y \\ A_z \end{bmatrix} \qquad \ldots (2.13)$$

The torque equation can be written as

$$\begin{bmatrix} t_x \\ t_y \\ t_z \end{bmatrix} = \begin{bmatrix} 0 & -\hat{z} & \hat{y} \\ \hat{z} & 0 & -\hat{x} \\ -\hat{y} & \hat{x} & 0 \end{bmatrix} \begin{bmatrix} A_x \\ A_y \\ A_z \end{bmatrix} dm + \begin{bmatrix} y\,\xi_z - z\,\xi_y \\ z\,\xi_x - x\,\xi_z \\ x\,\xi_y - y\,\xi_x \end{bmatrix} dm + \begin{bmatrix} \xi_y\,\xi_z - \xi_z\,\xi_y \\ \xi_z\,\xi_x - \xi_x\,\xi_z \\ \xi_x\,\xi_y - \xi_y\,\xi_x \end{bmatrix} dm$$

$$+\, 2\,dm \begin{bmatrix} \hat{z}\,\xi_z + \hat{y}\,\xi_y & -\hat{y}\,\xi_x & -\hat{z}\,\xi_x \\ -\hat{x}\,\xi_y & \hat{z}\,\xi_z + \hat{x}\,\xi_x & -\hat{z}\,\xi_y \\ -\hat{x}\,\xi_z & -\hat{y}\,\xi_z & \hat{y}\,\xi_y + \hat{x}\,\xi_x \end{bmatrix} \begin{bmatrix} p \\ q \\ r \end{bmatrix}$$

$$+\, dm \begin{bmatrix} \hat{z}^2 + \hat{y}^2 & -\hat{x}\hat{y} & -\hat{x}\hat{z} \\ -\hat{x}\hat{y} & \hat{z}^2 + \hat{x}^2 & -\hat{y}\hat{z} \\ -\hat{x}\hat{z} & -\hat{y}\hat{z} & \hat{y}^2 + \hat{x}^2 \end{bmatrix} \begin{bmatrix} p \\ q \\ r \end{bmatrix} + dm \begin{bmatrix} \hat{y}^2 - \hat{z}^2 & \hat{x}\hat{y} & -\hat{x}\hat{z} \\ -\hat{x}\hat{y} & \hat{z}^2 - \hat{x}^2 & -\hat{y}\hat{z} \\ \hat{x}\hat{z} & -\hat{y}\hat{z} & \hat{x}^2 - \hat{y}^2 \end{bmatrix} \begin{bmatrix} qr \\ rp \\ pq \end{bmatrix}$$

$$+\, dm \begin{bmatrix} 0 & \hat{y}\hat{z} & -\hat{y}\hat{z} \\ -\hat{x}\hat{z} & 0 & \hat{x}\hat{z} \\ \hat{x}\hat{y} & -\hat{x}\hat{y} & 0 \end{bmatrix} \begin{bmatrix} q^2 + r^2 \\ r^2 + p^2 \\ p^2 + q^2 \end{bmatrix} \qquad \ldots (2.14)$$

Now denoting

(i) $\hat{i}_{xx} = (\hat{y}^2 + \hat{z}^2)\,dm,\ \hat{i}_{yy} = (\hat{x}^2 + \hat{z}^2)\,dm,\ \hat{i}_{zz} = (\hat{x}^2 + \hat{y}^2)\,dm,$

$\hat{i}_{xy} = \hat{x}\hat{y}\,dm \qquad \hat{i}_{yz}\ \hat{y}\hat{z}\,dm \qquad \hat{i}_{zx} = \hat{z}\hat{x}\,dm,$ ... (2.15)

(ii) $\hat{i}_{zz} - \hat{i}_{yy}\ (\hat{y}^2 - \hat{z}^2)\,dm,$ since $(\hat{y}^2 - \hat{z}^2)\,dm\ (\hat{y}^2\ \hat{x}^2)\,dm - (\hat{x}^2\ \hat{z}^2)\,dm$

$\hat{i}_{xx} - \hat{i}_{zz}\ (\hat{z}^2 - \hat{x}^2)\,dm$

$\hat{i}_{yy} - \hat{i}_{zz}\ (\hat{z}^2 - \hat{y}^2)\,dm$ (2.16)

(iii) $\dfrac{d}{dt}(\hat{i}_{xx}) = \dfrac{d(\hat{y}^2 + \hat{z}^2)}{dt}\,dm = 2(\hat{y}\,\xi_y + \hat{z}\,\xi_z)\,dm$

$\dfrac{d}{dt}(\hat{i}_{yy}) = 2(\hat{x}\,\xi_x + \hat{z}\,\xi_z)\,dm$

$\dfrac{d}{dt}(\hat{i}_{zz}) = 2(\hat{x}\,\xi_x + \hat{y}\,\xi_y)\,dm$ ... (2.17)

$$(iv) \quad \frac{d}{dt}(\hat{i}_{xy}) = (\hat{x}\,\xi_y + \hat{y}\,\xi_x)\,dm \quad \cdot \quad \frac{d}{dt}(\hat{i}_{xyp}) \quad \frac{d}{dt}(\hat{i}_{xyq}) \qquad \ldots (2.18\ a)$$

Definition: $\dfrac{d}{dt}(\hat{i}_{xyp})$ is the component of $\dfrac{d}{dt}(\hat{i}_{xy})$ multiplying angular velocity p in the dynamic equation. As it is clear that any variation of x component along the p direction does not contribute to change in angular momentum and therefore differentiation should be done only for the coordinate normal to p. Similarly, differentiation of $\dfrac{d}{dt}(\hat{i}_{xyq})$ should be done for the coordinate normal to q.

$$\text{Similarly,} \qquad \frac{d}{dt}(\hat{i}_{yz}) = \frac{d}{dt}(\hat{i}_{yzq}) + \frac{d}{dt}(\hat{i}_{yzr}) = (\hat{y}\,\xi_z + \hat{z}\,\xi_y)\,dm \qquad .. (2.18\ b)$$

$$\frac{d}{dt}(\hat{i}_{xz}) = \frac{d}{dt}(\hat{i}_{xzp}) + \frac{d}{dt}(\hat{i}_{xzr}) = (\hat{x}\,\xi_z + \hat{z}\,\xi_x)\,dm \qquad \ldots (2.18\ c)$$

In order to understand the significance of the second and third term on right hand side of Equation 2.14, consider an elemental disc of the structure (Fig. 2.2) normal to x-axis.

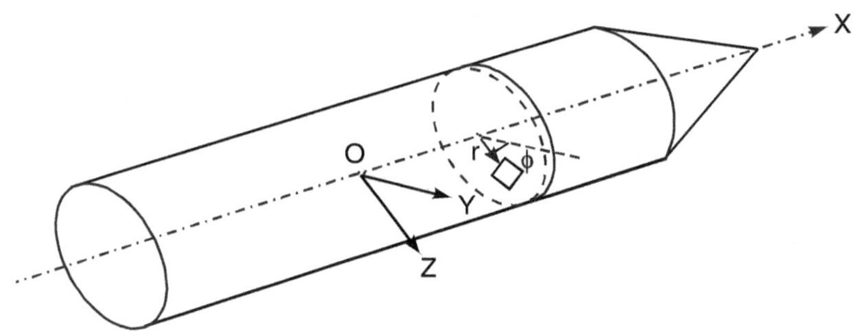

Fig. 2.2 Elemental disc normal to X-axis

The point (y, z) is given by:

$$y \quad r\cos\ ; \quad z \quad r\sin$$

$$dy \quad _y \quad -r\sin\ d\ ; \quad dz \quad _z \quad r\cos\ d$$

$$= -z\,d\phi \quad ; \quad = y\,d\phi \qquad \ldots (2.19)$$

Let $\quad _x \quad d \quad$ small angular perturbation

Then $\quad _y \quad -z\ _x$ and $\quad _z \quad y\ _x \qquad \ldots (2.20)$

Therefore, $\qquad (y\,\xi_z - z\,\xi_y)\,dm = (y^2 + z^2)\,dm\,\eta_x$

By integrating over the disc, one gets

$$\int_{disc} (y\,\xi_z - z\,\xi_y)\,dm = i_{xx}\,\eta_x \qquad \text{... (2.21 a)}$$

where i_{xx} indicates the inertia of the elemental disc. In a similar manner

$$\int_{slice} (z\,\xi_z - x\,\xi_z)\,dm = i_{yy}\,\eta_y \qquad \text{... (2.21 b)}$$

and

$$\int_{slice} (x\,\xi_y - y\,\xi_x)\,dm = i_{zz}\,\eta_z \qquad \text{... (2.21 c)}$$

Thus, the second term on RHS of equation 2.14 is given by:

$$\begin{bmatrix} i_{xx}\,\eta_x \\[6pt] i_{yy}\,\eta_y \\[6pt] i_{zz}\,\eta_z \end{bmatrix}$$

The third term on RHS of Eqn. 2.14 is given by

$$(\xi_y\,\xi_z - \xi_z\,\xi_y)\,dm = (-zy\eta_x\,\eta_x + yz\eta_x\,\eta_x)\,dm = 0 \qquad \text{... (2.22a)}$$

Similarly,
$$(\xi_z\,\xi_x - \xi_x\,\xi_z)\,dm = 0 \qquad \text{... (2.22b)}$$

And
$$(\xi_x\,\xi_Y - \xi_Y\,\xi_x)\,dm = 0 \qquad \text{... (2.22c)}$$

Thus, the third term on RHS of Equation 2.14 becomes zero.

It may be noted here that we have considered here angular perturbation as due to some sort of torsion in which case both y and z coordinates are affected by perturbation of about x-axis. Thus, the contribution to above terms due to structural torsion reduces to zero. This term will be selectively retained in the formulation to account for the gimballed engines as discussed in subsequent sections.

Now, let the vibration behaviour of the entire structure be represented by three translational vibrations (along x, y and z axes) and three torsional vibrations (about x, y and z axis), then

$$\overline{} = \sum_{i=1}^{n} \overline{}_i\, q_i \qquad \text{... (2.23)}$$

$$\overline{} = \sum_{i=1}^{n} \overline{}_i\, q_i \qquad \text{... (2.24)}$$

where
$$\overline{\Phi}_i = (\phi_{xi}\ \phi_{yi}\ \phi_{zi})^T$$

$$\overline{\Psi}_i = (\psi_{xi}\ \psi_{yi}\ \psi_{zi})^T \qquad \text{... (2.25)}$$

where ϕ_{xi} = Longitudinal vibration mode (pogo type)

ϕ_{yi} = Yaw plane bending mode

ϕ_{zi} = Pitch plane bending mode

ψ_{xi} = **Torsion mode about** x-axis

ψ_{yi} = Torsion mode about y-axis

ψ_{zi} = Torsion mode about z-axis

Now, denoting structural stress related force by $K_1\bar{\xi}_e$ and torque by $K_2\bar{\eta}_e$, the generalized equation of motion can be written as (K_1, K_2, K are appropriate stiffness matrices)

$$
\begin{bmatrix} f_x \\ f_y \\ f_z \\ t_x \\ t_y \\ t_z \end{bmatrix}
=
\begin{bmatrix}
dm & 0 & 0 & 0 & \hat{z}\,dm & -\hat{y}\,dm \\
0 & dm & 0 & -\hat{z}\,dm & 0 & \hat{x}\,dm \\
0 & 0 & dm & \hat{y}\,dm & -\hat{x}\,dm & 0 \\
0 & -\hat{z}\,dm & \hat{y}\,dm & \hat{i}_{xx} & -\hat{i}_{xy} & -\hat{i}_{xz} \\
\hat{z}\,dm & 0 & -\hat{x}\,dm & -\hat{i}_{xy} & \hat{i}_{yy} & -\hat{i}_{yz} \\
-\hat{y}\,dm & \hat{x}\,dm & 0 & -\hat{i}_{xz} & -\hat{i}_{yz} & \hat{i}_{zz}
\end{bmatrix}
\begin{bmatrix} A_x \\ A_y \\ A_z \\ p \\ q \\ r \end{bmatrix}
$$

$$
+
\begin{bmatrix}
0 & \hat{z}\,dm & \hat{y}\,dm & -\hat{x}\,dm & 0 & 0 \\
\hat{z}\,dm & 0 & \hat{x}\,dm & 0 & -\hat{y}\,dm & 0 \\
\hat{y}\,dm & \hat{x}\,dm & 0 & 0 & 0 & -\hat{z}\,dm \\
\hat{i}_{zz}-\hat{i}_{yy} & \hat{i}_{xy} & -\hat{i}_{xz} & 0 & \hat{i}_{yz} & -\hat{i}_{yz} \\
-\hat{i}_{xy} & \hat{i}_{xx}-\hat{i}_{zz} & \hat{i}_{yz} & -\hat{i}_{xz} & 0 & \hat{i}_{xz} \\
\hat{i}_{xz} & -\hat{i}_{yz} & \hat{i}_{yy}-\hat{i}_{xx} & \hat{i}_{xy} & -\hat{i}_{xy} & 0
\end{bmatrix}
\begin{bmatrix} qr \\ rp \\ pq \\ q^2+r^2 \\ r^2+p^2 \\ p^2+q^2 \end{bmatrix}
$$

$$
+2
\begin{bmatrix}
0 & \xi_z\,dm & -\xi_y\,dm \\
-\xi_z\,dm & 0 & \xi_x\,dm \\
\xi_y\,dm & -\xi_x\,dm & 0 \\
\frac{1}{2}\frac{d}{dt}(\hat{i}_{xx}) & -\frac{d}{dt}(\hat{i}_{xyq}) & -\frac{d}{dt}(\hat{i}_{xzr}) \\
-\frac{d}{dt}(\hat{i}_{xyp}) & \frac{1}{2}\frac{d}{dt}(\hat{i}_{yy}) & -\frac{d}{dt}(\hat{i}_{yzr}) \\
-\frac{d}{dt}(\hat{i}_{xzp}) & -\frac{d}{dt}(\hat{i}_{yzq}) & \frac{1}{2}\frac{d}{dt}(\hat{i}_{zz})
\end{bmatrix}
\begin{bmatrix} p \\ q \\ r \end{bmatrix}
+
\begin{bmatrix} \xi_x\,dm \\ \xi_y\,dm \\ \xi_z\,dm \\ \eta_x\,i_{xx} \\ \eta_y\,i_{yy} \\ \eta_z\,i_{zz} \end{bmatrix}
+ K
\begin{bmatrix} \xi_{xe} \\ \xi_{ye} \\ \xi_{ze} \\ \eta_{xe} \\ \eta_{ye} \\ \eta_{ze} \end{bmatrix}
$$

$$... (2.26)$$

2.3 A NOVEL METHOD OF INCORPORATING STRUCTURAL FLEXIBILITY, PROPELLANT SLOSHING AND GIMBALLED ENGINE DYNAMICS IN 6DOF EQUATIONS OF MOTION

The method of incorporating structural flexibility, propellant slosh and gimballed engine motion is illustrated here for a vehicle having two gimballed engines and flexible modes in pitch plane and yaw plane.

Fig. 2.3 shows the assumed convention for mode shapes and positive engine deflection. Positive angles give mass displacement towards positive body axes. The notations are:

$$\delta_{p_k} = k\text{th engine deflection in pitch plane}$$

$$\delta_{y_k} = k\text{th engine deflection in yaw plane}$$

$$\lambda_{p_j} = j\text{th slosh pendulum angle in pitch plane}$$

$$\lambda_{y_j} = j\text{th slosh pendulum angle in yaw plane}$$

$$M_{s_j} = j\text{th mode slosh mass}$$

$$L_{s_j} = \text{Pendulum length for } j\text{th mode}$$

$$x_{s_j} = \text{Location of } j\text{th slosh mass when } \lambda_{p_j} \text{ and } \lambda_{y_j} \text{ are zero}$$

$$L_c = \text{Distance of engine gimbal point along } x \text{ axis of body axes system}$$

$$L_a = \text{Distance of gimbal point from roll axis}$$

$$L_R = \text{Distance of engine CG from gimbal point}$$

$$I_R = \text{Engine moment of inertia about gimbal point}$$

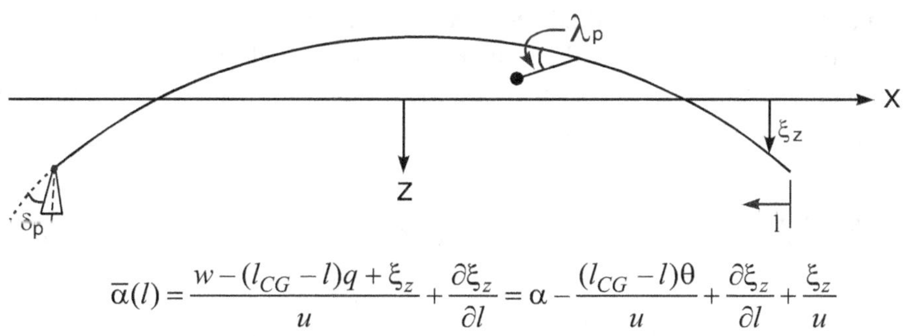

$$\bar{\alpha}(l) = \frac{w - (l_{CG} - l)q + \xi_z}{u} + \frac{\partial \xi_z}{\partial l} = \alpha - \frac{(l_{CG} - l)\theta}{u} + \frac{\partial \xi_z}{\partial l} + \frac{\xi_z}{u}$$

(a) Bending in pitch plane

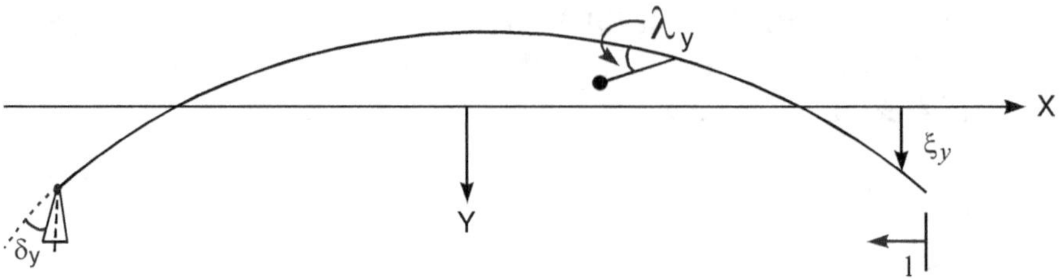

$$\bar{\beta}(l) = \frac{v + (l_{CG} - l)r + \xi_y}{u} + \frac{\partial \xi_y}{\partial l} = \beta + \frac{(l_{CG} - l)\psi}{u} + \frac{\partial \xi_y}{\partial l} + \frac{\xi_y}{u}$$

y yi q_i

(b) Bending in yaw plane

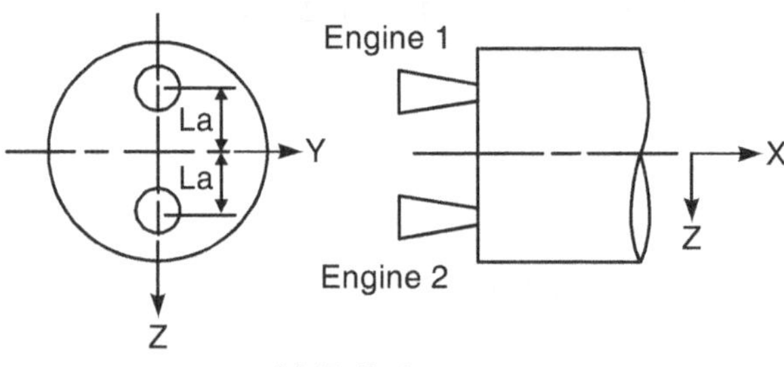

(c) Roll plane

Fig. 2.3

In equation 2.26, the perturbation in the position of any mass element is indicated by $\bar{}$ but it does not say anything about the reason for this perturbation.

It is now clarified that a mass element on the vehicle may be shifted from its nominal position due to following factors either alone or more than one factor operating simultaneously. Thus, the propellant slosh mass will be shifted due to vehicle bending and slosh motion simultaneously. Any element of mass of the gimballed engine will be shifted due to vehicle bending and engine motion occurring simultaneously. All other elemental masses of the vehicle will be perturbed by only vehicle bending.

Thus, $\xi_x(engine) = \sum\limits_{i=1}^{n} \phi_{xi}(e)q_i + l\left[1 - \cos\left(\sqrt{\delta_p^2 + \delta_y^2}\right)\right]$

$$_{xi}(e)q_i \quad l \quad \frac{_p^2 \quad _y^2}{2} \qquad l = \text{distance from gimbal point} \quad \dots (2.27)$$

$$_y(engine) \quad \sum\limits_{i\,1}^{n} \,_{yi}(e)\,q_i \quad l \quad _y \qquad l = \text{distance from hinge point} \quad \dots (2.28)$$

$$\xi_z(engine) = \sum_{i=1}^{n} \phi_{zi}(e)\, q_i + l\delta_p \qquad\qquad\qquad .. (2.29)$$

and slosh mass

$$\xi_x(s_j) = \sum_{i=1}^{n} \phi_{xi}(s_j) q_i + L_{s_j}\left(\frac{\lambda_{pj}^2 + \lambda_{yj}^2}{2}\right) \qquad ... (2.30)$$

$$\xi_y(s_j) = \sum_{i=1}^{n} \phi_{yi}(s_j) q_i + L_{s_j}\lambda_{yj} \qquad\qquad ... (2.31)$$

$$\xi_z(s_j) = \sum_{i=1}^{n} \phi_{zi}(s_j) q_i + L_{s_j}\lambda_{pj} \qquad\qquad ...(2.32)$$

If higher order terms of small quantities are ignored then contribution to $\quad_x$ due to second order engine and slosh motion can be ignored.

Thus, the same equations can be used to incorporate structural flexibility, propellant sloshing and/or gimballed engine dynamics without any extra effort in derivation. Further, the perturbation in the position of mass elements can be written down easily using the conventions adopted. There are no further changes in the model. Hence, validation of the model is very easy and the correctness of signs of various terms can be quickly assessed.

Rest of this chapter illustrates the applications of this generalized model for deriving various equations.

2.4 APPLICATIONS OF THE GENERALIZED MODEL

2.4.1 One Plane Vibrations of a Uniform Free-Free Beam

Consider vibrations in x-z plane and centre of gravity (CG) lying on x-axis. Then

$$\int \hat{y}\, dm = \int \hat{z}\, dm = \xi_x = \xi_x = p = r = 0 \qquad ... (2.33)$$

The third row of Equation 2.26 gives,

$$\xi_z\, dm + K\xi_z - \hat{x}q\, dm + A_z\, dm = f_z \qquad ... (2.34)$$

The elastic vibration equation for a slender beam is given by

$$\frac{\partial^2 \xi}{dt^2} + \frac{\partial^2}{\partial x^2}\left(EI\frac{\partial^2 \xi}{\partial x^2} \right) = 0$$

This has one zero frequency root and other infinite frequencies (Ref. 2). The modal shape for the zero frequency is given by

$$\quad_0 \quad a_0 \quad a_1 x \quad \text{if } x \text{ is measured from origin} \qquad ... (2.35)$$

or $\qquad\qquad\qquad _0 \quad b_1 \quad b_2 l \quad$ If l is measured from CG

This can be treated as two modes with zero frequency (rigid body modes)

$$\phi_{01} = b_1 \qquad \text{translational mode} \qquad\qquad \text{... (2.36)}$$

$$\phi_{02} = b_2 l \qquad \text{Rotational mode}$$

Then

$$\sum_{i\ 0}\ \phi_i q_i$$

$$\phi_{01} q_{01} \quad \phi_{02} q_{02} \quad \sum_{i\ 1} \phi_i q_i \qquad\qquad \text{... (2.37)}$$

Use Eqn. (2.37) in Eq. (2.35), multiply by ϕ_{01}, integrate over the beam length and use the orthogonality property *i.e.*,

$$\int \phi_i \phi_j \, dm \quad 0 \quad i \quad j \qquad\qquad \text{... (2.38)}$$

$$m_i \quad \text{for} \quad i \quad j$$

Then

$$\int b^2_{\ 1} q_{01} \, dm + \omega_{01} \int b_1^2 dm = \int f_z b_1$$

$$M b_1 q_{01} = F_z, \text{ since } \phi_{01} \quad 0$$

or

$$b_1 q_{01} = \frac{F_z}{M} = a_z \ \textit{i.e.}, \text{ acceleration of } CG \qquad\qquad \text{... (2.39)}$$

Similarly, for ϕ_{02} one gets

$$q_{02} b_2^2 \int l^2 dm + 0 = b_2 \int f_z l \qquad\qquad \phi_{02} \quad 0$$

or

$$b_2 q_{02} = \frac{-T_y}{I_y} = -q = -\theta \qquad\qquad \text{... (2.40)}$$

Now, putting equations 2.37, 2.39 and 2.40 in Equation 2.34, one gets

$$\xi dm + a_z \, dm - l\theta \, dm + \omega^2 \xi \, dm = f_z \qquad\qquad \text{... (2.41)}$$

where ϕ now consists of only non zero frequency flexible modes

If origin is considered any point other than *CG*, use

$$a_z - l\theta = A_z - x\theta$$

Then Eqn. 2.41 gives exactly the same Eqn. (2.34) as given by generalized equation illustrating that the generalized equation can be used to get the vibration equation. Further, the above derivation clarifies that the two zero frequency modes generally obtained in modal frequency estimation programs are the linear acceleration and angular accelerations of the vehicle. If we are not interested in the elastic modes of vibrations, only the zero frequency modes are retained which then give the normal rigid body 6DOF equations of motion.

2.4.2 Force Equations

Following configuration is assumed for the development of force equation. The equation can be extended to any other configuration quite easily.

1. Number of gimballed engines $= 2$ (See Fig. 2.3)
2. Number of slosh modes $= m$
3. Number of elastic modes $= n$

It is also assumed that the elastic modes are computed assuming the engine and propellant masses are frozen at their zero deflection position. In case, the engine mounting stiffnesses are assumed to be finite values, one may get branching of modes to indicate engine position. In such case, the branch corresponding to engine needs to be considered whenever engine mass position needs to be indicated by perturbed mass location.

The condition for natural vibration gives the following conditions:

$$
\begin{array}{lll}
\phi_{xi}\,dm & \phi_{yi}\,dm & \phi_{zi}\,dm \quad 0 \\[1em]
\phi_{yi}(x)\,x\,dm & \phi_{zi}(x)\,x\,dm \quad 0 \\[1em]
\phi_{xi}(y)\,y\,dm & \phi_{zi}(y)\,y\,dm \quad 0 \\[1em]
\phi_{xi}(z)\,z\,dm & \phi_{yi}(z)\,z\,dm \quad 0
\end{array}
\qquad \dots (2.42)
$$

for $i = 1, 2 \dots\dots n$

where ϕ_{xi}, ϕ_{yi} and ϕ_{zi} indicate the ith elastic mode shape.

Fig 2.4 illustrates various modes indicated in Eq. 2.42 in which a general structure may vibrate in fundamental mode.

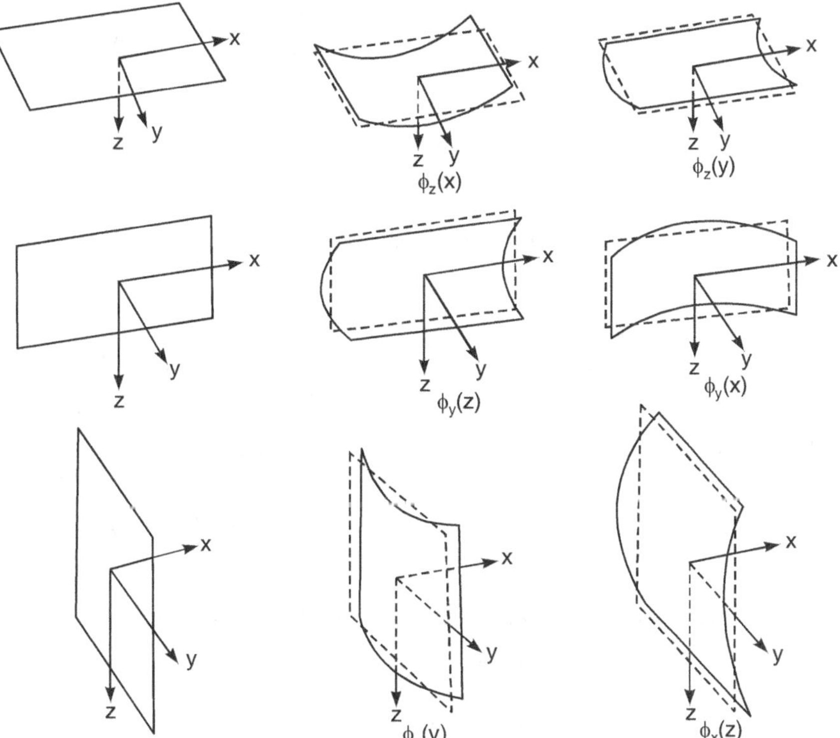

Fig. 2.4(a) Schematic of flexible modes

Force equation is then obtained by integrating Eqn. (2.8) [or the force equations in Eqn. 2.26] over the body and noting that

$$\int K_1 \begin{bmatrix} \xi_{xe} \\ \xi_{ye} \\ \xi_{ze} \end{bmatrix} = 0 \qquad \dots (2.43)$$

$$\begin{bmatrix} F_x \\ F_y \\ F_z \end{bmatrix} = \begin{bmatrix} A_x \\ A_y \\ A_z \end{bmatrix} M + \begin{bmatrix} -(q^2 + r^2) & -r + pq & q + pr \\ r + pq & -(r^2 + p^2) & -p + rq \\ -q + pr & p + rq & -(p^2 + q^2) \end{bmatrix} \begin{bmatrix} Mx_{cg} + \int \xi_x dm \\ My_{cg} + \int \xi_y\, dm \\ Mz_{cg} + \int \xi_z\, dm \end{bmatrix}$$

$$+ 2 \begin{bmatrix} 0 & -r & q \\ r & 0 & -p \\ -q & p & 0 \end{bmatrix} \begin{bmatrix} \int \xi_x dm \\ \int \xi_y\, dm \\ \int \xi_z\, dm \end{bmatrix} + \begin{bmatrix} \int \xi_x\, dm \\ \int \xi_y\, dm \\ \int \xi_z\, dm \end{bmatrix} \qquad \dots (2.44)$$

Using Eqn. 2.27 to 2.32 and Eqn. 2.42, one gets

$$\begin{bmatrix} F_x \\ F_y \\ F_z \end{bmatrix} = \begin{bmatrix} A_x \\ A_y \\ A_z \end{bmatrix} M + \begin{bmatrix} -(q^2 + r^2) & -r + pq & q + pr \\ r + pq & -(r^2 + p^2) & -p + rq \\ -q + pr & p + rq & -(p^2 + q^2) \end{bmatrix} \times$$

$$\begin{bmatrix} Mx_{cg} + \sum_{k=1}^{2} M_R L_R \left(\dfrac{\delta_{pk}^2 + \delta_{yk}^2}{2} \right) + \sum_{j=1}^{m} Ms_j L_{sj} \left(\dfrac{\lambda_{pj}^2 + \lambda_{yj}^2}{2} \right) \\ My_{cg} + \sum_{k=1}^{2} M_R L_R\, \delta_{yk} + \sum_{j=1}^{m} Ms_j L_{sj}\, \lambda_{yi} \\ Mz_{cg} + \sum_{k=1}^{2} M_R L_R \delta_{pk} + \sum_{j=1}^{m} Ms_j L_{sj} \lambda_{pj} \end{bmatrix}$$

$$+ 2 \begin{bmatrix} 0 & -r & q \\ r & 0 & -p \\ -q & p & 0 \end{bmatrix} \begin{bmatrix} \sum_{k=1}^{2} M_R L_R (\delta_{pk}\, \delta_{pk} + \delta_{yk}\, \delta_{yk}) + \sum_{j=1}^{m} M_{sj} L_{sj} (\lambda_{pj}\, \lambda_{pj} + \lambda_{yj} \lambda_{yj}) \\ \sum M_R L_R\, \delta_{yk} + \sum M_{sj} L_{sj}\, \lambda_{yj} \\ \sum M_R\, L_R\, \delta_{pk} + \sum M_{sj} L_{sj}\, \lambda_{pj} \end{bmatrix}$$

$$+ \begin{bmatrix} \sum M_R L_R (\delta_{pk}\delta_{pk} + \delta_{yk}\delta_{yk} + \delta_{pk}^2 + \delta_{yk}^2) + \sum M_{sj} L_{sj} (\lambda_{pj}\lambda_{pj} + \lambda_{yj}\, \lambda_{yj} + \lambda_{pj}^2 + \lambda_{yj}^2) \\ \sum M_R L_R\, \delta_{yk} + \sum M_{sj}\, L_{sj}\, \lambda_{yj} \\ \sum M_R L_R\, \delta_{pk} + \sum M_{sj}\, L_{sj}\, \lambda_{pj} \end{bmatrix}$$

$$\dots (2.45)$$

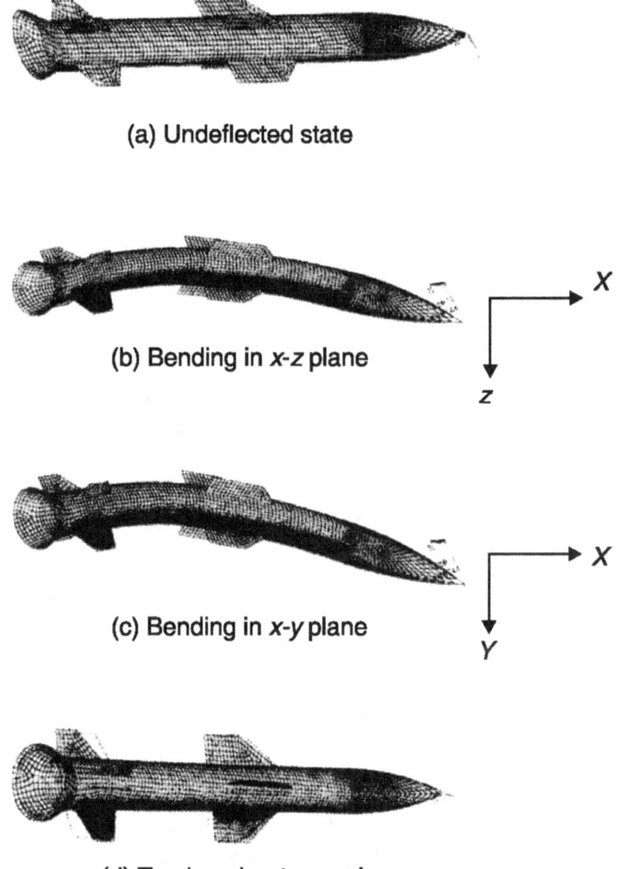

Fig. 2.4(b) Schematic of flexible modes

If one wishes to ignore the second order terms, the contribution of engine gimballing and slosh to ξ_x, ξ_x and ξ_x in F_x equation can be made zero which happens to be of second order. F_x, F_y and F_z on left hand side of Eq. 2.45 are components of the total external force acting on the body which consists of factors namely

(1) thrust (2) aerodynamics (3) gravity, and (4) control

As can be seen, Eq. 2.45 does not contain any term corresponding to structural flexibility on the right hand side. The structural flexibility will however, enter in these equations since some of the factors causing external forces such as thrust and aerodynamic forces will have components which will be functions of elastic mode displacements and velocities but none of them is a function of modal accelerations. Thus, it shows that the elastic modes and rigid body modes are totally decoupled and coupling occurs only due to external forces which are function of deformed shape of the body.

2.4.3   Torque Equations

Integrating the torque equations from the generalized Eqn. (2.26) one can write

$$
\begin{bmatrix} T_x \\ T_y \\ T_z \end{bmatrix} =
\begin{bmatrix} 0 & -\int \hat{z}\,dm & \int \hat{y}\,dm \\ \int \hat{z}\,dm & 0 & -\int \hat{x}\,dm \\ -\int \hat{y}\,dm & \int \hat{x}\,dm & 0 \end{bmatrix}
\begin{bmatrix} A_x \\ A_y \\ A_z \end{bmatrix} +
\begin{bmatrix} \hat{I}_{xx} & -\hat{I}_{xy} & -\hat{I}_{xz} \\ -\hat{I}_{xy} & \hat{I}_{yy} & -\hat{I}_{yz} \\ -\hat{I}_{xz} & -\hat{I}_{yz} & \hat{I}_{zz} \end{bmatrix}
\begin{bmatrix} p \\ q \\ r \end{bmatrix}
$$

$$
+ \begin{bmatrix} \hat{I}_{zz}-\hat{I}_{yy} & -\hat{I}_{xy} & -\hat{I}_{xz} \\ -\hat{I}_{xy} & \hat{I}_{xx}-\hat{I}_{zz} & +\hat{I}_{yz} \\ +\hat{I}_{xz} & -\hat{I}_{yz} & \hat{I}_{yy}-\hat{I}_{xx} \end{bmatrix}
\begin{bmatrix} qr \\ rp \\ pq \end{bmatrix} +
\begin{bmatrix} 0 & \hat{I}_{yz} & -\hat{I}_{yz} \\ -\hat{I}_{xz} & 0 & \hat{I}_{xz} \\ \hat{I}_{xy} & -\hat{I}_{xy} & 0 \end{bmatrix}
\begin{bmatrix} q^2+r^2 \\ r^2+p^2 \\ p^2+q^2 \end{bmatrix}
$$

$$
+ \begin{bmatrix} \dfrac{d}{dt}(\hat{I}_{xx}) & -2\dfrac{d}{dt}(\hat{I}_{xyq}) & -2\dfrac{d}{dt}(\hat{I}_{xzr}) \\[2mm] -2\dfrac{d}{dt}(\hat{I}_{xyp}) & \dfrac{d}{dt}(\hat{I}_{yy}) & -2\dfrac{d}{dt}(\hat{I}_{yzr}) \\[2mm] -2\dfrac{d}{dt}(\hat{I}_{xzp}) & -2\dfrac{d}{dt}(\hat{I}_{yzq}) & \dfrac{d}{dt}(\hat{I}_{zz}) \end{bmatrix}
\begin{bmatrix} p \\ q \\ r \end{bmatrix} +
\begin{bmatrix} \int_x \eta_x i_{xx} \\ \int_y \eta_y i_{yy} \\ \int_z \eta_z i_{zz} \end{bmatrix} +
\begin{bmatrix} \int(\xi_Y\xi_z - \xi_z\xi_y)\,dm \\ \int(\xi_z\xi_x - \xi_x\xi_z)\,dm \\ \int(\xi_x\xi_y - \xi_y\xi_x)\,dm \end{bmatrix}
$$

$$...(2.46)$$

where
$$
\begin{bmatrix} \int_x \eta_x i_{xx} \\ \int_y \eta_y i_{yy} \\ \int_z \eta_z i_{zz} \end{bmatrix} =
\begin{bmatrix} \int_x \eta_x i_{xx} \\ \int_y \eta_y i_{yy} \\ \int_z \eta_z i_{zz} \end{bmatrix}_{\text{for elastic vehicle}} +
\begin{bmatrix} \int_x (y\xi_z - z\xi_y)\,dm \\ \int_y (z\xi_x - x\xi_z)\,dm \\ \int_z (x\xi_y - y\xi_x)\,dm \end{bmatrix}_{\substack{\text{for engine and slosh} \\ \text{in deflected position}}}
\qquad ...(2.47)
$$

and
$$
\int k_2 \begin{bmatrix} \eta_{xe} \\ \eta_{ye} \\ \eta_{ze} \end{bmatrix} = 0
\qquad ... (2.48\ a)
$$

and
$$
\begin{bmatrix} \int(\xi_y\xi_z - \xi_z\xi_y)\,dm \\ \int(\xi_z\xi_x - \xi_x\xi_z)\,dm \\ \int(\xi_x\xi_y - \xi_y\xi_x)\,dm \end{bmatrix} =
\begin{bmatrix} 0 \\ 0 \\ 0 \end{bmatrix}_{\substack{\text{for elastic} \\ \text{structure} \\ \text{(eqn 2.22)}}} +
\begin{bmatrix} \int(\xi_y\xi_z - \xi_z\xi_y)\,dm \\ \int(\xi_z\xi_x - \xi_x\xi_z)\,dm \\ \int(\xi_x\xi_y - \xi_y\xi_x)\,dm \end{bmatrix}_{\substack{\text{for engine and} \\ \text{slosh in deflected} \\ \text{position}}}
\qquad ... (2.48\ b)
$$

Table 2.1 gives the expansion of various terms in the above equation. Expansion of few of the terms is illustrated on next page.

$$\int \hat{x}\, dm = \int (x + \xi_{xe})\, dm + \sum_{k=1}^{2} M_R L_R \left(\frac{\delta_{pk}^2 + \delta_{yk}^2}{2} \right) + \sum_{j=1}^{m} M_{sj} Ls_j \left(\frac{\lambda_{pj}^2 + \lambda_{yj}^2}{2} \right)$$

and $\int (x + \xi_{xe})\, dm = x_{cg} M + 0$

Thus, $$\int \hat{x}\, dm = x_{cg} M + \sum_{k=1}^{2} M_R L_R \left(\frac{\delta_{pk}^2 + \delta_{yk}^2}{2} \right) + \sum_{j=1}^{m} M_{sj} L_{sj} \left(\frac{\lambda_{pj}^2 + \lambda_{yj}^2}{2} \right) \qquad \text{...(2.49)}$$

$$\int \hat{y}\, dm = y_{cg} M + \sum_{k=1}^{2} M_R L_R \delta_{yk} + \sum_{j=1}^{m} M_{sj} L_{sj} \lambda_{yj} \qquad \text{...(2.50)}$$

$$\int \hat{z}\, dm = z_{cg} M + \sum_{k=1}^{2} M_R L_R \delta_{pk} + \sum_{j=1}^{m} M_{sj} L_{sj} \lambda_{pj} \qquad \text{...(2.51)}$$

Table 2.1: Values of Integrals

Notes:

(1) In this table, the summation ( ) is done for suffix

$i - 1$ to n (number of bending modes)

$k - 1$ to 2 (number of gimballed engines)

$j - 1$ to m (number of slosh modes)

(2) $\bar{Q}^T = \left[q_1, q_2, \ldots q_n \right]$

$$\int \hat{x}\, dm = x_{cg} M + \int \xi_x\, dm$$

$$\int \hat{y}\, dm = y_{cg} M + \int \xi_y\, dm$$

$$\int \hat{z}\, dm = z_{cg} M + \int \xi_z\, dm$$

$$\int \xi_x\, dm = \sum M_R L_R \left(\frac{\delta_{pk}^2 + \delta_{yk}^2}{2} \right) + \sum M_{sj} L_{sj} \left(\frac{\lambda_{pj}^2 + \lambda_{yj}^2}{2} \right)$$

$$\int \xi_y\, dm = \sum M_R L_R \delta_{yk} + \sum M_{sj} L_{sj} \lambda_{yj}$$

$$\int \xi_z\, dm = \sum M_R L_R \delta_{pk} + \sum M_{sj} L_{sj} \lambda_{pj}$$

The time derivatives $\ddot{}_x\, dm$ etc. can be easily obtained

3. $$\int x\xi_x dm = -\left(I_R + M_R L_R L_C\right)\sum\left(\frac{\delta_{pk}^2 + \delta_{yk}^2}{2}\right) + \sum M_{sj} x_{sj} L_{sj}\left(\frac{\lambda_{pj}^2 + \lambda_{yj}^2}{2}\right)$$

$$\int x\xi_y\, dm = -\left(I_R + M_R L_R L_C\right)\sum \delta_{yk} + \sum M_{sj} x_{sj} L_{sj}\lambda_{yj}$$

$$\int x\xi_z\, dm = -\left(I_R + M_R L_R L_C\right)\sum \delta_{pk} + \sum M_{sj} x_{sj} L_{sj}\lambda_{pj}$$

$$\int y\xi_x dm = \int y\xi_y dm = \int y\xi_z dm = 0$$

since $y = 0$ for engine and slosh masses for the given configuration. Also $z = 0$ for slosh masses.

$$\int z\xi_x dm = M_R L_R L_a\left[\frac{\delta_{p2}^2 + \delta_{y2}^2}{2} - \frac{\delta_{p1}^2 + \delta_{y1}^2}{2}\right]$$

$$\int z\xi_y dm = M_R L_R L_a\left[\delta_{y2} - \delta_{y1}\right]$$

$$\int z\xi_z dm = M_R L_R L_a\left[\delta_{p2} - \delta_{p1}\right]$$

$$\hat{I}_{xx} = \int\left[y^2 + z^2 + 2\left(y\xi_y + z\xi_z\right) + \xi_y^2 + \xi_z^2\right]\, dm$$

$$= I_{xx} + 2M_R L_R L_a\left[\delta_{p2} - \delta_{p1}\right] + \bar{Q}^T\left(M_{yy} + M_{zz}\right)\bar{Q}$$

$$+ \sum I_R\left(\delta_{yk}^2 + \delta_{pk}^2\right) + \sum M_{sj} L_{sj}^2\left(\lambda_{yj}^2 + \lambda_{pj}^2\right)$$

$$\hat{I}_{yy} = \int\left[x^2 + z^2 + 2\left(x\xi_x + z\xi_z\right) + \xi_x^2 + \xi_z^2\right].dm$$

$$= I_{yy} - 2\left[I_R + M_R L_R L_C\right]\sum\frac{\left(\delta_{pk}^2 + \delta_{yk}^2\right)}{2} + I_R\left[\frac{\left(\delta_{pk}^2 + \delta_{yk}^2\right)^2}{2} + \delta_{pk}^2\right]$$

$$+ 2\left[M_{sj} x_{sj} L_{sj}\frac{\left(\lambda_{pj}^2 + \lambda_{yj}^2\right)}{2}\right] + \sum M_{sj} L_{sj}^2\left[\left(\frac{\lambda_{pj}^2 + \lambda_{yj}^2}{2}\right)^2 + \lambda_{pj}^2\right] + \bar{Q}^T\left[M_{xx} + M_{zz}\right]\bar{Q}$$

$$\hat{I}_{zz} = \int\left[x^2 + y^2 + 2\left(x\xi_x + y\xi_y\right) + \xi_x^2 + \xi_y^2\right].dm$$

$$= I_{zz} - 2\left(I_R + M_R L_R L_C\right)\sum\frac{\left(\delta_{pk}^2 + \delta_{yk}^2\right)}{2} + I_R\left[\left(\frac{\delta_{pk}^2 + \delta_{yk}^2}{2}\right)^2 + \delta_{pk}^2\right]$$

$$+ 2\sum\left[M_{sj} x_{sj} L_{sj}\left[\frac{\left(\lambda_{pj}^2 + \lambda_{yj}^2\right)}{2}\right] + \sum M_{sj} L_{sj}^2\left[\left(\frac{\lambda_{pj}^2 + \lambda_{yj}^2}{2}\right)^2 + \lambda_{pj}^2\right]\right]$$

$$\bar{Q}^T \quad M_{xx} \quad M_{yy} \quad \bar{Q}$$

$$\hat{I}_{xy} = \int \left[xy + \left(x\xi_y + y\xi_x \right) + \xi_x \xi_y \right] dm$$

$$= I_{xy} - \left(I_R + M_R L_R L_C \right) \sum \delta_{yk} + \sum I_R \delta_{yk} \left(\frac{\delta_{pk}^2 + \delta_{yk}^2}{2} \right)$$

$$+ \sum \left[M_{sj} x_{sj} L_{sj} \lambda_{yj} + M_{sj} L_{sj}^2 \lambda_{yj} \left(\frac{\lambda_{pj}^2 + \lambda_{yj}^2}{2} \right) \right] + \overline{Q}^T M_{xy} \overline{Q}$$

$$\hat{I}_{yz} = \int \left[yz + \left(y\xi_z + z\xi_y \right) + \xi_y \xi_z \right] dm$$

$$= I_{yz} + M_R L_R L_a \left(\delta_{y2} - \delta_{y1} \right) + \sum I_R \delta_{yk} \delta_{pk} + \sum M_{sj} L_{sj} \lambda_{yj} \lambda_{pj}$$

$$\hat{I}_{xz} = \int \left[xz + \left(x\xi_z + z\xi_x \right) + \xi_x \xi_z \right] dm$$

$$= I_{xz} - \sum \left\{ \left(I_R + M_R L_R L_C \right) \delta_{pk} + M_R L_R L_a \left[\frac{\delta_{p2}^2 + \delta_{y2}^2}{2} - \frac{\delta_{p1}^2 + \delta_{y1}^2}{2} \right] \right.$$

$$\left. + I_R \sum \delta_{pk} \left(\frac{\delta_{pk}^2 + \delta_{yk}^2}{2} \right) \right\} + \sum \left[M_{sj} x_{sj} L_{sj} \lambda_{pj} + M_{sj} L_{sj}^2 \lambda_{pj} \left(\frac{\lambda_{pj}^2 + \lambda_{yj}^2}{2} \right) \right] + \overline{Q}^T M_{xz} \overline{Q}$$

$$\frac{d}{dt}\left(\hat{I}_{xyp} \right) = \int \left(x\xi_y + \xi_x \xi_y \right) dm$$

$$= \overline{Q}^T M_{xy} \overline{Q} - \sum \left(I_R + M_R L_R L_C \right) \delta_{yk} + \sum I_R \delta_{yk} \left(\frac{\delta_{pk}^2 + \delta_{yk}^2}{2} \right)$$

$$+ \sum \left\{ M_{sj} x_{sj} L_{sj} \lambda_{yj} + M_{sj} L_{sj}^2 \lambda_{yj} \left(\frac{\lambda_{pj}^2 + \lambda_{yj}^2}{2} \right) \right\}$$

$$\frac{d}{dt}\left(\hat{I}_{xyq} \right) = \int \left(y\xi_x + \xi_y \xi_x \right) dm$$

$$= \overline{Q}^T M_{xy} \overline{Q} + I_R \sum \delta_{yk} \left(\delta_{pk} \delta_{pk} + \delta_{yk} \delta_{yk} \right)$$

$$+ \sum M_{sj} L_{sj}^2 \lambda_{yj} \left[\lambda_{pj} \lambda_{pj} + \lambda_{yj} \lambda_{yj} \right]$$

$$\frac{d}{dt}\left(\hat{I}_{yzq} \right) = \int \left(y\xi_z + \xi_y \xi_z \right) dm$$

$$= \overline{Q}^T M_{yz} \overline{Q} + \sum I_R \delta_{yk} \delta_{pk} + \sum M_{sj} L_{sj}^2 \lambda_{yj} \lambda_{pj}$$

$$\frac{d}{dt}\left(\hat{I}_{yzr}\right) = \int\left(z\xi_y + \xi_z\xi_y\right)dm$$

$$= \bar{Q}^T M_{yz}\bar{Q} + M_R L_R L_a\left(\delta_{y2} - \delta_{y1}\right) + \sum I_R \delta_{pk}\delta_{yk} + \sum M_{sj}L_{sj}^2\lambda_{pj}\lambda_{yj}$$

$$\frac{d}{dt}\left(\hat{I}_{xzp}\right) = \int\left(x\xi_z + \xi_x\xi_z\right)dm$$

$$= \bar{Q}^T M_{xz}\bar{Q} - \sum\left(I_R + M_R L_R L_C\right)\delta_{pk} + I_R \delta_{pk}\left(\frac{\delta_{pk}^2 + \delta_{yk}^2}{2}\right)$$

$$M_{sj}X_{sj}L_{sj}\ \ _{pj}\qquad M_{sj}L_{sj}^2\ \ _{pj}\ \ \frac{^2_{pj}\quad ^2_{yj}}{2}$$

$$\frac{d}{dt}\left(\hat{I}_{xzr}\right) = \int\left(z\xi_x + \xi_z\xi_x\right)dm$$

$$= \bar{Q}^T M_{xz}\bar{Q} + M_R L_R L_a\left[\left(\delta_{p2}\delta_{p2} + \delta_{y2}\delta_{y2}\right) - \left(\delta_{p1}\delta_{p1} + \delta_{y1}\delta_{y1}\right)\right]$$

$$+ I_R\delta_{pk}\left[\delta_{pk}\delta_{pk} + \delta_{yk}\delta_{yk}\right] + \sum M_{sj}L_{sj}^2\lambda_{pj}\left[\lambda_{pj}\lambda_{pj} + \lambda_{yj}\lambda_{yj}\right]$$

$$\int\eta_x i_{xx} = \int\left[\hat{y}\xi_z - \hat{z}\xi_y\right]dm$$

$$= +M_R L_R L_a\left(\delta_{y1} - \delta_{y2}\right)$$

$$+ \sum I_R\left(\delta_{yk}\delta_{pk} - \delta_{pk}\delta_{yk}\right) + \sum M_{sj}L_{sj}^2\left(\lambda_{yj}\lambda_{pj} - \lambda_{pj}\lambda_{yj}\right)$$

$$\int\eta_y i_{yy} = \int\left[\hat{z}\xi_x - \hat{x}\xi_z\right]dm$$

$$= M_R L_R L_a\left(\delta_{x2} - \delta_{x1}\right) + \sum\left[\left(I_R + M_R L_R L_C\right)\delta_{pk} + I_R\left(\delta_{pk}\delta_{xk} - \delta_{xk}\delta_{pk}\right)\right]$$

$$- \sum\left[M_{sj}x_{sj}L_{sj}\lambda_{pj} + M_j L_{sj}^2\left[\lambda_{pj}\lambda_{xj} - \lambda_{xj}\lambda_{pj}\right]\right]$$

$$\text{where}\quad \delta_{x1} = \frac{\delta_{p1}^2 + \delta_{y1}^2}{2},\qquad \delta_{x2} = \frac{\delta_{p2}^2 + \delta_{y2}^2}{2},\qquad \lambda_{xj} = \frac{\lambda_{pj}^2 + \lambda_{yj}^2}{2}$$

$$\int\eta_z i_{zz} = \int\left[\hat{x}\xi_y - \hat{y}\xi_x\right]dm$$

$$= \sum\left[-\left(I_R + M_R L_R L_C\right)\delta_{yk} - I_R\left[\delta_{xk}\delta_{yk} - \delta_{yk}\delta_{xk}\right]\right]$$

$$+ \sum\left[M_{sj}X_{sj}L_{sj}\lambda_{yj} + M_{sj}L_{sj}^2\left(\lambda_{xj}\lambda_{yj} - \lambda_{yj}\lambda_{xj}\right)\right]$$

For the assumed configuration of the vehicle

  $y = 0$ for structural mass elements and gimballed engines in nominal position

  $z = 0$ for structural mass and slosh elements

    $= +L_a$ for engine 2 (see Fig 2.3(c).

    $= -L_a$ for engine 1

For a general configuration, when strapon stages are present as in case of GSLV and when the engine and slosh masses are present on strapon stages, their nominal location away from *x*-axis is to be appropriately specified by y and z as above. Thus, the same equations can be used to account for the sloshing and gimballing occurring on strapon stages also indicating the general nature of the equations of motion.

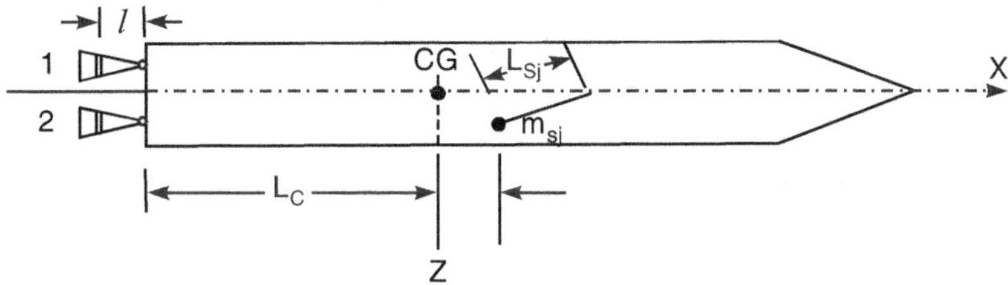

Fig. 2.5

As discussed above

$$\int \xi_x \, dm = \underset{\text{structure}}{\int \xi_x \, dm} + \underset{\substack{\text{engine} \\ \text{gimballing}}}{\int \xi_x \, dm} + \underset{\text{slosh}}{\int \xi_x \, dm}$$

$$= 0 + \sum_{k=1} M_R L_R \left(\frac{\delta_{pk}^2 + \delta_{yk}^2}{2} \right) + \sum_{j=1}^{m} M_{sj} L_{sj} \left(\frac{\lambda_{pj}^2 + \lambda_{yj}^2}{2} \right) \qquad \qquad ...(2.52)$$

$$\int x \xi_x \, dm = \int -(L_C + l) \, dm \left[l \left(\frac{\delta_{pk}^2 + \delta_{yk}^2}{2} \right) \right] + \sum_{j=1}^{m} M_{sj} \, x_{sj} \, L_{sj} \left(\frac{\lambda_{pj}^2 + \lambda_{yj}^2}{2} \right)$$

$$= - \sum_{k=1}^{2} (I_R + M_R L_R L_C) \left(\frac{\delta_{pk}^2 + \delta_{yk}^2}{2} \right) + \sum_{j=1}^{m} M_{sj} \, x_{sj} \, L_{sj} \left(\frac{\lambda_{pj}^2 + \lambda_{yj}^2}{2} \right) \qquad ...(2.53)$$

$$\int x \xi_x \, dm = - \sum_{k=1}^{2} (I_R + M_R L_R L_C) \left[\delta_{pk} \, \dot\delta_{pk} + \delta_{yk} \dot\delta_{yk} \right]$$

$$+ \sum_{j=1}^{m} M_{sj} \, x_{sj} \, L_{sj} (\lambda_{pj} \, \dot\lambda_{pj} + \lambda_{yj} \, \dot\lambda_{yj}) \qquad \qquad ... (2.54)$$

where L_R = Distance of engine *CG* from gimbal point.

Since $y = 0$ for both engine masses and slosh masses and $z = 0$ only for slosh masses

$$\int y \xi_x = \int y \xi_y \, dm = \int y \xi_z = 0$$

$$\int z \xi_z \, dm, \; \int x \xi_y \, dm, \; \int x \xi_z \, dm, \; \int z \xi_x \, dm, \; \int z \xi_y \, dm$$

are given in the table.

Now, let us consider the term $\int (\xi_y^2 + \xi_z^2) \, dm$ which occurs in

$$\hat{I}_{xx} = \int (y^2 + z^2 + 2(y\xi_y + z\xi_z) + \xi_y^2 + \xi_z^2) \, dm$$

Considering only elastic part of the structure

$$\xi_y = \sum_{i=1}^{n} \phi_{yi} q_i$$

$$\int \xi_y^2 \, dm = \int \left(\sum_{i=1}^{n} \phi_{yi} q_i \right)^2 dm = \overline{Q}_i^T (M_{yy}) \overline{Q}_i \qquad \ldots (2.55)$$

where $\qquad \overline{Q}_i = (q_1, q_2 \ldots q_n)^T$

and $\qquad M_{yy} = \begin{bmatrix} \int \phi_{y1}^2 dm & \int \phi_{y1} \phi_{y2} dm & \int \phi_{y1} \phi_{yn} dm \\ \int \phi_{y2} \phi_{y1} dm & \int \phi_{y2}^2 dm & \int \phi_{y2} \phi_{yn} dm \\ & & \\ \int \phi_{yn} \phi_{y1} dm & \int \phi_{yn} \phi_{y2} dm & \int \phi_{yn}^2 dm \end{bmatrix}$

Since the structural modes are orthogonal, $\int \phi_{yi} \phi_{yj} \, dm = 0$ for $i \neq j$

$$M_{yy} = \begin{bmatrix} \int \phi_{y1}^2 dm & 0 \\ & \\ 0 & \int \phi_{yn}^2 dm \end{bmatrix} \qquad \ldots (2.56)$$

Similarly, the elastic structural part of

$$\int \xi_y \xi_z \, dm = \int \left(\sum_{i=1}^{n} \phi_{yi} q_i \right) \left(\sum_{i=1}^{n} \phi_{zi} q_i \right) dm$$

$$\overline{Q}^T M_{yz} \overline{Q} \qquad \ldots (2.57)$$

where $\quad M_{yz} \quad$

$$\begin{bmatrix} y_1 \ z_1 \, dm & y_1 \ z_2 \, dm & & y_1 \ zn \, dm \\ y2 \ z_1 \, dm & y2 \ z2 \, dm & & \\ y_n \ z_1 \, dm & & & (\ y_{n1} \ z_n\,)dm \end{bmatrix}$$

Again noting that the modes are orthogonal

$$\int \phi_{y_i} \phi_{z_j}\, dm = 0 \ \text{ for } \ i \quad j$$

We get $\quad M_{yz} = \begin{bmatrix} \int\phi_{y_1}\phi_{z_1}\, dm & \cdots & 0 \\ & \int\phi_{y_2}\phi_{z_2}\, dm & \\ 0 & \cdots & \int(\phi_{y_n}\phi_{z_n})\, dm \end{bmatrix}$ $\qquad$ ... (2.58)

Using similar logic, all other terms in the torque equation can be written down. The gimballed engine and slosh mass contributions to the above integrals can be written down by inspection as below:

$$\int_{\substack{\text{engine} \\ \&\,\text{slosh}}} (\xi_y^2 + \xi_z^2)\, dm = \sum_{k=1}^{2} I_R(\delta_{yk}^2 + \delta_{pk}^2) + \sum_{j=1}^{m} M_{sj} L_{sj}^2 (\lambda_{yj}^2 + \lambda_{pj}^2) \qquad \text{... (2.59)}$$

$$\int_{\substack{\text{engine} \\ \&\,\text{slosh}}} \xi_y \xi_z\, dm = \sum_{k=1}^{2} I_R(\delta_{yk}\ \delta_{pk}) + \sum_{j=1}^{m} M_{sj} L_{sj}^2\ \lambda_{yj}\ \lambda_{pj} \qquad \text{... (2.60)}$$

Expansion of all other required terms has been given in Table 2.1.

Equation 2.47 can now be written as

$$\int \eta_x i_{xx} = \underbrace{\int \eta_{xe} i_{xx}}_{\substack{\text{elastic} \\ \text{structure}}} + \underbrace{\int (y\xi_z - z\xi_y)\, dm}_{\substack{\text{engine} \\ \&\,\text{slosh}}} = 0 \ + \ M_R L_R L_a (\delta_{y_1} - \delta_{y_2}) \qquad \text{... (2.61)}$$

$$y = 0 \text{ for engine and slosh}$$

$$z = 0 \text{ for slosh masses}$$

$$\int \eta_y i_{yy} = \underbrace{\int \eta_{ye} i_{yy}}_{\text{structure}} + \underbrace{\int (z\xi_x - x\xi_z)\, dm}_{\substack{\text{engine \&} \\ \text{slosh}}}$$

$$= M_R L_R L_a \left(\delta_{x_2} - \delta_{x_1}\right) + (I_R + M_R L_R L_C)\left(\delta_{P_1} + \delta_{P_2}\right) - \Sigma M_{sj} X_{sj} L_{sj} \lambda_{pj} \qquad \text{... (2.62)}$$

$$\text{where} \quad \delta x_k = \frac{\delta p_k^2 + \delta y_k^2}{2} \quad\quad K = 1, 2$$

$$\int \eta_z i_{zz} = \underbrace{\int \eta_{ze} i_{zz}}_{\text{structure}} + \underbrace{\int (x\xi_y - y\xi_x)\, dm}_{\text{engine \& slosh}}$$

$$= 0 - (I_R + M_R L_R L_C)(\delta_{y_1} + \delta_{y_2}) + \Sigma\, M_{sj} X_{sj} L_{sj} \lambda_{yj} \qquad \dots(2.63)$$

The torque equation is thus completely given without making simplifications. In applications such as rockets and missiles, the changes in moment of inertias due to elastic deflections may be quite less compared to changes in satellites and aircrafts due to large deflections in panels and wings. The equations are quite general and may be used to handle all such cases easily.

From the point of view of practical application of these moment equations in 6DOF trajectory simulation, there will be a significant increase in computational load if simplifying assumptions are not made. This is because an inversion of instantaneous inertia matrix will be needed at each integration step to solve for $(p, q, r)^T$. Computational load can be significantly reduced by neglecting off diagonal terms in the inertia matrix. The structures which are symmetric about any two of the x-y, y-z and x-z planes will have off diagonal terms $I_{xy} = I_{yz} = I_{zx} = 0$ for rigid body case. This is approximately true for most of the flight vehicles. The values will however, be disturbed to a small extent when the gimballed engine and propellant slosh masses are disturbed or deflected from their mean positions.

2.4.4   Slosh Equations

A pendulum analogy is used to represent propellant sloshing[2, 9, 10]. Ref. 2 gives the slosh equations for pitch plane and yaw plane assuming the motion occurs in one plane. Ref. 11 gives detailed derivation for slosh equations using Lagrangian approach when pitch and yaw motion simultaneously occurs and it also indicates the method of modifying the equations to suit spring mass model. It is assumed that the vehicle is having an inclination of θ w.r.t. horizontal and roll and yaw angles are zero. If the roll and yaw angles are present, the components of acceleration due to gravity need to be found along pitch, yaw and roll axes appropriately. The detailed equations are given as follows (see Appendix 2.1 for details).

$$L_{sj}\lambda_{pj} + \xi_{ze}(sj) + \left[w + pv - uq - g\cos\theta \right]$$

$$+ (-q + pr)\left[X_{sj} + L_{sj}\left(\frac{\lambda_{pj}^2 + \lambda_{yj}^2}{2} \right) + \xi_{xe}(sj) \right] + (p + qr)\left[L_{sj}\lambda_{yj} + \xi_{ye}(sj) \right]$$

$$- \left(p^2 + q^2 \right)\left(L_{sj}\lambda_{pj} + \xi_{ze}(sj) \right) + 2p\left(L_{sj}\lambda_{yj} + \xi_{ye}(sj) \right)$$

$$- 2q\left[L_{sj}\left(\lambda_{pj}\lambda_{pj} + \lambda_{yj}\lambda_{yj} \right) + \xi_{xe}(sj) \right] + \lambda_{pj} A = 0 \qquad \dots(2.64)$$

$$L_{sj}\lambda_{yj} + \xi_{ye}(sj) + (v + ru - pw) + (r + pq)\left[x_{sj} + L_{sj}\left(\frac{\lambda_{pj}^2 + \lambda_{yj}^2}{2}\right) + \xi_{xe}(sj)\right]$$

$$-(r^2 + p^2)(L_{sj}\lambda_{yj} + \xi_{ye}(sj)) + (-p + rq)\left(L_{sj}\lambda_{pj} + \xi_{ze}(sj)\right)$$

$$-2p\left(L_{sj}\lambda_{pj} + \xi_{ze}(sj)\right) + 2r(L_{sj}\left(\lambda_{pj}\lambda_{pj} + \lambda_{yj}\lambda_{yj}\right) + \xi_{xe}(sj)) + \lambda_{yj}A = 0 \qquad \text{... (2.65)}$$

where $A = (u + qw - rv + g\sin\theta)$

$$-\left(q^2 + r^2\right)\left[x_{sj} + L_{sj}\left(\frac{\lambda_{pj}^2 + \lambda_{yj}^2}{2}\right) + \xi_{xe}(sj)\right]$$

$$+(-r + pq)(L_{sj}\lambda_{yj} + \xi_{ye}(sj)) + (q + rp)(L_{sj}\lambda_{pj} + \xi_{ze}(sj))$$

$$-2r(L_{sj}\lambda_{yj} + \xi_{ye}(sj)) + 2q(L_{sj}\lambda_{pj} + \xi_{ze}(sj))$$

$$+L_{sj}(\lambda_{pj}\lambda_{pj} + \lambda_{yj}\lambda_{yj} + \lambda_{pj}^2 \; + \lambda_{yj}^2) \; + \xi_{xe}(sj)$$

$$\approx (u + qw - rv + g\sin\theta) \qquad \text{... (2.66a)}$$

or $\qquad A \quad L_{sj} \quad {}^{2}_{sj}$

Since, the slosh frequencies are given by

$$\omega_{sj}^2 = \frac{A}{L_{sj}} \qquad \text{... (2.66b)}$$

The final slosh equations can, therefore, be written by replacing acceleration 'A' in equation 2.64 and 2.65 by $L_{sj}\omega_{sj}^2$ and adding damping terms $2\zeta_{sj}\omega_{sj}\lambda_{pj}$ and $2\zeta_{sj}\omega_{sj}\lambda_{yj}$ respectively. These apparently complicated equations can be put in a very elegant form which makes it extremely easy to remember or rederive the expressions. To accomplish this, remember the following equations as explained earlier:

$$\xi_x(sj) = \xi_x(sj) + L_{sj}\left(\frac{\lambda_{pj}^2 + \lambda_{yj}^2}{2}\right)$$

$$\xi_y(sj) = \xi_{ye}(sj) + L_{sj}\lambda_{yj}$$

$$\xi_z(sj) = \xi_{ze}(sj) + L_{sj}\lambda_{pj} \qquad \text{... (2.67)}$$

Using Eq. (2.7b) and Eq. (2.67) in Eq. 2.64, 2.65 and 2.66, one can write:

Pitch plane

$$a_{z_{sj}} - g_z + \lambda_{pj}(a_{x_{sj}} - g_x) = 0 \quad \text{where } g_x \quad -g\sin \qquad \text{... (2.68a)}$$

OR

$$\boldsymbol{a_{z_{sj}} - g_z + L_{sj}\omega_{sj}^2\lambda_{pj} + 2\varsigma_{sj}\omega_{sj}\lambda_{pj} = 0} \qquad \text{... (2.68b)}$$

Yaw plane

$$a_{y_{sj}} - g_y + \lambda_{yj}(a_{x_{sj}} - g_x) = 0 \qquad \text{... (2.69a)}$$

OR
$$a_{y_{sj}} - g_y + L_{sj}\omega_{sj}^2\lambda_{yj} + 2\varsigma_{sj}\omega_{sj}\lambda_{yj} = 0 \qquad \text{... (2.69b)}$$

where g_x, g_y and g_z are components of acceleration due to gravity along body axes x, y and z. These equations can be easily written down by inspection of following Fig. 2.6.

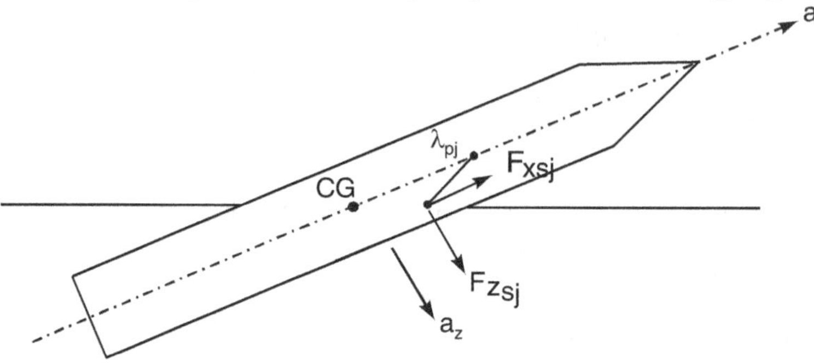

Fig. 2.6

Due to vehicle acceleration $a_{x_{sj}}$ and $a_{z_{sj}}$ at slosh location and gravity, the slosh pendulum will be stabilized by the following force balance:

$$F_{xsj} = -M_{sj}(a_{xsj} + g\sin\theta)$$

$$F_{zsj} = -M_{sj}(a_{zsj} - g\cos\theta)$$

Taking moments about slosh hinge point

$$F_{zsj}L_{sj}\cos(\lambda_{pj}) + F_{xsj}L_{sj}\sin(\lambda_{pj}) = 0$$

or
$$a_{z_{sj}} - g\cos\theta + \lambda_{pj}(a_{xsj} + g\sin\theta) = 0 \qquad \text{... (2.70)}$$

Similarly,
$$a_{y_{sj}} - g_y + \lambda_{yj}(a_{xsj} + g\sin\theta) = 0 \qquad \text{... (2.71)}$$

Eqns. 2.70 and 2.71 are the same as Eq. 2.68 and 2.69. If we substitute the expressions for a_{xsj}, a_{ysj} and a_{zsj} from Eq. 2.7(b) we get the same equations as 2.64 and 2.65.

The slosh equations can thus be very easily derived just by physical inspection of the slosh pendulum which is a great benefit of this method of derivation.

2.4.5   Gimballed Engine Dynamics and Actuator Equations

A procedure similar to the one adopted for slosh pendulum can be used for gimballed engines also. Since the engine is not a point mass, it can be considered as a compound pendulum hinged at the gimballed point. The acceleration at any point of the engine at a distance l from the gimbal point is given by eq. 7(b) where

$$x = -(L_c + l)$$

$y = L_{yk}$ = distance of gimbal point for kth engine from x-axis in y direction

$z = L_{zk}$ = distance of gimbal point for kth engine from x-axis in z direction and

$$\xi_x = \xi_x(G) + l\left(\frac{\delta_{pk}^2 + \delta_{yk}^2}{2}\right)$$

$$\xi_y = \xi_y(G) + l\left(\delta_{yk} + \frac{\partial \xi_{ye}(G)}{\partial l}\right)$$

$$\xi_z = \xi_z(G) + l\left(\delta_{pk} + \frac{\partial \xi_{ze}(G)}{\partial l}\right) \qquad \qquad \dots (2.72)$$

where G indicates gimbal point.

The inertial force acting on the engine is in a direction opposite to acceleration (*i.e.*, opposite to the force given by Eq. (2.8). The total inertial force on the engine is obtained by integration of Eq. (2.8) over the engine and setting negative sign and is given by

$$\begin{bmatrix} F_{Ix} \\ F_{Iy} \\ F_{Iz} \end{bmatrix} = -\begin{bmatrix} A_x \\ A_y \\ A_z \end{bmatrix} M_R - M_R \begin{bmatrix} -(q^2 + r^2) & -r + pq & q + rp \\ r + pq & -(r^2 + p^2) & -p + rq \\ -q + pr & p + qr & -(p^2 + q^2) \end{bmatrix} \begin{bmatrix} -(L_c + L_R) + \xi_{xe}(G) + L_R \dfrac{(\delta_{pk}^2 + d_{yk}^2)}{2} \\ L_{yk} + \xi_{yk}(G) + L_R\left(\delta_{yk} + \dfrac{\partial \xi_{ye}(G)}{\partial l}\right) \\ L_{zk} + \xi_{zk}(G) + L_R\left(\delta_{pk} + \dfrac{\partial \xi_{ze}(G)}{\partial l}\right) \end{bmatrix}$$

$$- 2M_R \begin{bmatrix} 0 & -r & q \\ r & 0 & -p \\ -q & p & 0 \end{bmatrix} \begin{bmatrix} \xi_{xe}(G) + L_R(\delta_{pk}\delta_{pk} + \delta_{yk}\delta_{yk}) \\ \xi_{ye}(G) + L_R\left(\delta_{yk} + \dfrac{\partial \xi_{ye}(G)}{\partial l}\right) \\ \xi_{ze}(G) + L_R\left(\delta_{pk} + \dfrac{\partial \xi_{ze}(G)}{\partial l}\right) \end{bmatrix}$$

$$- M_R \begin{bmatrix} \xi_{xe}(G) + L_R(\delta_{pk}\delta_{pk} + \delta_{pk}^2 + \delta_{yk}\delta_{yk} + \delta_{yk}^2) \\ \xi_{ye}(G) + L_R\left(\delta_{yk} + \dfrac{\partial \xi_{ye}(G)}{\partial l}\right) \\ \xi_{ze}(G) + L_R\left(\delta_{pk} + \dfrac{\partial \xi_{Ze}(G)}{\partial l}\right) \end{bmatrix} \qquad \dots (2.73)$$

The gravity force acting on the engine in pitch plane is given by

$$F_{Gx} \quad -M_R g \sin$$

$$F_{Gy} \quad 0$$

$$F_{Gz} \quad M_R g \cos \qquad\qquad \text{... (2.74)}$$

This assumes that the yaw and roll angles are zero. For a general case, one must find gravity force in body axes frame.

Let M_{ay} and M_{az} be the torque supplied by the actuator in x-y plane and x-z plane respectively at any instant of time. Since, the gimbal point cannot supply any torque, the sum of all torques about gimbal point must be zero. (M_{ay} and M_{az} are assumed positive in the direction of increasing $_p$ and $_y$ respectively). Then for pitch actuator.

$$M_{ay} \quad (F_{Ix} \quad F_{Gx})L_R \quad _{pk} \quad \frac{_{ze}(G)}{l} \quad (F_{Iz} \quad F_{Gz})L_R \quad 0$$

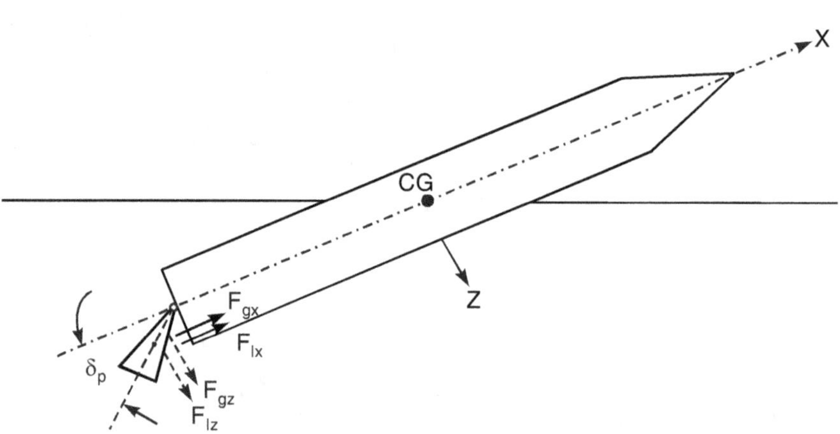

Fig. 2.7

$$M_{ay} + (F_{Iz} + F_{Gz})L_R + (F_{Ix} + F_{Gx})L_R\left(\delta_{pk} + \frac{\partial \xi_{ze}(G)}{\partial l}\right) = 0 \qquad \text{... (2.75)}$$

Let the actuator motion in the absence of vehicle motion be represented by second order equation

$$\ddot{\delta}_p + 2\zeta_n \omega_n \dot{\delta}_p + \omega_n^2 \delta_p = \omega_n^2 \delta_{pc}$$

The torque supplied by the actuator (in absence of vehicle motion) is given by

$$I_R \ddot{\delta}_p = I_R(\omega_n^2 \delta_{pc} - 2\zeta_n \omega_n \dot{\delta}_p - \omega_n^2 \delta_p) \qquad \text{... (2.76)}$$

RHS gives the torque given by actuator in terms of natural frequency. The response of actuator in the presence of inertial load due to vehicle motion is then given by equating the RHS to M_{ay}.

Using $I_R = M_R L_R^2$ and transposing, the dynamic equation for pitch actuator can be written as:

$$-I_R \delta_{pk} = M_R L_R \left\{ (w + pv - qu - g\cos\theta) \right.$$

$$-(-q + pr)\left[L_c + L_R - \xi_{xe}(G) - \frac{L_R(\delta_{pk}^2 + \delta_{yk}^2)}{2} \right]$$

$$+(p + qr)\left[L_{yk} + \xi_{ye}(G) + L_R\left(\frac{\partial \xi_{ye}(G)}{\partial l} + \delta_{yk} \right) \right]$$

$$-(p^2 + q^2)\left[L_{zk} + \xi_{ze}(G) + L_R\left(\frac{\partial \xi_{ze}(G)}{\partial l} + \delta_{pk} \right) \right]$$

$$-2q\left[\xi_{xe}(G) + L_R(\delta_{pk}\delta_{pk} + \delta_{yk}\delta_{yk}) \right]$$

$$+2p\left[\xi_{ye}(G) + L_R\left(\frac{\partial \xi_{ye}(G)}{\partial l} + \delta_{yk} \right) \right] + \left[\xi_{ze}(G) + L_R \frac{\partial \xi_{ze}(G)}{\partial l} \right] \right\}$$

$$+M_R L_R \left(\delta_{pk} + \frac{\partial \xi_{ze}(G)}{dl} \right)\left\{ u + qw - rv + g\sin\theta \right.$$

$$+\left(q^2 + r^2\right)\left[L_c + L_R - \xi_{xe}(G) - L_R\left(\frac{\delta_{pk}^2 + \delta_{yk}^2}{2} \right) \right]$$

$$+(-r + pq)\left[L_{yk} + \xi_{ye}(G) + L_R\left(\frac{\partial \xi_{ye}}{\partial l} + \delta_{yk} \right) \right]$$

$$+(q + rp)\left[L_{zk} + \xi_{ze}(G) + L_R\left(\frac{\partial \xi_{ze}}{\partial l} + \delta_{pk} \right) \right]$$

$$-2r\left[\xi_{ye}(G) + L_R\left(\frac{\partial \xi_{ye}(G)}{\partial l} + \delta_{yk} \right) \right] + 2q\left[\xi_{ze}(G) + L_R\left(\frac{\partial \xi_{ze}(G)}{\partial l} + \delta_{pk} \right) \right]$$

$$+\left[\xi_{xe}(G) + L_R\left(\delta_{pk}\delta_{pk} + \delta_{pk}^2 + \delta_{yk}\delta_{yk} + \delta_{yk}^2 \right) \right] \right\}$$

$$+I_R\left[2\zeta_n \omega_n \delta_{pk} + \omega_n^2 \delta_{pk} - \omega_n^2 \delta_{pkc} \right] \qquad \dots (2.77)$$

The equation incorporates the exact load torque expression. For ease of implementation one may neglect the third and higher order terms from the expression for load torque.

The short form of equation (2.77) is given as

$$I_R \delta_{pk} = I_R \left[-2\zeta_n \omega_n \delta_{pk} - \omega_n^2 \delta_{pk} + \omega_n^2 \delta_{pc} \right] - M_R L_R \left[a_z'(ecg) - g_z \right]$$

$$- M_R L_R \left(\delta_{pk} + \frac{\partial \xi_{ze}(G)}{\partial l} \right) (a_x(ecg) - g_x) \qquad ...(2.78)$$

where $g_x \quad -g\sin$

and $a_z(ecg) \quad$ acceleration at engine CG due to vehicle motion and flexibility when δ_{pk} is made zero.

Following a similar procedure, the yaw equation is given by

$$I_R \delta_{yk} = I_R (-2\zeta_n \omega_n \delta_{yk} - \omega_n^2 \delta_{yk} + \omega_n^2 \delta_{yc}) - M_R L_R \left[a_y'(ecg) - g_y \right]$$

$$- M_R L_R \left(\delta_{yk} + \frac{\partial \xi_{ye}(G)}{\partial l} \right) (a_x(ecg) - g_x) \qquad ... (2.79)$$

where $a_y(ecg) \quad$ acceleration at engine CG due to vehicle motion and flexibility when δ_{yk} is made zero

2.4.6 Equation for Generalized Coordinate for Structural Mode

Three different approaches can be used to get the modal coordinate equations:

1. One plane at a time

The required number of modes can be considered in each plane. The approach is similar to the one discussed in section 2.3.1 by considering one of the linear acceleration or force equation. This approach will give the simplest equations.

2. Multiple plane linear vibrations modes

Consider the translational mode to consist of 3 components *i.e.*, $\overline{\Phi} = \left[\phi_{xi}, \phi_{yi}, \phi_{zi}, \right]^T$

Premultiply the force equations with $\overline{\Phi}^T$ and integrate over the whole vehicle and use orthogonality property to get the equations for modal coordinates.

Similarly, torsional modes will consist of 3 components $\overline{\psi} = \left[\psi_{xi}, \psi_{yi}, \psi_{zi} \right]^T$. Premultiply the moment equations by $^{-T}$ and integrate over the whole body and use orthogonality property to get the equation for modal coordinates.

The required number of 3 component translational and 3 component rotational (torsional) modes may be considered to represent the vehicle elastic deflection.

3. Simultaneous linear and torsional vibrations

Consider each mode to consist of 6 components as $\overline{\Phi}^T = \left[\phi_{xi}, \phi_{yi}, \phi_{zi}, \psi_{xi}, \psi_{yi}, \psi_{zi} \right]^T$. Premultiply the generalized set of equations of eq. 2.26, integrate over the whole vehicle and use orthogonality property to get the equation of modal coordinate. The required number of 6 component modes may then be considered to represent the vehicle behaviour completely.

The three approaches will give equations of increasing complexity and generality but many of the matrices will be filled with zeros. For example, symmetric missile will have same pitch and yaw plane first mode frequency and mode shape but for general approach they will be treated as two modes. For the first mode only entries corresponding to pitch plane will have non-zero values and for the second mode only entries corresponding to yaw plane will have non-zero values.

The second and third approaches will be useful when multiple component modes or coupled vibrations exist in the structure. The second approach (with 3 component mode shape) is illustrated below. Following equations are used:

$$\int \phi_{xi}^2 \, dm = m_{xxi}, \quad \int \phi_{yi}^2 \, dm = m_{yyi}, \quad \int \phi_{zi}^2 \, dm = m_{zzi},$$

$$\int \phi_{xi} \phi_{yi} \, dm = m_{xyi}, \quad \int \phi_{xi} \phi_{zi} \, dm = m_{xzi}, \quad \int \phi_{yi} \phi_{zi} \, dm = m_{yzi},$$

$$\int \phi_{xi} \phi_{xj} \, dm = \int \phi_{xi} \phi_{yi} \, dm = \int \phi_{xi} \phi_{zj} \, dm = \int \phi_{yi} \phi_{zj} \, dm = 0$$

$$\int \phi_{yi} \phi_{yj} \, dm = \int \phi_{zi} \phi_{zj} \, dm = 0 \qquad \qquad \dots (2.80)$$

etc. all product integrals are zero for $i \neq j$

Also $\displaystyle\int x\phi_{yi} \, dm = \int x\phi_{zi} \, dm = \int y\phi_{xi} \, dm = \int y\phi_{zi} \, dm = \int z\phi_{xi} \, dm = \int z\phi_{yi} \, dm = 0$ $\quad \dots (2.81)$

and $\displaystyle\int x\phi_{xi} \, dm = 0, \ \int y\phi_{yi} \, dm = 0, \ \int z\phi_{zi} \, dm = 0,$ $\qquad \dots (2.82)$

Let $\quad \phi_{xi}(ecg), \phi_{yi}(ecg), \phi_{zi}(ecg), \qquad - \qquad$ ith mode value at engine CG

$\qquad \phi_{xi}(s_j), \phi_{yi}(s_j), \phi_{zi}(s_j), \qquad - \qquad$ ith mode value at jth slosh mass location

and Denote by

$$E_k = \begin{bmatrix} E_{xk} \\ E_{yk} \\ E_{zk} \end{bmatrix} = \begin{bmatrix} M_R L_R \left(\dfrac{\delta_{pk}^2 + \delta_{yk}^2}{2} \right) \\ M_R L_R \delta_{yk} \\ M_R L_R \delta_{pk} \end{bmatrix} \qquad \dots(2.83)$$

$$S_j = \begin{bmatrix} S_{xj} \\ S_{yj} \\ S_{zj} \end{bmatrix} = \begin{bmatrix} M_{sj}L_{sj}\left(\dfrac{\lambda_{pj}^2 + \lambda_{yj}^2}{2}\right) \\ M_{sj}L_{sj}\lambda_{yj} \\ M_R L_R \lambda_{pj} \end{bmatrix} \qquad \ldots (2.84)$$

$$W = \begin{bmatrix} 0 & -r & q \\ r & 0 & -p \\ -q & p & 0 \end{bmatrix} \qquad \ldots (2.85)$$

$$R = \begin{bmatrix} -(q^2 + r^2) & -r + pq & q + pr \\ r + pq & -(r^2 + p^2) & -p + rq \\ -q + pr & p + rq & -(p^2 + q^2) \end{bmatrix} \qquad \ldots (2.86)$$

Now, remembering that $\bar{\xi}$ consists of deflection due to three sources *i.e.*, elastic deflection $\bar{\xi}_e$, engine deflection and slosh deflection, the equation for generalized coordinate is obtained as follows:

Rewrite the elemental force equation from Eq. 2.26 (or Eqn. 2.8)

$$\begin{bmatrix} \ddot{\xi}_x \\ \ddot{\xi}_y \\ \ddot{\xi}_z \end{bmatrix} dm + K \begin{bmatrix} \xi_{xe} \\ \xi_{ye} \\ \xi_{ze} \end{bmatrix} + \begin{bmatrix} A_x \\ A_y \\ A_z \end{bmatrix} dm + \begin{bmatrix} R \end{bmatrix} \begin{bmatrix} x + \xi_x \\ y + \xi_y \\ z + \xi_y \end{bmatrix} dm + 2 \begin{bmatrix} W \end{bmatrix} \begin{bmatrix} \dot{\xi}_x \\ \dot{\xi}_y \\ \dot{\xi}_z \end{bmatrix} dm = \begin{bmatrix} f_x \\ f_y \\ f_z \end{bmatrix}$$

$$\ldots (2.87)$$

Premultiplying by $\bar{\Phi}_i^T$ and integrating over the whole body, we get

$$\left[m_{xx_i} + m_{yy_i} + m_{zz_i} \right] \ddot{q}_i + \bar{\Phi}_i^T(ecg)\sum_{k=1}^{m} E_k + \sum_{j=1}^{m}\bar{\Phi}_i^T(sj)S_j$$

$$+ \left[\int \bar{\Phi}_i^T K \bar{\Phi}_i dm \right] q_i - (q^2 + r^2)\left[m_{xx_i} + \int x\phi_{x_i} dm \right] - (r^2 + p^2)\left[m_{yy_i} + \int y\phi_{y_i} dm \right]$$

$$- (p^2 + q^2)\left[m_{zz_i} + \int z\phi_{z_i} dm \right] + \bar{\Phi}_i^T(ecg)R\sum_{k=1}^{2} E_k + \sum_{j=1}^{m}\bar{\Phi}_i(sj)RS_j$$

$$+ 2\bar{\Phi}_i^T(ecg)W\sum_{k=1}^{2} E_k + 2\bar{\Phi}_i^T(sj)WS_j = \int \bar{\Phi}_i^T \bar{F} dl \qquad \ldots (2.88)$$

This gives the equation for the generalized coordinate assuming 3 component planar elastic mode.

Torsional modal equation can be similarly written by premultiplying the last three equations from the set (2.26) by $^{-T}$ and integrating over the whole body.

2.4.7 External Forces And Moments

For the structures under consideration such as launch vehicles and missiles, following external forces and moments are considered:

   (*i*) Propulsion   (*ii*) Aerodynamic

(*iii*) Gravitational   (*iv*) Control

Section 1.4 of chapter I gives these external forces for a rigid body vehicle. When the vehicle structure is considered flexible and experiences deformation of shape it will give rise to additional components of forces and moments. Formulation of such modified forces is discussed in the present section.

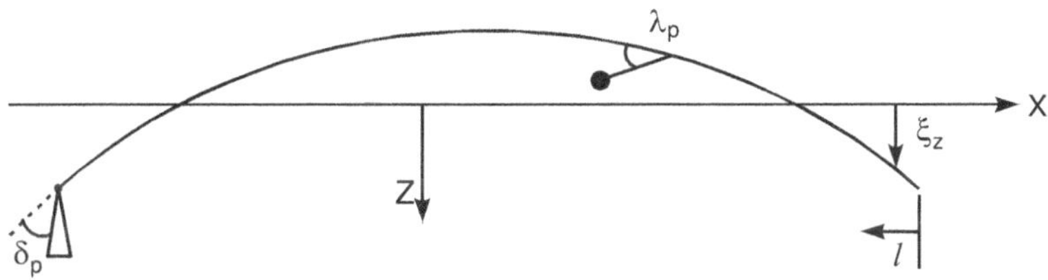

Fig. 2.8

The thrust is assumed to act at gimballing point for the gimballed engine or the throat of the nozzle for the solid propellant rockets.

The aerodynamic force is distributed over the entire structure. For the rigid body case, vehicle is assumed to have a single angle of attack. However, for flexed vehicle the angle of attack will vary along the length of the vehicle and will depend on the local slope of the flexed vehicle and also on the local speed experienced by the vehicle element due to vibration.

The gravity force is distributed over the whole vehicle structure and the resultant at any time acts at the instantaneous centre of gravity of the structure.

Control forces act at the control force location and will have components due to deformed shape of the vehicle. However, in case of aerodynamic control, the control surface deflection will be much larger than the slope due to flexibility and hence, contribution due to flexibility may be ignored.

The external forces are then given by following expressions for a vehicle with two gimballed engine configuration.

$$\overline{F}_T = \left[\sum_{i=1}^{2} T_i \cos \sqrt{\left(\delta_p + \frac{\partial \xi_z}{\partial l}(G) \right)^2 + \left(\delta_{yi} + \frac{\partial \xi_y}{\partial l}(G) \right)^2} \right] i$$

$$- \left[\sum_{i=1}^{2} T_i \sin \left(\delta_{yi} + \frac{\partial \xi_y}{\partial l}(G) \right) \right] j - \left[\sum_{i=1}^{2} T_i \sin \left(\delta_{zi} + \frac{\partial \xi_z}{\partial l}(G) \right) \right] k \qquad \text{... (2.89)}$$

$$\overline{F}_A = \left[-QS \int C_A(\overline{r}) \right] i + \left[-QS \int \frac{\partial C_{y\beta}(\overline{r})}{\partial l} \beta(\overline{r}) \, \partial l \right] j + \left[-QS \int \frac{\partial C_{N\alpha}(\overline{r})}{\partial l} \alpha(\overline{r}) \, dl \right] k$$

$$\text{... (2.90)}$$

where $\overline{\alpha}(\overline{r}), \beta(\overline{r})$ are local angle of attack and sideslip angles.

$\dfrac{\partial C_{y\beta}(\overline{r})}{\partial l}$ and $\dfrac{\partial C_{N\alpha}(\overline{r})}{\partial l}$ are local slopes of $C_{N\alpha}$ and $C_{y\beta}$ distribution on the vehicle and l is the length measured from vehicle nose tip.

$$l = l_{cg} - x$$

The parameter $\overline{r}$ indicates the vehicle station at which α, β or slopes of normal forces are considered. A general station $\overline{r}$ is considered instead of x or l to take care of the more complicated configuration such as with strapons.

$$\overline{F}_G = F_{gx} i + F_{gy} j + F_{gz} k \qquad \text{... (2.91)}$$

$$\overline{F}_C = F_{cx} i + F_{cy} j + F_{cz} k \qquad \text{... (2.92)}$$

$$\approx F_{cy} j + F_{cz} k$$

The torque acting on the vehicle due to external forces is then given by

$$\overline{T} = T_x i + T_y j + T_z k$$

$$= \hat{r}_{cg} \times \overline{F}_G + \overline{r}_T \times \overline{F}_T + \int \overline{r}_A \times \overline{F}_A + \overline{r}_C \times \overline{F}_C$$

$$= \left[\hat{y}_{cg}F_{gz} - \hat{z}_{cg}F_{gy} + \sum_{I=1}^{2}\left(\hat{y}_{Ti}F_{Tz} - \hat{z}_{Ti}F_{Ty} \right) + \hat{y}_{c}F_{cz} - \hat{z}_{c}F_{cy} \right.$$

$$\left. + \left\{ -\int_{Y}\left(QS\int_{X}\frac{\partial C_{N\alpha}}{\partial x}\bar{\alpha}(x)\hat{y}dx \right)dy + \int_{z}\left(QS\int\frac{\partial c_{y\beta}}{\partial x}\bar{\beta}(x)\hat{z}dx \right)dz \right\} \right] i$$

$$+ \left[\hat{z}_{cg}F_{gx} - \hat{x}_{cg}F_{gz} + \sum_{i=1}^{2}\left(z_{Ti}F_{Tix} - x_{Ti}F_{Tiz} \right) + \left(\hat{z}_{c}F_{cx} - \hat{x}_{cz}F_{cz} \right) \right.$$

$$\left. + \left\{ -\int_{z}\left(QS\int_{X}\frac{\partial C_{A(\bar{r})}}{\partial x}\hat{z}dx \right)dz + \int_{x}\left(QS\frac{\partial c_{N\alpha}}{\partial x}\bar{\alpha}(x)\hat{x} \right)dx \right\} \right] j$$

$$+ \left[\left(\hat{x}_{cg}F_{gy} - \hat{y}_{cg}F_{gx} + \sum_{i=1}^{2}\left(x_{Ti}F_{Tiy} - y_{Ti}F_{Tx} \right) + \left(\hat{x}_{c}F_{cy} - \hat{y}_{c}F_{cx} \right) \right) \right.$$

$$\left. + \left\{ -\int_{x}\left\{ QS\frac{\partial C_{y\beta}}{\partial x}\bar{\beta}(x)\hat{x}dx + \int_{y}\left(QS\frac{\partial C_{A(\bar{r})}}{\partial x}\hat{y}dx \right)dy \right\} \right] k \right. \qquad \dots (2.93)$$

(**Note:** Second integrals in aerodynamic terms in '*i*' and '*j*' components of torque need to be considered only when the aerodynamic force distribution on multiple boosters is separately considered.)

The position vectors for the point of application of different forces in the above equations are given as follows:

(*i*) $\hat{r}_{cg}$ is given by

$$M\hat{r}_{cg} = \left[\int (x + \xi_x)dm \right]i + \left[\int (y + \xi_y)dm \right]j + \left[\int (z + \xi_z)dm \right]k \qquad (2.94)$$

If the mode shapes are computed for the total vehicle

$$\int \xi_{xe}\,dm = \int \xi_{ye}\,dm = \int \xi_{ze}\,dm = 0$$

i.e., the structural component of deflection will go to zero and only gimballing and slosh components will remain (Ref. Eqs. 2.27 to 2.32)

$$M\hat{r}_{cg} = \left[MX_{cg} + \sum_{i=1}^{2} M_R L_R \left(\frac{\delta_{pk}^2 + \delta_{yk}^2}{2} \right) + \sum_{j=1}^{m} M_{sj} L_{sj} \left(\frac{\lambda_{pj}^2 + \lambda_{yj}^2}{2} \right) \right] i$$

$$+ \left[MY_{cg} + \sum_{i=1}^{2} M_R L_R \delta_{yi} + \sum_{j=1}^{m} M_{sj} L_{sj} \lambda_{yj} \right] j$$

$$+ \left[MZ_{cg} + \sum_{i=1}^{2} M_R L_R \delta_{pi} + \sum_{j=1}^{m} M_{sj} L_{sj} \lambda_{pj} \right] k \qquad \ldots (2.95)$$

$(ii) \quad \hat{r}_{T1} = \left[-L_C \right] i + \left[\xi_y(G) \right] j + \left[-L_a + \xi_z(G) \right] k \qquad \ldots (2.96)$

$\quad\quad \hat{r}_{T2} = \left[-L_C \right] i + \left[\xi_y(G) \right] j + \left[L_a + \xi_z(G) \right] k \qquad \ldots (2.97)$

$(iii) \quad \hat{\bar{r}}_A = (x + \xi_{xe}) i + (y + \xi_{ye}) j + (z + \xi_{ze}) k \qquad \ldots (2.98)$

$(iv) \quad \hat{\bar{r}}_c = (x_c + \xi_{xe}(c)) i + (y_c + \xi_{ye}(c)) j + (z_c + \xi_{ze}(c)) k \qquad \ldots (2.99)$

For illustration, the components of moment are written below for an axis symmetric missile (with no strapons) and having two gimballed engines in pitch plane.

For central body $\hat{y} = y + \xi_y = \xi_y$ and $\hat{z} = \xi_z$ as $y = z = 0$.

Then,

$$T_x = \left[MY_{cg} + \sum_{i=1}^{2} M_R L_R \delta_{yi} + \sum_{j=1}^{m} M_{sj} L_{sj} \lambda_{yj} \right] g_z$$

$$- \left[MZ_{cg} + \sum_{i=1}^{2} M_R L_R \delta_{pi} + \sum_{j=1}^{m} M_{sj} L_{sj} \lambda_{pj} \right] g_y - \sum_{i=1}^{2} \xi_y(G) T_i \left(\delta_{pi} + \frac{\partial \xi_z(G)}{\partial l} \right)$$

$$+ (-L_a + \xi_z(G)) T_1 \left(\delta_{y1} + \frac{\partial \xi_y(G)}{\partial l} \right) + (L_a + \xi_z(G)) T_2 \left(\delta_{y2} + \frac{\partial \xi_y(G)}{\partial l} \right)$$

$$- QS \int_x \frac{\partial C_{N_\alpha}(x)}{\partial l} \bar{\alpha}(x) \xi_y \, dx + QS \int_x \frac{\partial C_{y\beta}(x)}{\partial l} \bar{\beta}(x) \xi_z dx \qquad \ldots (2.100)$$

$$T_y = \left[Mz_{cg} + \sum_{i=1}^{2} M_R L_R \delta_{pi} + \sum_{j=1}^{m} M_{sj} L_{sj} \lambda_{pj} \right] g_x$$

$$- \left[Mx_{cg} + \sum_{i=1}^{2} M_R L_R \left(\frac{\delta_{pk}^2 + \delta_{yk}^2}{2} \right) + \sum_{j=1}^{m} M_{sj} L_{sj} \left(\frac{\lambda_{pj}^2 + \lambda_{yj}^2}{2} \right) \right] g_z$$

$$+ \left(-L_a + \xi_z(G) \right) T_1 \left(1 - \left(\frac{\delta_{p1} + \partial \xi_z(G)}{2} \right)^2 - \left(\frac{\delta_{y1} + \partial \xi_y(G)}{2} \right)^2 \right)$$

$$+ (L_a + \xi_z(G)) T_2 \left(1 - \left(\frac{\delta_{p2} + \dfrac{\partial \xi_z(G)}{\partial l}}{2} \right)^2 - \left(\frac{\delta_{y2} + \dfrac{\partial \xi_y(G)}{\partial l}}{2} \right)^2 \right)$$

$$- L_c \sum_{i=1}^{2} T_i \left(\delta_{pi} + \frac{\partial \xi_z(G)}{\delta l} \right) - QS \int_x \frac{\partial C_A(\overline{r})}{\partial l} \xi_z dx + QS \int_x \frac{\partial C_{N_\alpha}(\overline{r})}{\partial l} \overline{\alpha}(x) \, x dx$$

$$\dots (2.101)$$

Here the component F_{cx} and $_x$ in $\hat{x}$ are neglected.

$$T_z = \left[Mx_{cg} + \sum_{i=1}^{2} M_R L_R \left(\frac{\delta_{pk}^2 + \delta_{yk}^2}{2} \right) + \sum_{j=1}^{m} M_{sj} L_{sj} \left(\frac{\lambda_{pj}^2 + \lambda_{yj}^2}{2} \right) \right] g_y$$

$$- \left[My_{cg} + \sum_{i=1}^{2} M_R L_R \delta_{yi} + \sum_{j=1}^{m} M_{sj} L_{sj} \lambda_{yj} \right] g_x + L_c \sum_{i=1}^{2} T_i \left(\delta_{yi} + \frac{\partial \xi_y(G)}{\partial l} \right)$$

$$- \xi_y(G) \sum_{i=1}^{2} T_i \left(1 - \frac{\left(\delta_{pi} + \dfrac{\partial \xi_z(G)}{\partial l} \right)^2}{2} - \frac{\left(\delta_{yi} + \dfrac{\partial \xi_y(G)}{\partial l} \right)^2}{2} \right)$$

$$- QS \int \frac{\partial c_{y\beta}}{\partial l} \overline{\beta}(x) \, x dx + QS \int \frac{\partial c_A(\overline{r})}{\partial x} \xi_y \, dx \qquad \dots (2.102)$$

The external forces and moments, thus have been completely given for the given configuration. For any other configuration, the forces and moments can be similarly written without much difficulty.

2.4.8   Linear Model for Aero-Structure-Control-Slosh Interaction Studies

A detailed model consisting of force, torque, slosh modes, engine gimballing and generalized coordinates of flexible modes has been given in Sections 2.4.2 to 2.4.7 without making simplifications. This model will be useful for trajectory simulation incorporating these features for validating the control system performance. It is easy to see that the right hand side (RHS) of equation 2.45 does not contain structural mode terms. Thus, the rigid body modes are decoupled from the elastic body bending modes. (The coupling, however, enters through the force terms appearing on left hand side as seen from Eqs. 2.89 and 2.90). This is the effect of using mode shapes computed for the total vehicle in which gimballed masses and slosh masses are assumed to be frozen at their nominal position. These equations are different from Greensite's[2] equations in respect of modal coupling. The derivation of Greensite's[2] equations from the generalized equations will be shown in subsequent section.

The design of control systems is done using a linearised model. Hence, this section gives the linear model useful for control system stability analysis and design of suitable filters.

Following simplifications have been made to arrive at the linear model:

1. Only the perturbation model is considered. The missile is assumed to be aligned with forward velocity. Hence the components of velocity, v and w are assumed small quantities.

2. Products and squares of small quantities are neglected.

3. Product inertias *i.e.* I_{xy}, I_{yz}, I_{xz} and their derivatives are neglected.

4. Variation of moment of inertia due to flexibility, engine gimballing and slosh motion is neglected which gives $\hat{I}_{xx} \approx I_{xx}$, $\hat{I}_{yy} \approx I_{yy}$, and $\hat{I}_{zz} \approx I_{zz}$ and their time derivatives are therefore zero.

5. Longitudinal flexibility $(\ _x)$ is neglected.

6. The thrust of two engines are assumed equal.

   Therefore, $T_1 = T_2 = T_E =$ Thrust of each engine

7. The pitch, yaw and roll motions are assumed decoupled which implies that $x_{cg}\ y_{cg}\ z_{cg}\ 0$ (Origin coincides with *CG* of the rigid body vehicle) and body rate $p = 0$.

8. $\delta_{p1} = \delta_{p2} = \delta_p$ $\qquad\qquad\qquad \delta_{p1} + \delta_{p2} = 2\delta_p$

   $$\delta_{y1} + \delta_{y2} = 2\delta_y$$

   $$\delta_{y2} - \delta_{y1} = 2\delta_R$$

where, δ_p, δ_y and δ_R are engine deflection for pitch, yaw and roll respectively.

The acceleration equations for the missile are then obtained from Eqs. 2.45, 2.89, 2.90, 2.91, 2.92 and 2.13 and can be written as:

$$
M \begin{bmatrix} \dot{u} \\ \dot{v} + ur \\ \dot{w} - uq \end{bmatrix} =
\begin{bmatrix}
0 \\[2ex]
-\sum_{k=1}^{2} M_R L_R \delta_{yk} - \sum_{j=1}^{m} M_{sj} L_{sj} \lambda_{yj} \\[3ex]
-\sum_{k=1}^{2} M_R L_R \delta_{pk} - \sum_{j=1}^{m} M_{sj} L_{sj} \lambda_{pj}
\end{bmatrix}
$$

$$
+ \begin{bmatrix}
F_{gx} + 2T_E - C_D QS \\[2ex]
F_{gy} - \sum_{k=1}^{2} T_E \left(\delta_{yk} + \dfrac{\partial \xi_y(G)}{\partial l} \right) - QS \int \dfrac{\partial C_{y\beta}}{dl} \overline{\beta}(l)dl \\[3ex]
F_{gz} - \sum_{k=1}^{2} T_E \left(\delta_{pk} + \dfrac{\partial \xi_z(G)}{\partial l} \right) - QS \int \dfrac{\partial C_{N\alpha}}{dl} \overline{\alpha}(l)dl
\end{bmatrix} \quad \dots (2.103)
$$

The angular equations for the missile are obtained using Eq. 2.46, 2.47, 2.48. 2.49, 2.50, 2.51, 2.54, 2.100, 2.101, 2.102 and given below:

$$
\begin{bmatrix} I_{xx}\dot{p} \\ I_{yy}\dot{q} \\ I_{zz}\dot{r} \end{bmatrix} =
\begin{bmatrix}
A_y(+\Sigma M_R L_R \delta_{pi} + \Sigma M_{sj} L_{sj} \lambda_{pj}) - A_z(\Sigma M_R L_R \delta_{yi} + \Sigma M_{sj} L_{sj} \lambda_{yj}) \\[2ex]
-A_x(\Sigma M_R L_R \delta_{pi} + \Sigma M_{sj} L_{sj} \lambda_{pj}) \\[2ex]
A_x(\Sigma M_R L_R \delta_{yi} + \Sigma M_{sj} L_{sj} \lambda_{yj})
\end{bmatrix}
$$

$$
+ \begin{bmatrix}
2M_R L_R L_a \delta_R \\[2ex]
-2(I_R + M_R L_R L_C)\,\delta_p + \sum M_{sj} x_{sj} L_{sj} \lambda_{pj} \\[2ex]
2(I_R + M_R L_R L_C)\delta_y - \sum M_{sj} x_{sj} L_{sj} \lambda_{yj}
\end{bmatrix}
$$

$$
+ \begin{bmatrix}
g_y \left[-2M_R L_R \delta_p - \sum M_{sj} L_{sj} \lambda_{pj} \right] + g_z \left[2M_R L_R \delta_y + \sum M_{sj} L_{sj} \lambda_{yj} \right] \\[2ex]
g_x \left(2M_R L_R \delta_p + \sum M_{sj} L_{sj} \lambda_{pj} \right) \\[2ex]
-g_x \left(2M_R L_R \delta_y + \sum M_{sj} L_{sj} \lambda_{yj} \right)
\end{bmatrix}
$$

(eq. contd.)

$$+\begin{bmatrix} 2T_E L_a \delta_R + 2T_E \xi_z(G)\left[\delta_y + \dfrac{\partial \xi_y(G)}{\partial l}\right] - 2T_E \xi_y(G)\left[\delta_p + \dfrac{\partial \xi_z(G)}{\partial l}\right] \\[2ex] +2T_E \xi_z(G) - 2L_c T_E\left[\delta_p + \dfrac{\partial \xi_z(G)}{\partial l}\right] \\[2ex] -2T_E \xi_y(G) + 2L_c T_E\left[\delta_y + \dfrac{\partial \xi_y(G)}{\partial l}\right] \end{bmatrix}$$

$$+\begin{bmatrix} C_L QSd \\[2ex] QS\displaystyle\int \dfrac{\partial C_{N_\alpha}}{\partial l}\,\overline{\alpha}(x)\; x\; dx \\[2ex] -QS\displaystyle\int \dfrac{\partial C_{y_\beta}}{\partial l}\,\overline{\beta}(x)\; x\; dx \end{bmatrix} \qquad\qquad \dots (2.104)$$

The linearised slosh equations are obtained from Eqs. 2.64 to 2.69 and are given below:

$$L_{sj}\lambda_{pj} = -\xi_z(sj) - \left(w - uq - g\cos\theta\right) + qx_{sj} - L_{sj}\omega_{sj}^2\lambda_{pj} - 2\varsigma_i\omega_{sj}L_{sj}\lambda_{pj} \qquad \dots (2.105)$$

$$L_{sj}\lambda_{yj} = -\xi_y(sj) - (v + ur) - rx_{sj} - L_{sj}\omega_{sj}^2\lambda_{yj} - 2\varsigma_{sj}\omega_{sj}L_{sj}\lambda_{yj} \qquad \dots (2.106)$$

The linearised actuator equations are obtained from Eqs. 2.78 and 2.79 and are given below:

$$I_R\delta_p = -M_R L_R \left\{ w - uq - g\cos\theta + q(L_c + L_R) \right.$$

$$+ \left(\xi_{ze}(G) + L_R\dfrac{\partial \xi_{ze}(G)}{\partial l}\right) + \left(u + g\sin\theta\right)\left. \left(\delta_p + \dfrac{\partial \xi_{ze}(G)}{\partial l}\right)\right\}$$

$$+ I_R\left(\omega_n^2\delta_{pc} - 2\zeta_n\omega_n\delta_p - \omega_n^2\delta_p\right) \qquad\qquad \dots (2.107)$$

$$I_R\delta_y = -M_R L_R \left\{ v + ur - r(L_R + L_C) + \left(\xi_{ye}(G) + L_R\dfrac{\partial \xi_{ye}(G)}{\partial l}\right)\right.$$

$$+ (u + g\sin\theta)\left. \left(\delta_y + \dfrac{\partial \xi_{ye}(G)}{\partial l}\right)\right\} + I_R\left(\omega_n^2\delta_{yc} - 2\zeta_n\omega_n\delta_y - \omega_n^2\delta_y\right)$$

$$\dots (2.108)$$

Control system stability analysis is carried out considering vehicle motion in one plane. Hence, the generalized mode equation is obtained from Eq. 2.88 by suppressing motions in other planes and are given below:

$$m_{yy}q_{yi} + \omega_i^2 m_{yyi}q_{yi} + 2\phi_{yi}(ecg)M_R L_R \delta_y + \sum \phi_{yi}(sj)M_{sj}L_{sj}\lambda_{yj} = \int f_y(l)\phi_{yi}(l)dl$$

$$\text{... (2.109)}$$

$$m_{zz}q_{zi} + \omega_i^2 m_{zzi}q_{zi} + 2\phi_{zi}(ecg)M_R L_R \delta_p + \sum \phi_{zi}(sj)M_{sj}L_{sj}\lambda_{pj} = \int f_z(l)\phi_{zi}(l)dl$$

$$\text{... (2.110)}$$

To write the equations in the form of final state equations, the following information is used:

1. Lateral acceleration (pitch plane)

$$A_z = w - uq = u\alpha + u\alpha - uq \qquad \text{... (2.111)}$$

   Lateral acceleration (yaw plane)

$$A_y = v + ur = u\beta + u\beta + ur \qquad \text{... (2.112)}$$

2. x $l_{CG} - l$ where l, l_{CG} are measured from nose tip.
3. Local angle of attack and side slip angles:

$$\bar{\alpha}(l) = \alpha - \frac{(l_{CG} - l)\theta}{u} + \frac{\partial \xi_{ze}(l)}{\partial l} + \frac{\xi_{ze}(l)}{u} \qquad \text{... (2.113)}$$

$$\bar{\beta}(l) = \beta + \frac{(l_{CG} - l)\psi}{u} + \frac{\partial \xi_{ye}(l)}{\partial l} + \frac{\xi_{ye}(l)}{u} \qquad \text{... (2.114)}$$

4. Aerodynamic forces and moments

$$F_{AZ} = -QS \int \frac{\partial C_{N\alpha}(l)}{\partial l} \bar{\alpha}(l)\,dl \qquad \text{(2.115)}$$

$$F_{AY} = -QS \int \frac{\partial C_{y\beta}(l)}{\partial l} \bar{\beta}(l)\,dl \qquad \text{... (2.116)}$$

$$M_{AY} = QS \int \frac{\partial C_{N_a}(l)}{\partial l} \bar{\alpha}(l)(l_{cg} - l)\,dl \qquad \text{... (2.117)}$$

$$M_{AZ} = -QS \int \frac{\partial C_{y\beta}(l)}{\partial l} \bar{\beta}(l)(l_{cg} - l)\,dl \qquad \text{... (2.118)}$$

5. The total pitch attitude angle $\theta = \gamma_0 + d\theta$

   Then $g\cos\theta = g\cos\gamma_0 - g\sin\gamma_0 d\theta$

$$\lambda_{pj}\, g\sin\theta = \lambda_{pj}\, g\sin\gamma_0 + \lambda_{pj}\, g\cos\gamma_0 d\theta = \lambda_{pj}\, g\sin\gamma_0 \qquad \text{... (2.119)}$$

   Similarly, $\delta_p\, g\sin\theta = \delta_p\, g\sin\gamma_0$

The constant term $g\cos\gamma_0$ and non-linear terms $g\cos\gamma_0\lambda_{pj}d\theta$ and $g\cos\gamma_0\delta_p\,d\theta$ are neglected. (The constant term will be balanced by other steady components of forces and moments and not considered for perturbation model which is used for control system stability analysis). The variable $d\theta$ is reassigned symbol θ for simplicity.

6. The actuator deflections for pitch, yaw and roll are given by:

$$\text{Pitch} \quad : \quad \delta_{p1} + \delta_{p2} = 2\delta_p$$

$$\text{Yaw} \quad : \quad \delta_{y1} + \delta_{y2} = 2\delta_y$$

$$\text{Roll} \quad : \quad \delta_{y2} - \delta_{y1} = 2\delta_R$$

7. While considering roll motion, lateral components of acceleration *i.e.*, A_y, A_z, g_y, g_z are neglected to avoid coupling.

8. Structural damping term $2\zeta_i\omega_i q_i$ and slosh damping terms $2\zeta_s\omega_s\lambda_{pj}$, $2\zeta_s\omega_s\lambda_{yj}$ are added in the generalized mode equations and slosh equations.

The linearized pitch plane equations are then given as follows:

$$\ddot{\alpha} = \left[-\frac{\dot{u}}{u} - \frac{QS}{Mu}\int\frac{\partial C_{N\alpha}}{\partial l}(l)dl\right]\alpha + \left[-\frac{g\sin\gamma_0}{u}\right]\theta$$

$$+ \left[1 + \frac{QS}{Mu^2}\int\frac{\partial C_{N\alpha}(l)}{\partial l}(l_{cg}-l)dl\right]\dot{\theta} + \left[-\frac{2T_E}{Mu}\right]\delta_p + \left[-\frac{2M_R L_R}{Mu}\right]\dot{\delta}_p$$

$$+ \sum_{j=1}^{m}\left[-\frac{M_{sj}L_{sj}}{Mu}\right]\ddot{\lambda}_{pj} + \sum_{i=1}^{n}\left[-\frac{2T_E\dfrac{\partial\phi_{zi}(G)}{\partial l}}{Mu} - \frac{QS}{Mu}\int\frac{\partial C_{N\alpha}}{\partial l}(l)\frac{\partial\phi_{zi}}{\partial l}(l)dl\right]q_i$$

$$+ \sum_{i=1}^{n}\left[-\frac{QS}{Mu^2}\int\frac{\partial C_{N\alpha}}{\partial l}\phi_{zi}(l)dl\right]\ddot{q}_i \qquad\qquad \text{... (2.120)}$$

$$\ddot{\theta} = \left[\frac{QS}{I_{yy}}\int\frac{\partial C_{N\alpha}(l)}{\partial l}(l_{CG}-l)dl\right]\alpha + \left[-\frac{QS}{I_{yy}u}\int\frac{\partial C_{N\alpha}(l)}{\partial l}(l_{CG}-l)^2\,dl\right]\dot{\theta}$$

$$+ \left[-\frac{2T_E L_C + 2M_R L_R(u + g\sin\gamma_0)}{I_{yy}}\right]\delta_p + \left[-\frac{2(I_R + M_R L_R L_C)}{I_{yy}}\right]\dot{\delta}_p$$

$$+ \sum_{j=1}^{m}\left[-\frac{M_{sj}L_{sj}(u + g\sin\gamma_0)}{I_{yy}}\right]\ddot{\lambda}_{pj} + \sum_{j=1}^{m}\left[\frac{M_{sj}X_{sj}L_{sj}}{I_{yy}}\right]\ddot{\lambda}_{pj}$$

(eq. contd.)

$$+\sum_{i=1}^{n}\left[\frac{2T_E\left[\phi_{zi}(G)-L_C\dfrac{\partial\phi_{zi}(G)}{\partial l}\right]}{I_{yy}}+\frac{QS}{I_{yy}}\int\frac{\partial C_{N\alpha}(l)}{\partial l}(l_{cg}-l)\frac{\partial\phi_{zi}(l)}{\partial l}dl\right]q_i$$

$$+\sum_{i=1}^{n}\left[\frac{QS}{I_{yy}u}\int_{0}^{L}\frac{\partial C_{N\alpha}}{\partial l}(l_{cg}-l)\phi_{zi}(l)dl\right]q_i \qquad \ldots (2.121)$$

$$\delta_p=\left[\frac{-M_RL_Ru}{I_R}\right]\alpha+\left[\frac{-M_RL_Ru}{I_R}\right]\alpha$$

$$+\left[\frac{-M_RL_Rg\sin\gamma_o}{I_R}\right]\theta+\left[\frac{M_RL_Ru}{I_R}\right]\theta+\left[-1-\frac{M_RL_RL_C}{I_R}\right]\theta$$

$$+\left[-\omega_n^2-\frac{M_RL_R(u+g\sin\gamma_0)}{I_R}\right]\delta_p-2\zeta_n\omega_n\delta_p$$

$$+\sum_{i=1}^{n}\left[-\frac{M_RL_R(u+g\sin\gamma_0)}{I_R}\frac{\partial\phi_{zi}(G)}{\partial l}\right]qi$$

$$+\sum_{i=1}^{n}\left[-\frac{\partial\phi_{zi}(G)}{\partial l}-\frac{M_RL_R\phi_{zi}(G)}{I_R}\right]q_i+\omega_n^2\delta_{pc} \qquad \ldots (2.122)$$

$$\lambda_{pj}=\left[-\frac{u}{L_{sj}}\right]\alpha+\left[-\frac{u}{L_{sj}}\right]\alpha+\left[\frac{-g\sin\gamma_0}{L_{sj}}\right]\theta+\left[\frac{u}{L_{sj}}\right]\theta+\left[\frac{X_{sj}}{L_{sj}}\right]\theta$$

$$+\left[-\omega_{pj}^2\right]\lambda_{pj}+\left[-2\zeta_{pj}\omega_{pj}\right]\lambda_{pj}+\sum_{i=1}^{n}\left[-\frac{\phi_i(sj)}{L_{sj}}\right]q_i \qquad \ldots (2.123)$$

where
$$\omega_{pj}^2=\frac{u+g\sin\gamma_0}{L_{sj}} \qquad \ldots (2.124)$$

and damping term $(-2\zeta_{pj}\omega_{pj})\lambda_{pj}$ has been added

$$q_i=\left[\frac{-QS}{m_i}\int\frac{\partial C_{N\alpha}(l)}{\partial l}\phi_{zi}(l)dl\right]\alpha+\left[\frac{QS}{m_iu}\int\frac{\partial C_{N\alpha}(l)}{\partial l}(l_{cg}-l)\phi_{zi}(l)dl\right]\theta$$

(eq. contd.)

$$+\left[-\frac{2T_E\phi_{zi}(G)}{m_i}\right]\delta + \left[-\frac{2M_RL_R\left(\phi_{zi}(G)+L_R\dfrac{\partial\phi_{zi}(G)}{\partial l}\right)}{m_i}\right]\delta$$

$$+\sum_{j=1}^{m}\left[-\frac{M_{sj}L_{sj}\phi_{zi}(sj)}{m_i}\right]\lambda_{pj}$$

$$+(-\omega_i^2)q_i + \sum_{j=1}^{n}\left[-\frac{2T_E\phi_{zi}(G)\dfrac{\partial\phi_{zi}(G)}{\partial l}}{m_i} - \frac{QS}{m_i}\int\frac{\partial C_{N\alpha}(l)}{\partial l}\frac{\partial\phi_{zj}(l)}{\partial l}\phi_{zi}(l)dl\right]q_j$$

$$+(-2\zeta_i\omega_i)q_i + \sum_{j=1}^{n}\left[-\frac{QS}{m_iu}\int\frac{\partial C_{N\alpha}(l)}{\partial l}\phi_{zj}(l)\phi_{zi}(l)dl\right]q_j \qquad \text{... (2.125)}$$

The damping term $(-2\zeta_i\omega_iq_i)$ has been added to account for structural damping. Now define the vehicle state as

$$\bar{X} = \begin{bmatrix} \alpha & \theta & \dot\theta & \delta & \dot\delta & \lambda_{p1} & \dot\lambda_{p1} & ...... & \lambda_{pm} & \dot\lambda_{pm} & q_i & \dot q_1 & ....\dot q_n & q_n \end{bmatrix} \qquad \text{... (2.126)}$$

The above linearized equations can be summarily written as:

$$\dot{\bar{X}} = A\bar{X} + D\dot{\bar{X}} \qquad\qquad \text{... (2.127)}$$

or

$$\dot{\bar{X}} = F\bar{X} \qquad\qquad \text{... (2.128)}$$

where

$$F \quad (I-D)^{-1}A \qquad\qquad \text{... (2.129)}$$

Most of the launch vehicles and missiles will be axis symmetric. Hence, the stability analysis carried out for pitch plane will be used for yaw plane also and the design in pitch and yaw plane will be the same. However, in some cases such as ASLV the vehicle is not symmetric. It is also observed that there is a sideways load distribution $\dfrac{\partial C_{y\alpha}(l)}{\partial l}$ in the presence of   . Hence, yaw plane analysis needs to be carried out separately.

2.4.9 Derivation of Greensite's Model Using Generalised Model

We now derive the Greensite's[2] model and show that the model is valid only under following conditions (though Greensite does not explicitly say so):

1. The origin of the body coordinate system is taken at the centre of gravity of 'reduced' vehicle (*i.e.*, the vehicle excluding gimbaled engine and sloshing masses).

2. Mode shapes and frequencies are obtained for the reduced vehicle.

3. Nonlinear terms are neglected.

Greensite's conventions for vehicle deflection are shown in Fig. 2.9.

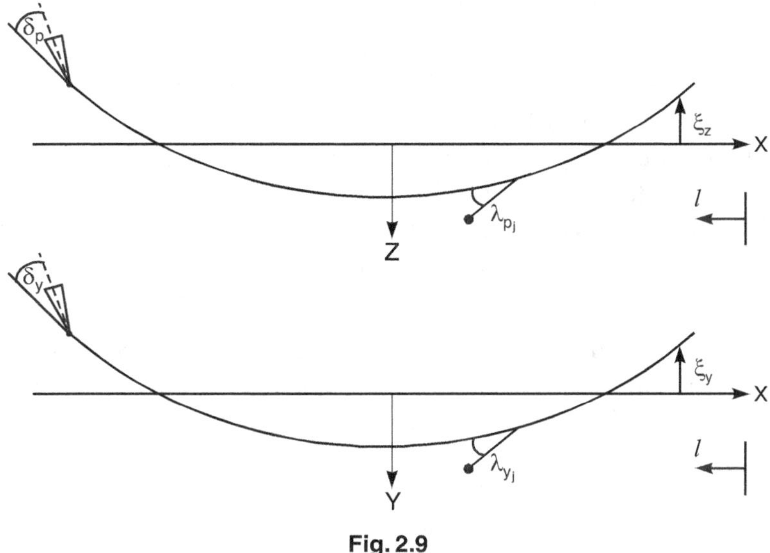

Fig. 2.9

Comparison of Fig. 2.9 with Fig. 2.3 shows appropriate signs for various variables to be used to get Greensite's state variables from this author's model. The relation is given in Table 2.2.

TABLE 2.2

Variable	Authors' model	To be used for Greensite's model
1. Rigid vehicle:	θ	θ
	α	α
2. Flexible vehicle:	q_i	$-q_i$
ξ_z	$\Sigma\phi_{zi}q_i$	$-\Sigma\phi_{zi}q_i$
ξ_y	$\Sigma\phi_{yi}q_i$	$-\Sigma\phi_{yi}q_i$
3. Engine deflection:	δ_p	$-\delta_p$
	δ_y	$-\delta_y$
4. Slosh pendulum angle:	λ_{pj}	λ_{pj}
	λ_{yj}	λ_{yj}

Greensite[2] has derived this model using Lagrangian equation. It is felt that this model and the derivation methodology has following limitations.

1. The model shows coupling between modes and the rigid body modes are also coupled with flexible modes.

2. It is not clear from derivation what modifications need to be made if one wishes to use the modes computed for the total vehicle.

It is felt that though the mode shape data is worked out for the total vehicle, it is being erroneously used by control designers under the tacit assumption that the model and data are compatible. The generalized model given in this chapter enables one to appreciate the difference between structural modes computed for reduced vehicle as well as those for the total vehicle and also enables one to derive any one of them without any extra complications. Though the numerical values of modal frequencies for reduced vehicle and total vehicle may not show great difference, it gives great technical satisfaction to see that the mathematical model really shows the decoupling of the orthogonal modes.

The derivation of pitch plane model with parameters of reduced vehicle is illustrated here summarily by using author's conventions for variables. Greensite's equations can then be written down easily using Table 2.2. For this derivation use:

$$p \quad r \quad 0$$

$$p = r = 0$$

$$_x(l) \quad _y(l) \quad 0$$

Also, product inertias (i_{xy}, i_{xz}, i_{yz}) and their time derivatives and higher order terms are assumed zero.

Fig. 2.10 shows the schematic diagram of the vehicle with engine and slosh masses and centre of gravity of reduced vehicle.

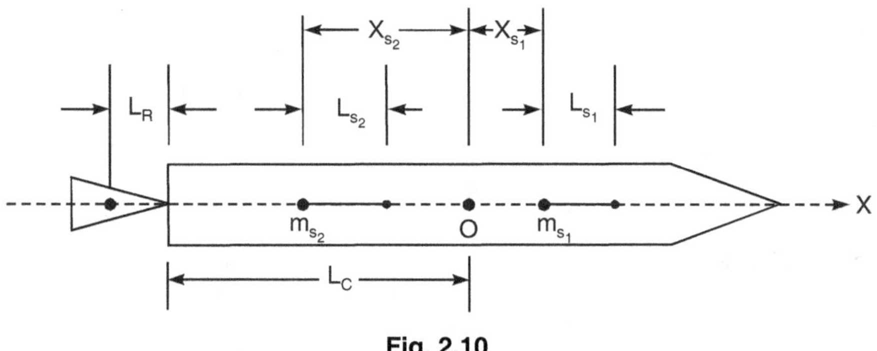

Fig. 2.10

The *CG* of total vehicle w.r.t. origin (*i.e.*, *CG* of reduced vehicle) is given by

$$Mx_{cg} = -(L_C + L_R)M_R + \sum_{j=1}^{k} M_{sj}x_{sj} \qquad \ldots (2.130)$$

$$M = M_o + M_R + \sum_{j=i}^{m} m_{sj}, \quad M_o = \text{mass of reduced vehicle} \qquad \dots (2.131)$$

The force $\left(\int f_z(l)\right)$ and torque equations $\left(\int t_y\right)$ (eqn. 2.26) after integrating over the whole vehicle are then given by:

$$\int f_z(l) = A_z M - q\, x_{cg}\, M + \int \xi_z dm + K \int \xi_{ze} \qquad \dots (2.132)$$

$$\int t_y = A_x \int \hat{z} dm - A_z \int \hat{x} dm + q \int \hat{i}_{yy} + \int \eta_y \hat{i}_{yy} + K \int \eta_{ye} + q \int \frac{d\hat{i}_{yy}}{dt} \qquad \dots (2.133)$$

Force equation:

Now, following the procedure of Section 2.3.

$$\int \xi_z dm = \int_R \xi_{ze} dm + \int_{engine} \left[\xi_{ze}(G) + l\left(\delta_p + \frac{\partial \xi_{(G)}}{\partial l} \right) \right] dm + \sum_{j=1}^{m} \left(\xi_{ze}(sj) + L_{sj}\lambda_{pj} \right) m_{sj} \qquad \dots (2.134)$$

$\int_R$ indicates integration over reduced vehicle and $\xi_{ze}(G)$ indicates deflection at gimbal point.

We now state that the Greensite's model can be obtained only if we assume that the elastic modes are computed for reduced vehicle, in which case:

$$\int_R \xi_{ze} dm = \sum q_i \int \phi_{zi}(l) dm = 0 \qquad \dots (2.135)$$

also
$$K \int \xi_{ze} = 0 \qquad \dots (2.136)$$

Using Eqs. 2.134, 2.135 and 2.136 in 2.132 gives

$$MA_z = MX_{cg}q - M_R \left[L_R\delta_p + \sum_{i=1}^{n}\left(\phi_{zei}(G) + L_R \frac{\partial \phi_{zei}(G)}{\partial l} \right) q_i \right]$$

$$- \sum_{j=1}^{m}\left[\left(\xi_{ze}(sj) + L_{sj}\lambda_{pj} \right) M_{sj} + \int f_z(l) dl \right] \qquad \dots (2.137)$$

From Eqs. 2.64 and 2.66, on simplification, one gets

$$L_{sj}\lambda_{pj} + \xi_{ze}(sj) = x_{sj}q - (w - uq - g\cos\theta) - \lambda_{pj}(u + g\sin\theta) \qquad \dots (2.138)$$

and
$$\int f_z(l) dl = Mg\cos\theta - T\left(\delta_p + \frac{\partial \xi_{ze}(G)}{\partial l} \right) - QS \int \frac{\partial C_{N\alpha}(l)}{\partial l} \bar{\alpha}(l) dl \qquad \dots (2.139)$$

Using Eqs. 2.138 and 2.139 in Eq. 2.137, one gets,

$$(w - uq)\left(M - \sum_{j=1}^{m} M_{sj} \right) = \left(M - \sum_{j=1}^{m} M_{sj} \right) g \cos\theta - T\left(\delta_p + \frac{\partial \xi_{ze}(G)}{\partial l} \right)$$

$$-QS \int \frac{\partial C_{N\alpha}(l)}{\partial l} \bar{\alpha}(l) dl - M_R \left(L_R + L_C \right) q - M_R L_R \delta_P$$

$$+ \sum_{j=1}^{m} M_{sj} (u + g \sin\theta) \lambda_{pj} - M_R \sum_{j=1}^{n} \left(\phi_{zi}(G) + L_R \frac{\partial \xi_{ze}(G)}{\partial l} \right) q_i \quad \ldots (2.140)$$

This is same as the Greensite's force equation. It may be noted that Greensite's equation needs vehicle mass excluding slosh masses, which is also indicated by Eq. (2.140).

The torque eq. (2.133) is expanded as follows:

LHS = External torque $\quad ty$

$$= g_x \left\{ Mz_{cg} + M_R \left[L_R \delta_P + \xi_{ze}(G) + L_R \frac{\partial \xi_{ze}(G)}{\partial l} \right] + \sum_{j=1}^{m} \left(\xi_{ze}(sj) + L_{sj} \lambda_{pj} \right) \right\}$$

$$- g_z \left\{ -M_R \left(L_R + L_C \right) + \sum_{j=1}^{m} x_{sj} M_{sj} \right\} + T \left[\xi_{ze}(G) - L_c \left(\frac{\partial \xi_{ze}(G)}{\partial l} + \delta_P \right) \right]$$

$$+ QS \int_0^l \frac{\partial C_{N\alpha}(l)}{\partial l} \bar{\alpha}(l_{cg} - l) dl \qquad\qquad \ldots (2.141)$$

RHS terms of eq. 2.133 are expanded as:

$$A_x \int \hat{z} dm = A_x \left[\int_R \xi_z dm + M_R \left(\xi_{ze}(G) + L_R \left(\frac{\partial \xi_{ze}(G)}{\partial l} + \delta_P \right) \right) + \sum_{j=1}^{m} \left(\xi_{ze}(sj) + L_{sj} \lambda_{pj} \right) \right]$$

$$\ldots (2.142)$$

$$A_z \int \hat{x} dm = A_z \left[-M_R \left(L_R + L_C \right) + \sum_{j=1}^{m} x_{sj} M_{sj} \right] \qquad\qquad .. (2.143)$$

$$q \int \hat{i}_{yy} = q I_{yy}; \quad q \int \frac{d\hat{i}_{yy}}{dt} \approx 0 \qquad\qquad \ldots (2.144)$$

$$\int \eta_y i_{yy} = \int (z \xi_x - x \xi_z) dm = - \int x \xi_z \, dm$$

$$= \left(I_R + M_R L_R L_C\right)\delta_P - M_R\left(L_R + L_C\right)\xi_{ze}(G) - \left(I_R + M_R L_R L_C\right)\frac{\partial \xi_z(G)}{\partial l}$$

$$-\sum_{j=1}^{m} x_{sj}\left(\xi_{ze}(sj) + L_{sj}\lambda_{pj}\right)M_{sj} \qquad \ldots (2.145)$$

$$K\int \eta_{ye} = 0 \qquad \ldots (2.146)$$

Also $\qquad z_{cg} = 0, \qquad \int_R \xi_z\, dm = 0 \qquad \ldots (2.146)$

Torque Equation:

Using Eq. (2.138) and eqns. 2.141 to 2.146, the torque equation is given by:

$$\left(I_{yy} - \sum M_{sj} x_{sj}^2\right)q = -TL_C\left(\delta_P + \frac{\partial \xi_{ze}(G)}{\partial l}\right) + T\xi_{ze}(G) + QS\int \frac{\partial C_{N\alpha}(l)}{\partial l}\overline{\alpha}(l_{cg} - l)dl$$

$$-M_R L_R\left(A_x + g\sin\theta\right)\delta_P - M_R\left(L_R + L_C\right)\left(A_z - g\cos\theta\right)$$

$$-\left(I_R + M_R L_R L_C\right)\delta_P - M_R\left(A_x + g\sin\theta\right)\left(\underline{\xi_{ze}(G)} + L_R\frac{\partial \xi_{ze}(G)}{\partial l}\right)$$

$$-\sum_{j=1}^{m} M_{sj}\left(A_x + g\sin\theta\right)\left(X_{hj}\lambda_{pj} + \underline{\xi_{ze}(sj)}\right)$$

$$-M_R\left(L_R + L_C\right)\xi_{ze}(G) - \left(I_R + M_R L_R L_C\right)\frac{\partial \xi_{ze}(G)}{\partial l} \qquad \ldots (2.147)$$

where $X_{hj} = x_{sj} + L_{sj} = $ Location of hinge point for slosh pendulum. $\qquad \ldots (2.148)$

Underlined terms are missing in Greensite's model. Greensite remarks that the slosh mass, being free to move inside the propellant tank, does not contribute to moment of inertia. Hence, the moment of inertia in his equation needs to be obtained by excluding the slosh masses.

This fact has been automatically obtained in Eq. 2.147

Greensite uses following expressions for force and moment due to slosh.

$$\text{Force} = \sum_{i=1}^{m} m_{sj}\, u\, \lambda_{pj} \qquad \ldots (2.149)$$

$$\text{Moment} = -\sum_{i=1}^{m} m_{sj} x_{hj}\, u\, \lambda_{pj} \qquad \ldots (2.150)$$

These expressions are obviously inadequate (even after ignoring the correction for potential energy which he has neglected) as they consider the effect of only forward acceleration and do not account for the effect of lateral acceleration (A_z) and angular acceleration (q).

However, surprisingly enough, these are the terms finally appearing in the dynamic equations and the other terms after rearrangement suggest that the magnitudes of the mass and moment of inertia to be used in the dynamic equation should be

$$M' = M - \sum_{j=1}^{m} M_{sj} \qquad \qquad \text{... (2.151)}$$

$$I'_{yy} = I_{yy} - \sum_{j=1}^{m} M_{sj} x_{sj}^2 \qquad \qquad \text{... (2.152)}$$

which is also the assumption made by Greensite. (Also see Eq. 4.20 and 4.21 of chapter IV).

Generalized Coordinate Equation

Consider the force equation f_z from Eq. (2.26) and ignoring nonlinear and coupling terms, one gets

$$f_z = A_z\, dm - \hat{x}\, dmq + \xi_z\, dm + K\xi_{ze} \qquad \qquad \text{... (2.153)}$$

OR $\qquad \xi_z\, dm + K\xi_{ze} = -A_z\, dm + \hat{x}q\, dm + f_z$

Let $\qquad \qquad \xi_z = \sum_{i=1}^{n} \phi_{zi}\, q_i$

To get the equation for ith generalized coordinate, multiply both sides by $\phi_{zi}(l)$ and integrate over the whole vehicle.

Remembering that the whole vehicle consists of:

  (*i*) Reduced vehicle

  (*ii*) Gimballed engine and

(*iii*) Slosh masses

and since the mode shapes (for Greensite's model) are calculated for reduced vehicle,

$$\int_R \phi_{zi}(x)\, dm = 0, \quad \int_R \phi_{zi}(x)\, x\, dm = 0, \qquad \qquad \text{... (2.155)}$$

$$\int_R \phi_{zi}(x)\phi_{zj}(x)\, dm = m_i \qquad \text{for } i = j$$

$$= 0 \qquad \text{for } i \ne j \qquad \qquad (2.156)$$

where m_i is the generalised mass for ith mode.

$$\int K \ \phi_{zi}(x)\sum_{j=1}^{n}\phi_{zj}(x)q_j = \omega_i^2 q_i m_i \qquad\qquad \ldots (2.157)$$

1. $\displaystyle\int \phi_{zi}\xi_z \, dm = \int \phi_{zi}\sum_{j=1}^{n}\phi_{zj}q_j \, dm = \int_{R} + \int_{Engine} + \int_{slosh}$

$$= m_i q_i + M_R\left[\phi_{zi}(G) + L_R\frac{\partial\phi_{zi}(G)}{\partial l}\right]\left[L_R\delta_P + \sum_{j=1}^{n}\left(\phi_{zj}(G) + L_R\frac{\partial\phi_{zj}(G)}{\partial l}\right)q_j\right]$$

$$+\sum M_{sj}\phi_{zi}(sj)\left(\xi_{zj}(sj) + L_{sj}\lambda_{pj}\right) \qquad\qquad .. (2.158)$$

2. $\displaystyle\int \phi_{zi}K\xi_{ze} = \omega_i^2 m_i q_i \qquad\qquad \ldots (2.159)$

3. $\displaystyle -\int A_z\phi_{zi}(l)dm = -\left[\int_{R}\ldots = 0\right] - A_z M_R\left[\phi_{zi}(G) + L_R\frac{\partial\phi_{zi}(G)}{\partial l}\right] - A_z\sum_{j=1}^{m}M_{sj}\phi_{zi}(sj)$

$$\ldots (2.160)$$

4. $\displaystyle\int \phi_{zi}(l)\hat{x}\ q\ dm = [0] - M_R q[L_C + L_R]\left[\phi_{zi}(G) + L_R\frac{\partial\phi_{zi}(G)}{\partial l}\right]$

$$+q\sum_{j=1}^{m}\phi_{zi}(sj)x_{sj}M_{sj} \qquad\qquad \ldots (2.161)$$

5. $\displaystyle\int f_z\phi_{zi} = -T\left(\delta_p + \frac{\partial\xi_{zi}(G)}{\partial l}\right)\phi_{zi}(G) - QS\int\frac{\partial C_{N\alpha}(l)}{\partial l}\bar{\alpha}(l)\phi_{zi}(l)dl$

$$+\left(g\cos\theta\int_{R}\phi_{zi}\,dm\,dl = 0\right) + M_R g\cos\theta\left[\phi_{zi}(G) + L_R\frac{\partial\phi_{zi}(G)}{\partial l}\right]$$

$$+g\cos\theta\sum M_{sj}\phi_{zi}(sj) \qquad\qquad \ldots (2.162)$$

Then collecting various terms and using slosh equation 2.138 and equations 2.158 to 2.162, the generalized coordinate equation can be written as:

$$m_i q_i + \omega_i^2 m_i q_i = -T\phi_{zi}(G)\delta_P - T\phi_{zi}(G)\sum_{j=1}^{n}\frac{\partial\phi_{zj}(G)}{\partial l}q_j - QS\int\frac{\partial C_{N\alpha}}{\partial l}(l)\bar{\alpha}(l)\phi_{zi}(l)dl$$

$$-M_R\left(A_z - g\cos\theta\right)\left[\phi_{zi}(G) + L_R\frac{\partial\phi_{zi}(G)}{\partial l}\right] - M_R L_R\left[\phi_{zi}(G) + L_R\frac{\partial\phi_{zi}(G)}{\partial l}\right]\delta_p$$

$$-M_R(L_C + L_R)\left[\phi_{zi}(G) + L_R \frac{\partial \phi_{zi}(G)}{\partial l}\right]\theta$$

$$-M_R\left[\underline{\phi_{zi}(G) + L_R \frac{\partial \phi_{zi}(G)}{\partial l}}\right]\sum_{k=1}^{n}\left(\phi_{zk}(G) + L_R \frac{\partial \phi_{zk}(G)}{\partial l}\right)q_k$$

$$+(u + g\sin\theta)\sum_{j=1}^{m} M_{sj}\,\underline{\underline{\phi_{zi}(sj)}}\lambda_{pj} \qquad\qquad \dots (2.163)$$

In the above equation, $q \equiv \theta$ has been used.

This is the generalized coordinate equation when the mode shapes and frequencies are obtained for 'reduced' vehicle for the sign conventions assumed by this author for the state variables. The Greensite's model for his conventions can be simply obtained using Table 2.2. The model exactly matches with the Greensite's model except for the underlined terms. The terms underlined by single line (_______) are absent in Greensite's model implying that the engine mass is considered to act at gimbal point instead of at its CG. The term underlined by double line (=========) is taken by Greensite at the hinge point of slosh pendulum instead of at CG of slosh mass.

Equations 2.137, 2.138, 2.147 and 2.162 when modified using Table 2.2 exactly matches with Greensite's model. This fact confirms that the Greensite's model requires mode shapes for the 'reduced' vehicle and the motion is also considered about the CG of the reduced vehicle. These two factors are primarily responsible for making this model a coupled model in the sense that:

1. Linear acceleration equation has θ and q_i terms on right hand side (RHS)
2. Angular acceleration equation has A_z and θ_i on RHS.
3. q_i equation has A_z, θ and q_j for $j = 1$ to n on RHS.

This coupling is present even when external forces and moments are neglected. If the modes are orthogonal, one mode should not be affected by the other mode. The above derivation makes it clear that this coupling is occurring due to the fact that the 'reduced' vehicle parameters have been used for this model. The Lagrangian approach adopted by Greensite does not give any clue as to how to avoid this coupling.

The generalised model described in this chapter not only enables one to obtain the decoupled model using total vehicle parameters but also enables one to obtain the coupled model using the 'reduced' vehicle parameters. The model equations 2.120, 2.121 and 2.125 do not have the coupling described above if the external forces are neglected. The coupling between the modes however occurs due to the fact that the aerodynamic and propulsion forces depend on the net deflected shape of the vehicle which is the effect of several modes present simultaneously.

The fact that the coupled model requires 'reduced' vehicle parameters and mode shapes also for 'reduced' vehicle is not generally realized by many users of this model and total vehicle parameters and mode shapes themselves are (erroneously) used under tacit assumption that those are compatible.

We have not made any study to see the difference in stability margins due to erroneous use of data. It is felt that the difference may not be significant. However, it gives a great satisfaction to have a decoupled model when all modes are really orthogonal.

2.5 CONCLUDING REMARKS

This chapter gives the derivation of the generalized equations of motion. The format is similar to 6DOF equations with few additional terms. It then gives a simple procedure which enables one to incorporate any number of modes of flexibility, slosh and gimballed engine dynamics very easily without having to go for a fresh derivation.

It also gives models for slosh and gimballed engine dynamics which can be written down by inspection considering it just like a pendulum. There is no need to remember the expressions. This is a very attractive feature of this methodology of derivation.

The chapter then gives the complete model which is useful for 6DOF simulation and also a linear model which is suitable for stability analysis of control system.

It has been shown that the model gives decoupling between the orthogonal modes when the modes are computed for total vehicle. It also shows how the Greensite's model which indicates coupling between the modes can be derived using the generalized model bringing out clearly that the coupling is due to the fact that the Greensite's model assumes that the structural modes are computed for reduced vehicle and the motion is considered about the CG of reduced vehicle. In the process it brings out clearly what needs to be done to avoid the coupling.

REFERENCES

1. **V.L. Alley, A.M. Gerringer:** A matrix method for the determination of natural vibrations of free-free unsymmetrical beams with applications to launch vehicles NASA TN D 1247, April 1962.

2. **A.L. Greensite:** Control theory, Vol. II – Analysis and design of space vehicle flight control systems, Spartan books, 1970 (Chapter 2)

3. **R.L. Swaim:** Control system synthesis for launch vehicles with severe mode ineractions, IEEE Trans. on automatic control, Vol. 14, Oct 1969, pp. 517 – 523

4. **Harold C. Lester, Dennis F. Collins:** Determination of loads on a flexible launch vehicle during ascent through winds. NASA TN D – 2590, Feb. 1965

5. **K.L. Handoo, B. Saseendran, K. Prabhakaran:** Flexible vehicle response analysis of PSLV – Details of modeling and mathematical formulation – Revision 1 VSSC.SEG:SDD:TR(72)/1990, June 1990.

6. **N.V. Kadam:** A generalized model for the dynamics of a flexible structure and its application to modelling the control-structure-slosh interactions in missiles, DRDL.6100.1024.000, April 1995.

7. **N.V. Kadam:** A generalized model for structure-control-slosh interactions in missiles. Proceedings of National Systems Confe.rence NSC-97 (Jan 23-24, 1998) at RCI, Hyderabad

8. **N.V. Kadam:** A model for simultaneous pitch and yaw slosh motions, DRDL.6100.1013.000, March 1993.

9. **J.E. Roberts, E.R. Basurto, Pei–Ying Chen:** Slosh design handbook I, NASA CR 406, May 1966.

10. **H.N. Abramson:** The dynamic behaviour of liquids in moving containers, NASA SP 106, 1966.

11. **N.V. Kadam:** A model for simultaneous pitch and yaw slosh motions, DRDL.6160.1013.000, Mar 1993.

12. **N.V. Kadam, K.L. Handoo, E.M. George, T.B. Haridas:** A short period model for ASLV including vehicle flexibility for use in the design and analysis of attitude control system, VSSC-CGD-CSA-5-84, May 1984.

APPENDIX 2.1	A MODEL FOR SIMULTANEOUS PITCH AND YAW SLOSH MOTIONS

1. INTRODUCTION

This appendix gives the derivation of slosh dynamics simultaneously in pitch and yaw planes using Langrangian approach. The derivation uses the usual body coordinate system and assumes pendulum anology for the slosh motion. A table is given in the end to modify the equations for the spring mass model.

2. DERIVATION OF EQUATIONS

Let the missile vibrations be represented as

$$\overline{\xi}(x,t) = \xi_x(x,t)\ \overline{i} + \xi_y(x,t)\ \overline{j} + \xi_z(x,t)\ \overline{k}$$

$$= \sum_{i=1}^{n} \left[\phi_{x_i}(x)\ i + \phi_{y_i}(x)\ \overline{j} + \phi_{z_i}(x)\ \overline{k} \right] q_i$$

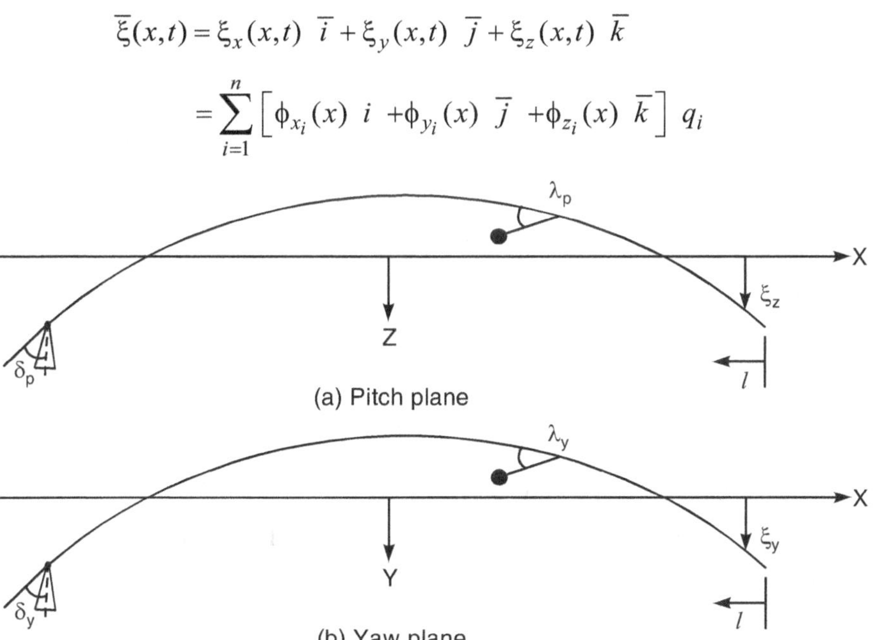

Fig. A2.1.1 Gives the convention for positive structural deflection

Let O' be the origin of body coordinate system

$l_0{}'$: distance of origin from nose tip

l_{hi} : hinge point from nose tip

pi : Angle of instantaneous ith pendulum line with missile axis in pitch plane

$_{yi}$	: Angle of instantaneous ith pendulum line with missile axis in yaw plane
$\bar{\rho}_i$	: Position vector for ith slosh mass
m_{si}	: ith slosh mass
L_{si}	: ith slosh pendulum length
u, v, w	: Missile velocity components in body axes
p, q, r	: Missile angular velocity in body axes system
$_x(si),\ _y(si),\ _z(si)$	: Structural deflection at ith slosh mass location

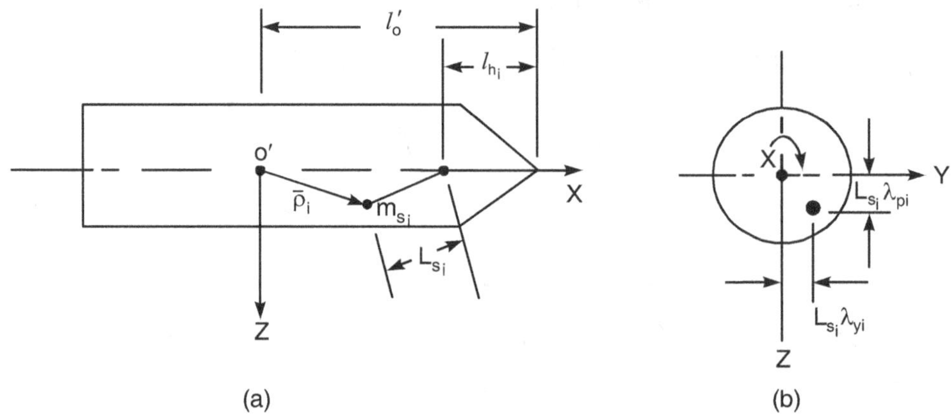

Fig. A2.1.2

Fig. A2.1.2 gives a sketch of instantaneous ith pendulum position.

Each slosh mass is independent of other slosh masses and the equation for the individual modes can be written independently.

$$\bar{\rho}_i = \left[l_{0'} - l_{hi} - L_{si}\cos(\sqrt{\lambda_{pi}^2 + \lambda_{yi}^2}) + \xi_x(si) \right]\bar{i}$$

$$+ \left[L_{si}\lambda_{yi} + \xi_y(si) \right]\bar{j} + \left[Lsi\lambda_{pi} + \xi_z(si) \right]\bar{k}$$

where (si) indicates vehicle deflection at ith slosh mass position.

The velocity of slosh mass is given by

$$\bar{\mu}_i = \bar{\mu}_{0'} + \frac{\partial \bar{\rho}_i}{\partial t} + \bar{\omega} \times \bar{\rho}_i$$

$$= \left[u + L_{si}(\lambda_{pi}\lambda_{pi} + \lambda_{yi}\lambda_{yi}) + \xi_x(si) + q(L_{si}\lambda_{pi} + \xi_z(si)) - r(L_{si}\lambda_{yi} + \xi_y(si)) \right] \bar{i}$$

$$+ \left[v + L_{si}\lambda_{yi} + \xi_y(si) + r\left(l_{0'} - l_{hi} - L_{si} + L_{si}\left(\frac{\lambda_{pi}^2 + \lambda_{yi}^2}{2} \right) + \xi_x(si) \right) - p(L_{si}\lambda_{pi} + \xi_z(si)) \right] \bar{j}$$

$$+ \left[w + L_{si}\lambda_{pi} + \xi_z(si) + p\left(L_{si}\lambda_{yi} + \xi_y(si) \right) - q\left(l_{0'} - l_{hi} - L_{si} + L_{si}\left(\frac{\lambda_{pi}^2 + \lambda_{yi}^2}{2} \right) + \xi_x(si) \right) \right] \bar{k}$$

Kinetic energy:
$$T = \frac{1}{2} m_{si} \bar{\mu}_i \cdot \bar{\mu}_i$$

Potential energy (V): $\qquad h_i \qquad x_i \sin \qquad z_i \cos$

Potential energy V is given by

$$V \quad m_{si} g \quad h_i$$

$$= m_{si} g \left[L_{si}\left(\frac{\lambda_{pi}^2 + \lambda_{yi}^2}{2} \right) + \xi_x(si) \right] \sin\theta - \left(L_{si}\lambda_{pi} + \xi_z(si) \right)\cos\theta$$

$$\therefore \qquad \frac{\partial V}{\partial \lambda_{pi}} = m_{si} g L_{si}\left[\lambda_{pi} \sin\theta - \cos\theta \right]$$

$$\frac{\partial V}{\partial \lambda_{yi}} = m_{si} g L_{si}\lambda_{yi} \sin\theta$$

Damping term $2\xi_i\omega_i L_{si}\lambda_{pi}$ and $2\xi_i\omega_i L_{si}\lambda_{yi}$ will be added in the final equations (otherwise a dissipation function can be considered in the beginning itself)

The slosh equation will then be given by

$$\frac{d}{dt}\frac{\partial T}{\partial \lambda_{pi}} - \frac{\partial T}{\partial \lambda_{pi}} + \frac{\partial V}{\partial \lambda_{pi}} = 0$$

$$\frac{d}{dt}\frac{\partial T}{\partial \lambda_{yi}} - \frac{\partial T}{\partial \lambda_{yi}} + \frac{\partial V}{\partial \lambda_{yi}} = 0$$

On putting the values of each term and simplifying these equations will lead to (after incorporating damping)

$$L_{si}\lambda_{pi} + \xi_z(si) + 2\xi_i\omega_i L_{si}\lambda_{pi} + w + pv - qu - g\cos\theta$$

$$+(-q + pr)\left\{ l_{o'} - l_{hi} - L_{si} + L_{si}\left(\frac{\lambda_{pi}^2 + \lambda_{yi}^2}{2} \right) + \xi_x(si) \right\}$$

$$+(p + qr)\left(L_{si}\lambda_{yi} + \xi_y(si) \right) - (p^2 + q^2)(L_{si}\lambda_{pi} + \xi_z(si))$$

$$+2p\left(L_{si}\lambda_{yi}+\xi_y(si)\right)-2q\left\{L_{si}(\lambda_{pi}\lambda_{pi}+\lambda_{yi}\lambda_{yi})+\xi_x(si)\right\}$$

$$+\lambda_{pi}[u+qw-rv+g\sin\theta-(q^2+r^2)(l_{o'}-l_{hi}-L_{si}-L_{si}\left(\frac{\lambda_{pi}^2+\lambda_{yi}^2}{2}\right)+\xi_x(si))$$

$$-(r+pq)\left(L_{si}\lambda_{yi}+\xi_y(si)\right)+(q+rp)(L_{si}\lambda_{pi}+\xi_z(si))$$

$$-2r\left(L_{si}\lambda_{yi}+\xi_y(si)\right)+2q\left(L_{si}(\lambda_{pi}+\xi_z(si)\right)$$

$$+L_{si}\left(\lambda_{si}^2+\lambda_{yi}^2+\lambda_{pi}\lambda_{pi}+\lambda_{yi}\lambda_{yi}\right)+\xi_x(si)]=0$$

For simplicity, neglect higher order terms in square bracket which are negligible in comparison with the forward acceleration of the missile $(u+qw-rv)$ and use

$$\omega_{si}^2=(u+qw-rv+g\sin\theta)/L_{si}$$

Then the pitch plane slosh motion is given by:

(Also neglect vehicle deflection along x *i.e.*, $\xi_x(si)$ for the purpose of present 6DOF program)

$$L_{si}\lambda_{pi}+2\zeta_i\omega_{si}L_{si}\lambda_{pi}+\omega_{si}^2 L_{si}\lambda_{pi}+\xi_z(si)+(w+pv-qu-g\cos\theta)$$

$$+(-q+pr)\left[l_{o'}-l_{hi}-L_{si}+L_{si}\left(\frac{\lambda_{pi}^2+\lambda_{yi}^2}{2}\right)\right]$$

$$+(p+qr)\left(L_{si}\lambda_{yi}+\xi_y(si)\right)-(p^2+q^2)(L_{si}\lambda_{pi}+\xi_z(si))$$

$$+2p(L_{si}\lambda_{yi}+\xi_y(si))-2q\left[L_{si}\left(\lambda_{pi}\lambda_{pi}+\lambda_{yi}\lambda_{yi}\right)\right]=0$$

Equation for yaw plane slosh motion is similarly obtained and is given as:

$$L_{si}\lambda_{yi}+2\zeta_i\omega_{si}L_{si}\lambda_{yi}+\omega_{si}^2 L_{si}\lambda_{yi}+\xi_y(si)$$

$$+(v+ru-pw)+(r+pq)\left[l_{o'}-l_{hi}-L_{si}+L_{si}\left(\frac{\lambda_{pi}^2+\lambda_{yi}^2}{2}\right)\right]$$

$$-(r^2+p^2)\left(L_{si}\lambda_{yi}+\xi_y(si)\right)+(-p+rq)(L_{si}\lambda_{pi}+\xi_z(si))$$

$$+2r[L_{si}(\lambda_{pi}\lambda_{pi}+\lambda_{yi}\lambda_{yi})]-2p(L_{si}\lambda_{pi}+\xi_z(si))=0$$

3. CONVERSION OF PENDULUM MODEL TO SPRING MASS MODEL

The above equations are given for pendulum model. These can be converted to spring mass model for using in 6DOF program by identifying following quantities:

Qty.	Pendulum Model	Spring mass model
1. Slosh mass displacement in pitch plane	$L_{si}\lambda_{pi}$	z_{si}
2. Slosh mass displacement in yaw lane	$L_{si}\lambda_{yi}$	y_{si}
3. Velocity in pitch plane	$L_{si}\lambda_{pi}$	z_{si}
4. Velocity in yaw plane	$L_{si}\lambda_{yi}$	y_{si}
5. Acc^n in pitch plane	$L_{si}\lambda_{yi}$	z_{si}
6. Acc^n in yaw plane	$L_{si}\lambda_{yi}$	y_{si}
7. Position of slosh mass along x = axis for $\lambda_{pi}=\lambda_{yi}=0$	$(l_o - l_{hi} - L_{si})$	X_{si} if X is measured +ve towards nose
8. Position of slosh mass along x-axis for non zero $\lambda_{pi}, \lambda_{yi}$	$l_o - l_{hi} - L_{si} + L_{si}\left(\dfrac{\lambda_{pi}^2 + \lambda_{yi}^2}{2}\right)$	No provision in spring mass model*
9. Velocity of slosh mass along x-axis	$L_{si}(\lambda_{pi}\lambda_{pi} + \lambda_{yi}\lambda_{yi})$	No provision in spring mass model*

*To be ignored

CHAPTER 3

Control Systems Design–1: Configuration and Sizing

3.1 INTRODUCTION

Control System design for both the launch vehicles and missiles consists of two parts. One part covers the configuration of the control system feedback loops, detailed stability analysis and design of control system gains, filters or compensators for the desired stability margins and speed of response etc. Other part of control system design addresses the more general aspects of the problem such as resolution of errors to generate the required signals for control system, planning the control systems for various segments of trajectory, sizing of the control system which implies determination of maximum control force required, control impulse required to be stored, acceptable bounds on attitude errors, roll manoeuvring to orient the vehicle axes in the required orientation, static margin requirement, actuator requirement, sensor requirements etc. This chapter concentrates on these general aspects of control system design and subsequent chapter gives the detailed loop design aspects.

3.2   RESOLUTION OF ATTITUDE ERRORS IN BODY AXES FRAME

The launch vehicles mainly adopt attitude control law. Fig 3.1. shows a schematic block diagram of attitude control systems for pitch, yaw and roll axes.

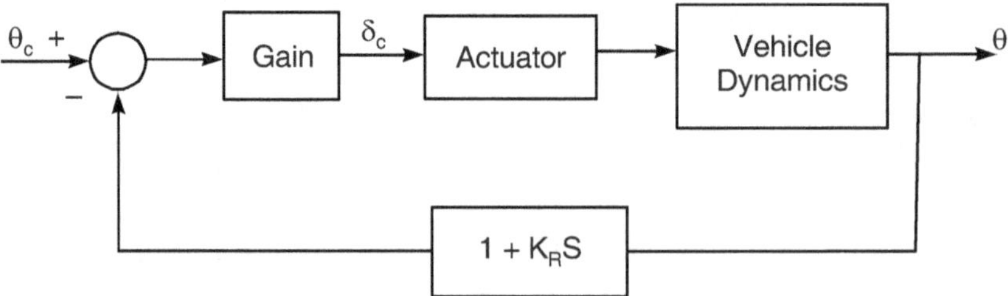

Fig. 3.1 Schematic Diagram of Attitude Control System

The vehicle attitude with reference to inertial frame of reference is sensed by inertial measurement unit. This unit basically senses linear accelerations and angular rates which are further integrated by processors to get the complete navigation information consisting of vehicle position (x, y, z), velocity (V_x, V_y, V_z) and vehicle attitude θ, ψ, ϕ. There are two standard methods for describing the vehicle attitudes *i.e.*, using the Euler angles (θ, ψ, ϕ) or quaternions, $Q = (q_0, q_1, q_2, q_3)$. Depending on the scheme, the preprogrammed time variation of the vehicle attitude is stored in the computer which serves as the commands for the controls system.

Mainly, the launch vehicles use the pitch program to achieve the given orbit but many times yaw program is also used to have a desirable ground trace so that the separated stages do not have a possibility of impacting on populated areas. Thus, a general stored trajectory is given in the form of θ_d, ψ_d, ϕ_d as a function of time (as a table) or in the form of desired quaternion history Q_d which means four components of quaternion $(q_{d0}, q_{d1}, q_{d2}, q_{d3})$ as a function of time.

Control actuators and sensors are mounted on the vehicle and thus fixed with reference to body axes frame. For influencing the vehicle attitude the appropriate actuators need to be given the control signals. Hence, it is essential that the attitude errors which are obtained by comparing the vehicle attitude sensed by the inertial reference unit with the stored attitude at a particular time are resolved in body coordinate system for proper functioning of the control system.

We describe here both the Euler angle as well as quaternion version of attitude sensing scheme.

3.2.1 Error Resolution in Terms of Euler Angles

Let X_R, Y_R, Z_R denote the reference frame

X_b, Y_b, Z_b denote the vehicle body frame

θ_d, ψ_d, ϕ_d give the desired orientation of the vehicle at a time

(in that sequence of Euler angles)

and θ, ψ, ϕ give the actual orientation of the vehicle. Where θ is rotation about y-axis, ψ is rotation about z-axis and ϕ is rotation about x-axis.

Then,

$$\begin{bmatrix} X_{bd} \\ Y_{bd} \\ Z_{bd} \end{bmatrix} = \begin{bmatrix} 1 & 0 & 0 \\ 0 & \cos\phi_d & \sin\phi_d \\ 0 & -\sin\phi_d & \cos\phi_d \end{bmatrix} \begin{bmatrix} \cos\psi_d & \sin\psi_d & 0 \\ -\sin\psi_d & \cos\psi_d & 0 \\ 0 & 0 & 1 \end{bmatrix} \begin{bmatrix} \cos\theta_d & 0 & -\sin\theta_d \\ 0 & 1 & 0 \\ \sin\theta_d & 0 & \cos\theta_d \end{bmatrix} \begin{bmatrix} X_R \\ Y_R \\ Z_R \end{bmatrix}$$

$$\dots (3.1)$$

It is assumed that the reference and body frame x,y,z axes coincide for $\theta = \psi = \phi = 0$. In case, the inertial reference axes system is different, an appropriate transformation needs to be used to relate inertial frame with the assumed reference frame. Actual orientation is given by

$$\begin{bmatrix} X_{ba} \\ Y_{ba} \\ Z_{ba} \end{bmatrix} = \begin{bmatrix} 1 & 0 & 0 \\ 0 & \cos\phi_a & \sin\phi_a \\ 0 & -\sin\phi_a & \cos\phi_a \end{bmatrix} \begin{bmatrix} \cos\psi_a & \sin\psi_a & 0 \\ -\sin\psi_a & \cos\psi_a & 0 \\ 0 & 0 & 1 \end{bmatrix} \begin{bmatrix} \cos\theta_a & 0 & -\sin\theta_a \\ 0 & 1 & 0 \\ \sin\theta_a & 0 & \cos\theta_a \end{bmatrix} \begin{bmatrix} X_R \\ Y_R \\ Z_R \end{bmatrix}$$

$$...(3.2)$$

Denote the transformation matrix as $[\theta_d], [\psi_d]$ etc. Then the desired body axis orientation is obtained in terms of actual body axis orientation as follows:

$$\begin{bmatrix} X_{bd} \\ Y_{bd} \\ Z_{bd} \end{bmatrix} = [\phi_d][\psi_d][\theta_d][\theta_a]^T[\psi_a]^T[\phi_a]^T \begin{bmatrix} X_{ba} \\ Y_{ba} \\ Z_{ba} \end{bmatrix} = [D][\phi_a]^T \begin{bmatrix} X_{ba} \\ Y_{ba} \\ Z_{ba} \end{bmatrix} \qquad ... (3.3)$$

where D matrix is given by

$$D_{11} = \cos(\theta_d - \theta_a)\cos\psi_d\cos\psi_a + \sin\psi_d\sin\psi_a \qquad\qquad ... (3.4)$$

$$D_{21} = -\cos(\theta_d - \theta_a)\cos\psi_a\cos\phi_d\sin\psi_d + \sin\psi_a\cos\phi_d\cos\psi_d$$
$$+ \sin(\theta_d - \theta_a)\cos\psi_a\sin\phi_d \quad ... (3.5)$$

$$D_{31} = \cos(\theta_d - \theta_a)\cos\psi_a\sin\phi_d\sin\psi_d - \sin\psi_a\sin\phi_d\cos\psi_d$$
$$+ \sin(\theta_c - \theta_a)\cos\psi_a\cos\phi_d \quad ... (3.6)$$

$$D_{12} = -\cos(\theta_d - \theta_a)\sin\psi_a\cos\psi_d + \cos\psi_a\sin\psi_d \qquad\qquad ... (3.7)$$

$$D_{22} = \cos(\theta_d - \theta_a)\sin\psi_a\cos\phi_d\sin\psi_d + \cos\psi_a\cos\phi_d\cos\psi_d$$
$$- \sin(\theta_d - \theta_a)\sin\psi_a\sin\phi_d \quad ... (3.8)$$

$$D_{32} = -\cos(\theta_d - \theta_a)\sin\psi_a\sin\phi_d\sin\psi_d - \cos\psi_a\sin\phi_d\cos\psi_d$$
$$- \sin(\theta_d - \theta_a)\sin\psi_a\cos\phi_d \quad ... (3.9)$$

$$D_{13} = -\sin(\theta_d - \theta_a)\cos\psi_d \qquad\qquad ... (3.10)$$

$$D_{23} = \sin(\theta_d - \theta_a)\cos\phi_d\sin\psi_d + \cos(\theta_d - \theta_a)\sin\phi_d \qquad\qquad ... (3.11)$$

$$D_{33} = \sin(\theta_d - \theta_a)\sin\phi_d\sin\psi_d + \cos(\theta_d - \theta_a)\cos\phi_d \qquad\qquad ... (3.12)$$

Then
$$\begin{bmatrix} X_{bd} \\ Y_{bd} \\ Z_{bd} \end{bmatrix} = \begin{bmatrix} D_{11} & D_{12}\cos\phi_a + D_{13}\sin\phi_a & -D_{12}\sin\phi_a + D_{13}\cos\phi_a \\ D_{21} & D_{22}\cos\phi_a + D_{13}\sin\phi_a & -D_{22}\sin\phi_a + D_{23}\cos\phi_a \\ D_{31} & D_{32}\cos\phi_a + D_{33}\sin\phi_a & -D_{32}\sin\phi_a + D_{33}\cos\phi_a \end{bmatrix} \begin{bmatrix} X_{ba} \\ Y_{ba} \\ Z_{ba} \end{bmatrix} \quad \ldots (3.13)$$

Now, let θ_e, ψ_e, ϕ_e be the angular rotations about the actual body axes required to align them with the desired body axes system. Being the control errors, these are assumed to be small angles. Then

$$\begin{bmatrix} X_{bd} \\ Y_{bd} \\ Z_{bd} \end{bmatrix} = \begin{bmatrix} 1 & 0 & 0 \\ 0 & 1 & \phi_e \\ 0 & -\phi_e & 1 \end{bmatrix} \begin{bmatrix} 1 & \psi_e & 0 \\ -\psi_e & 1 & 0 \\ 0 & 0 & 1 \end{bmatrix} \begin{bmatrix} 1 & 0 & -\theta_e \\ 0 & 1 & 0 \\ \theta_e & 0 & 1 \end{bmatrix} \begin{bmatrix} X_{ba} \\ Y_{ba} \\ Z_{ba} \end{bmatrix}$$

$$= \begin{bmatrix} 1 & \psi_e & -\theta_e \\ -\psi_e + \theta_e\phi_e & 1 & \psi_e\theta_e + \phi_e \\ \phi_e\psi_e + \theta_e & -\phi_e & -\theta_e\phi_e\psi_e + 1 \end{bmatrix} \begin{bmatrix} X_{ba} \\ Y_{ba} \\ Z_{ba} \end{bmatrix} \quad \ldots (3.14)$$

Equating the corresponding elements of matrices from Eq. 3.13 and 3.14, one gets

$$\theta_e = D_{12}\sin\phi_a - D_{13}\cos\phi_a \qquad\qquad \ldots (3.15)$$

$$\psi_e = D_{12}\cos\phi_a + D_{13}\sin\phi_a \qquad\qquad \ldots (3.16)$$

$$\phi_e = -D_{32}\cos\phi_a - D_{33}\sin\phi_a \qquad\qquad \ldots (3.17)$$

Using eqs. 3.4 to 3.13 and assuming $(\theta_d - \theta_a)$ and $(\psi_d - \psi_a)$ as small angles, one gets:

$$\theta_e = (\theta_d - \theta_a)\cos\psi_d \cos\phi_a + (\psi_d - \psi_a)\sin\phi_a \qquad\qquad \ldots (3.18)$$

$$\psi_e = -(\theta_d - \theta_a)\cos\psi_d \sin\phi_a + (\psi_d - \psi_a)\cos\phi_a \qquad\qquad \ldots (3.19)$$

$$\phi_e = (\phi_c - \phi_a) + (\theta_d - \theta_a)\sin\psi_d \qquad\qquad \ldots (3.20)$$

[we have used here

$$\sin\psi_a \cos\phi_d \cos\phi_a + \sin\psi_d \sin\phi_d \sin\phi_a \cong \sin\psi_a \cos(\phi_d - \phi_a) \approx \sin\psi_a \approx \sin\psi_d]$$

Eqs. 3.18, 3.19 and 3.20 give the required errors in the body axes frame which can be used as control system errors.

Alternate Derivation

The above error equations can also be derived as follows:

Fig. 3.2 gives the body axes orientation in relation to reference frame as obtained after θ, ψ and ϕ rotation.

The vehicle angular rates can be described by p,q,r in body frame as well as θ, ψ, θ as shown in the Fig. 3.2. Then

$$p = \phi + \theta \cos(90 - \psi) \qquad\qquad\qquad \ldots (3.21)$$

$$q = \theta \cos\psi \cos\phi + \psi \cos(90 - \phi) \qquad\qquad \ldots (3.22)$$

$$r = \psi \cos\phi + \theta \cos\psi \cos(90 + \phi) \qquad\qquad \ldots (3.23)$$

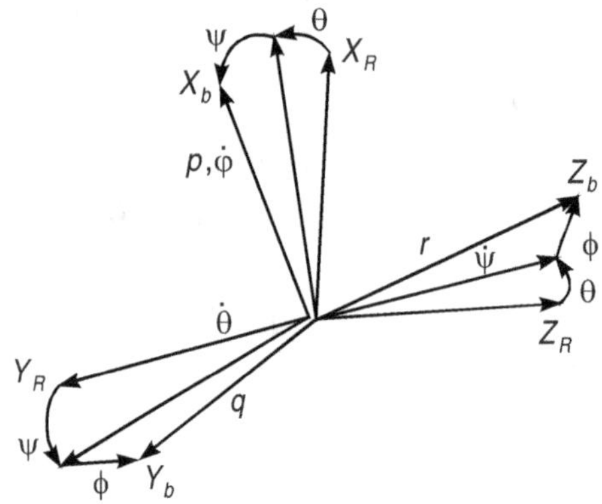

Fig. 3.2

Now, noting that,

$$p = \frac{d\phi_b}{dt}, q = \frac{d\theta_b}{dt}, r = \frac{d\psi_b}{dt} \qquad\qquad \ldots (3.24)$$

and

$$\phi = \frac{d\phi_R}{dt}, \theta = \frac{d\theta_R}{dt}, \psi = \frac{d\psi_R}{dt} \qquad\qquad \ldots (3.25)$$

we get (multiplying both sides by *dt*)

$$d\theta_b = d\theta_R \cos\psi \cos\phi + d\psi_R \sin\phi \qquad\qquad \ldots (3.26)$$

$$d\psi_b = -d\theta_R \cos\psi \sin\phi + d\psi_R \cos\phi \qquad\qquad \ldots (3.27)$$

$$d\phi_b = d\phi_R + d\theta_R \sin\psi \qquad\qquad\qquad \ldots (3.28)$$

Eqs. 3.26, 3.27 and 3.28 are same as Eqs. 3.18, 3.19 and 3.20 respectively, except that they use $\cos\psi_d$ instead of $\cos\psi_a$. If the error $(\psi_d - \psi_a)$ is small, any one of them would be usable. However, if the errors are not so small, it is preferable to use $\cos\psi_a$ instead of $\cos\psi_d$ since it accounts for actual missile orientation required for resolution. This is made clear by Eqns. 3.26 and 3.27. When the mission profile does not require yaw programming or yaw manoeuvre, the values $\psi_d \approx \psi_a \approx 0$ or are very small values. Then, the error equations reduce to

$$\theta_e = (\theta_d - \theta_a)\cos\phi_a + (\psi_d - \psi_a)\sin\phi_a \qquad \text{... (3.29)}$$

$$\psi_e = -(\theta_d - \theta_a)\sin\phi_a + (\psi_d - \psi_a)\cos\phi_a \qquad \text{... (3.30)}$$

$$\text{and } \phi_e = \phi_c - \phi \qquad \text{... (3.31)}$$

These are the expressions normally used for resolving errors in body frame accounting for vehicle developing some roll angles.

Fig. 3.3 shows a schematic diagram indicating the error resolution.

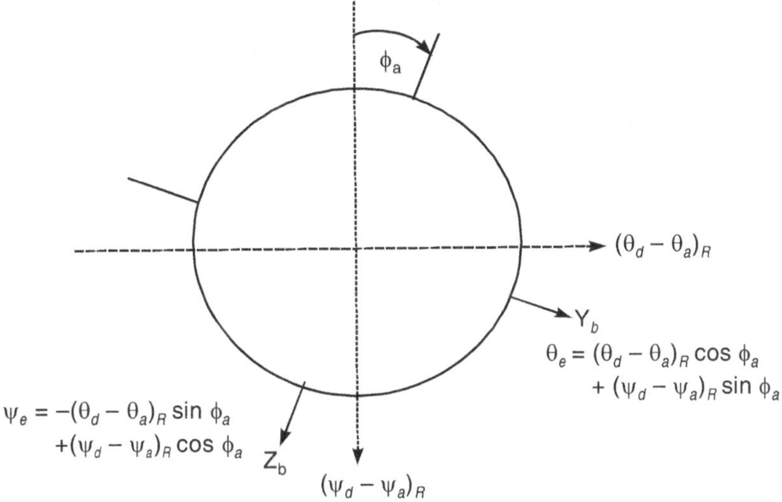

Fig. 3.3 Resolution of errors in Body Frame

If we solve equations 3.21, 3.22, 3.23 for θ, ψ, ϕ we get

$$\theta = (q\cos\phi - r\sin\phi)/\cos\psi \qquad \text{... (3.32)}$$

$$\psi = q\sin\phi + r\cos\phi \qquad \text{... (3.33)}$$

$$\phi = p - \tan\psi \,(q\cos\phi - r\sin\phi) \qquad \text{... (3.34)}$$

These equations bring out the singularity aspect of Euler angle approach as the middle rotation (in this case ψ) approaches 90°. The magnitudes of θ and ϕ tend to infinity.

3.2.2   Error Angles in Terms of Quaternions

3.2.2.1   *Brief Introduction about Quaternions*

A Quaternion gives a single rotation by which one coordinate frame is aligned with another coordinate frame. It consists of four elements (q_0, q_1, q_2, q_3) such that $(q_0^2 + q_1^2 + q_2^2 + q_3^2) = 1$. The first element q_0 is called the scalar (or real) part and gives the magnitude of the rotation. The next three elements are called the vector (or imaginary) part and give the direction cosines of the axis of rotation along the reference frame axes. The quaternions in such cases is also written as:

$$Q = (q_0 + q_1 i + q_2 j + q_3 k) \qquad \qquad \text{... (3.35)}$$

where i, j, k are the unit vectors along the reference axes system. The conjugate of the quaternion is then given by

$$Q^* = (q_0 - q_1 i - q_2 j - q_3 k) \qquad \qquad \text{... (3.36)}$$

And
$$\|Q\| = \sqrt{QQ^*} = \sqrt{q_0^2 + q_1^2 + q_2^2 + q_3^2} \qquad \qquad \text{... (3.37)}$$

The quantities i, j, k have the properties as
$$i^2 = j^2 = k^2 = -1, \quad ij = -ji = k, \quad jk = -kj = i \text{ and } ki = -ik = j \quad \text{... (3.38)}$$

Now, let a velocity vector in reference coordinate frame be given by

$$\overrightarrow{V}_R = u_R i + v_R j + w_R k \qquad \qquad \text{... (3.39)}$$

Then the product $Q^*\overline{V}_R Q$ is given by

$$Q^*\overrightarrow{V}_R Q = \left\{ \left(q_0^2 + q_1^2 - q_2^2 - q_3^2 \right) u_R + 2(q_1 q_2 + q_0 q_3) v_R + 2(q_1 q_3 - q_0 q_2) w_R \right\} i$$

$$+ \left\{ 2(q_1 q_2 - q_0 q_3) u_R + \left(q_0^2 - q_1^2 + q_2^2 - q_3^2 \right) v_R + 2(q_2 q_3 + q_0 q_1) w_R \right\} j$$

$$+ \left\{ 2(q_1 q_3 + q_0 q_2) u_R + 2(q_2 q_3 - q_0 q_1) v_R + \left(q_0^2 - q_1^2 - q_2^2 + q_3^2 \right) w_R \right\} k \quad \text{...(3.40)}$$

This can be written as

$$\overrightarrow{V}_B = Q^*\overrightarrow{V}_R Q = C\underset{R}{\overset{B}{\uparrow}}\overrightarrow{V}_R \qquad \qquad \text{... (3.41)}$$

$$= \begin{bmatrix} \left(q_0^2 + q_1^2 - q_2^2 - q_3^2 \right) & 2(q_1 q_2 + q_0 q_3) & 2(q_1 q_3 - q_0 q_2) \\ 2(q_1 q_2 - q_0 q_3) & q_0^2 - q_1^2 + q_2^2 - q_3^2 & 2(q_2 q_3 + q_0 q_1) \\ 2(q_1 q_3 + q_0 q_2) & 2(q_2 q_3 - q_0 q_1) & q_0^2 - q_1^2 - q_2^2 + q_3^2 \end{bmatrix} \begin{bmatrix} u_R \\ v_R \\ w_R \end{bmatrix} \quad \text{... (3.42a)}$$

$$= 2 \begin{bmatrix} (q_0^2 + q_1^2 - \frac{1}{2}) & q_1q_2 + q_0q_3 & q_1q_3 - q_0q_2 \\ q_1q_2 - q_0q_3 & (q_0^2 + q_2^2 - \frac{1}{2}) & q_2q_3 + q_0q_1 \\ q_1q_3 + q_0q_2 & q_2q_3 - q_0q_1 & (q_0^2 + q_3^2 - \frac{1}{2}) \end{bmatrix} \begin{bmatrix} u_R \\ v_R \\ w_R \end{bmatrix} \qquad \text{... (3.42b)}$$

This is similar to transforming a vector from one coordinate frame to another coordinate frame by using transformation matrix. $C\!\uparrow_R^B$, where $C\!\uparrow_R^B$ is easily identifiable from Eq. (3.42a)

Thus, if a quaternion q is given which rotates the reference frame to a body frame, the direction cosine matrix $C\!\uparrow_R^B$ can be written down as in Eq. (3.42a).

The axis of single rotation and the rotation angle α which aligns the reference frame to body frame is then given as follows[3,4] :

$$\cos \alpha = \frac{1}{2}(C_{11} + C_{22} + C_{33} - 1) \qquad \text{... (3.43)}$$

$$E_x = (C_{23} - C_{32})/2\sin\alpha \qquad \text{... (3.44)}$$

$$E_y = (C_{31} - C_{13})/2\sin\alpha \qquad \text{... (3.45)}$$

$$E_z = (C_{12} - C_{21})/2\sin\alpha$$

Where C_{ij} are the elements of direction cosine matrix $C\!\uparrow_R^B$ and

$$\overline{E} = (E_x i + E_y j + E_z k) \qquad \text{... (3.46)}$$

which on simplification gives

$$q_0 = \cos(\alpha/2)$$

$$q_1 = E_x \sin(\alpha/2)$$

$$q_2 = E_y \sin(\alpha/2)$$

$$q_3 = E_z \sin(\alpha/2) \qquad \text{... (3.47)}$$

Thus, if the angle of rotation and the direction cosines (E_x, E_y and E_z) are known, the quaternion can be easily written down.

3.2.2.2   Error Angles in Terms of Quaternions

The concept of single rotation is useful for control of the missile, since the control torque simultaneously acts along pitch, yaw and roll axes. Considering small angle α as a vector quantity, its components αE_x, αE_y and αE_z along the x, y and z axes serve as the error angles for control systems and taking $\sin \alpha / 2 \approx \alpha / 2$, one gets angular errors along various axes as

$$\phi_e = \alpha\, E_x = 2q_1$$

$$\theta_e = \alpha\, E_y = 2q_2$$

$$\psi_e = \alpha\, E_z = 2q_3 \qquad\qquad \text{... (3.48)}$$

For large angles

$$\phi_e = 2E_x \sin(\alpha/2) = 2q_1$$

$$\theta_e = 2E_y \sin(\alpha/2) = 2q_2$$

$$\psi_e = 2E_z \sin(\alpha/2) = 2q_3 \qquad\qquad \text{... (3.49)}$$

Now, let A give a quaternion which rotates X_0 frame to X_1 and B give another quaternion which rotates X_1 frame to X_2 frame. Then, the resultant quaternion C which is the product of A and B gives the quaternion which rotates the frame X_0 to X_2 where the product is defined as

$$C = AB = \begin{bmatrix} a_0 \\ a_1 \\ a_2 \\ a_3 \end{bmatrix} \cdot \begin{bmatrix} b_0 \\ b_1 \\ b_2 \\ b_3 \end{bmatrix} = \begin{bmatrix} a_0 b_0 - \overline{A}\cdot\overline{B} \\ a_0\overline{B} + b_0\overline{A} + \overline{A}\times\overline{B} \end{bmatrix} \qquad \text{... (3.50a)}$$

$$= \begin{bmatrix} a_0 b_0 - & (a_1 b_1 + & a_2 b_2 + & a_3 b_3) \\ a_0 b_1 + & b_0 a_1 + & a_2 b_3 - & a_3 b_2 \\ a_0 b_2 + & b_0 a_2 + & a_3 b_1 - & a_1 b_3 \\ a_0 b_3 + & b_0 a_3 + & a_1 b_2 - & a_2 b_1 \end{bmatrix} \qquad \text{... (3.50b)}$$

Further, if $A\,B = C$

Then $B = A^{-1} C = A^{*}\, C$ and $A = C\,B^{-1} = CB^{*}$ ... (3.51)

Now, consider that the rotation axis for A and B is same and A gives rotation through θ_A and B gives rotation through θ_B. Then

$$
C = AB =
\begin{bmatrix}
\cos\dfrac{\theta_A}{2} \\[2ex]
E_x \sin\dfrac{\theta_A}{2} \\[2ex]
E_y \sin\dfrac{\theta_A}{2} \\[2ex]
E_z \sin\dfrac{\theta_A}{2}
\end{bmatrix}
\begin{bmatrix}
\cos\dfrac{\theta_B}{2} \\[2ex]
E_x \sin\dfrac{\theta_B}{2} \\[2ex]
E_y \sin\dfrac{\theta_B}{2} \\[2ex]
E_z \sin\dfrac{\theta_B}{2}
\end{bmatrix}
$$

$$
=
\begin{bmatrix}
\cos\dfrac{\theta_A}{2}\cos\dfrac{\theta_B}{2} - (Ex^2 + Ey^2 + Ez^2)\sin\dfrac{\theta_A}{2}\cos\dfrac{\theta_B}{2} \\[3ex]
E_x\left(\cos\dfrac{\theta_A}{2}\sin\dfrac{\theta_B}{2} + \sin\dfrac{\theta_A}{2}\cos\dfrac{\theta_B}{2}\right) + 0 \\[3ex]
E_y\left(\cos\dfrac{\theta_A}{2}\sin\dfrac{\theta_B}{2} + \sin\dfrac{\theta_A}{2}\cos\dfrac{\theta_B}{2}\right) + 0 \\[3ex]
E_z\left(\cos\dfrac{\theta_A}{2}\sin\dfrac{\theta_B}{2} + \sin\dfrac{\theta_A}{2}\cos\dfrac{\theta_B}{2}\right) + 0
\end{bmatrix}
$$

$$
=
\begin{bmatrix}
\cos\dfrac{(\theta_A + \theta_B)}{2} \\[2ex]
E_x \sin\dfrac{(\theta_A + \theta_B)}{2} \\[2ex]
E_y \sin\dfrac{(\theta_A + \theta_B)}{2} \\[2ex]
E_z \sin\dfrac{(\theta_A + \theta_B)}{2}
\end{bmatrix}
\qquad \text{... (3.52)}
$$

Thus, if the axis of rotation is same, the multiplication of quaternions is same as addition of the angles, then Eq. 3.51 in terms of angles would be

$$
\theta_A + \theta_B = \theta_C
$$

Then

$$
\theta_B = -\theta_A + \theta_C = \theta_C - \theta_A
$$

$$
\theta_A = \theta_C + (-\theta_B) = \theta_C - \theta_B \qquad \text{... (3.53)}
$$

Now to obtain control errors in terms of quaternions,

Let Q_D be the desired quaternion which defines the desired orientation of body axes frame relative to reference frame and Q_a be the actual quaternion describing the actual orientation of the vehicle body axes relative to reference frame.

Let q_e be the error quaternion which gives the desired orientation of the body axes frame relative to the actual orientation of the reference frame.

Then
$$Q_a q_e = Q_D$$

or
$$q_e = Q_a^{-1} Q_D = Q_a^* Q_D \qquad \text{... (3.54)}$$

The control errors are then obtained using Eq. 3.48 or 3.49, *i.e.*,

$$\phi_e = 2q_{e1}$$

$$\theta_e = 2q_{e2}$$

$$\psi_e = 2q_{e3} \qquad \text{... (3.55)}$$

3.2.2.3   *Comparative Features of Euler Angles and Quaternions*

Following description of comparative features of angles and quaternions are important for a control engineer as it has a potential to create catastrophic situation due to positive feedback if appropriate care is not taken in the design software.

Everyone is familiar with the scheme of measuring angles. The angle can vary from 0 to 360° or 0 to ±180°. Similar possibility in terms of quaternion is not quite familiar to many. Hence, it is described here in somewhat detail.

Let us consider a single axis rotation about X-axis indicated by an angle ϕ. Then the corresponding quaternion is given by

$$Q = \begin{bmatrix} \cos\dfrac{\phi}{2} \\[2mm] \sin\dfrac{\phi}{2} \\[2mm] 0 \\[1mm] 0 \end{bmatrix} \qquad \text{... (3.56)}$$

Fig 3.4 gives the signs of scalar component and vector component of the quaternion in different quadrants if the angle ϕ varies from 0 to +360° or 0 to −360°.

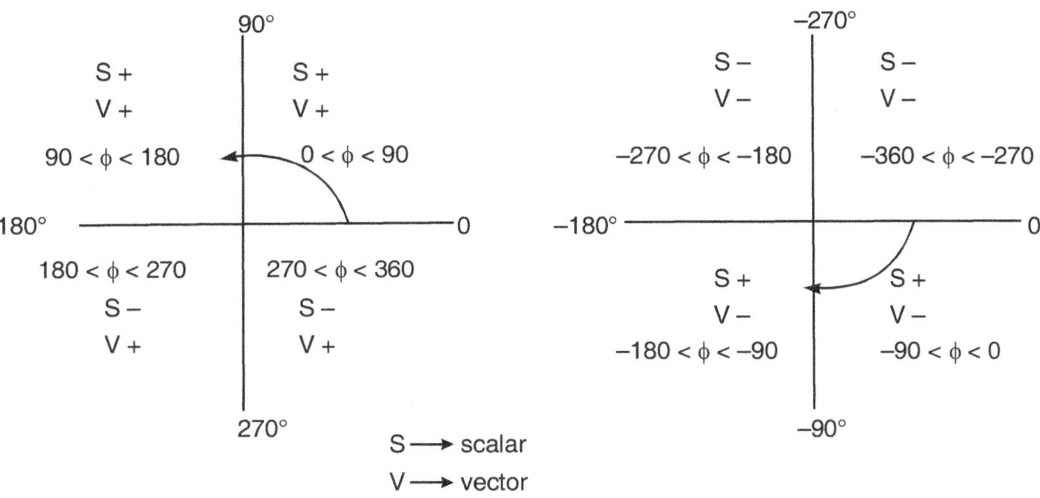

Fig. 3.4. Signs of scalar and vector components for +ve and −ve rotation

Thus, the same physical orientation can be described by a +ve angle (between 0 and 360°) or a negative angle (between 0 and −360°). Similar to that the position can be obtained by positive rotation quaternion or negative rotation quaternion. To bring uniqueness, it is better to define angles between 0 to ±180°. However, this creates a discontinuous change or jump scenario in the angle. For example, if there is a constant rate and angle continuously increases and crosses 180°, a jump will be experienced. (For example 181° will be sensed as −179°). Fig. 3.5 shows the variation of positive rate and negative rate scenario. [jump discontinuity also occurs for angles measured between 0 to ±360° unless the angles are continuously incremented above 360°].

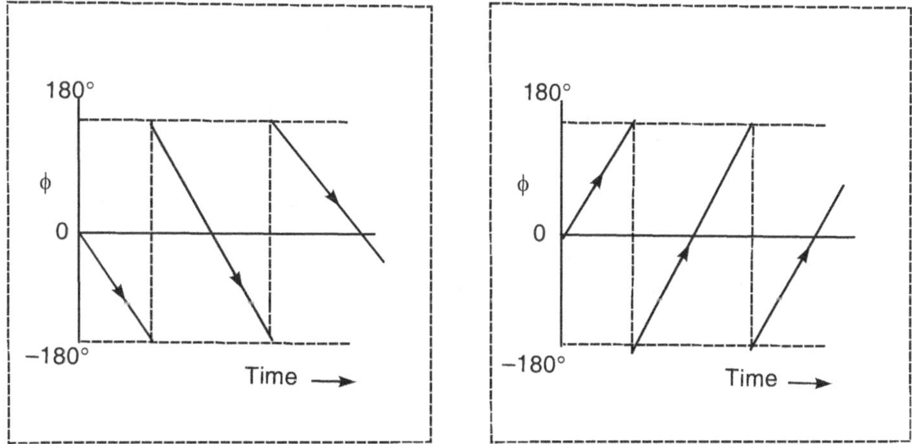

Fig. 3.5 Variation of angle from 0 to ±180°

It is known that this scenario is handled in software as follows:

$$\phi = \phi - 360° \qquad \text{if} \qquad \phi > 180°$$

$$= \phi + 360° \qquad \text{if} \qquad \phi < -180° \qquad \qquad \text{... (3.57)}$$

Not many are familiar with similar provision in case of quaternions. Hence this is described here in detail. As we have already seen that addition of angle is similar to multiplication of quaternion. Then, the effect of 360° addition or subtraction is given by multiplication by a quaternion.

$$\begin{bmatrix} \cos\left(\dfrac{360}{2}\right) \\ \sin\left(\dfrac{360}{2}\right) \\ 0 \\ 0 \end{bmatrix} = \begin{bmatrix} -1 \\ 0 \\ 0 \\ 0 \end{bmatrix} = \begin{bmatrix} \cos\left(\dfrac{-360}{2}\right) \\ \sin\left(\dfrac{-360}{2}\right) \\ 0 \\ 0 \end{bmatrix} \qquad \text{... (3.58)}$$

Thus, the addition of 360° or subtraction of 360° is same as multiplication of the quaternion by a quaternion (–1 0 0 0). Now

$$\begin{bmatrix} q_0 \\ q_1 \\ q_2 \\ q_3 \end{bmatrix} \cdot \begin{bmatrix} -1 \\ 0 \\ 0 \\ 0 \end{bmatrix} = \begin{bmatrix} -q_0 \\ -q_1 \\ -q_2 \\ -q_3 \end{bmatrix} \qquad \text{... (3.59)}$$

Thus, addition or subtraction of 360° is similar to changing the sign (or multiplication by –1) of all the four components of the quaternion.

It is seen from Fig. 3.4 that the scalar part of quaternion changes sign whenever it crosses 180° either in clockwise direction or in anticlockwise direction. This fact is clear from the scalar part given by $\cos\left(\dfrac{\theta}{2}\right)$ or $\cos\left(\dfrac{-\theta}{2}\right)$. The sign is positive for $\left|\dfrac{\theta}{2}\right| < 90°$ but changes sign for $\left|\dfrac{\theta}{2}\right| > 90°$ and it is zero for $\left|\dfrac{\theta}{2}\right| = 90°$. This is utilized here to restrict the quaternion to a scenario where the scalar part is always maintained positive between 0 to 1. Thus, maintaining scalar part of quaternion always positive is similar to restricting angle between 0 and ±180° and the provision

$$\text{if } q_0 < 0 \text{ make } Q = -Q \qquad \qquad \text{... (3.60)}$$

is similar to the provision of adding or subtracting 360° to the angle if $|\phi| > 180°$ given by Eq. 3.57.

3.2.2.4 Protection Logics for Euler Angles and Quaternions

We have described above the need to restrict the angles within 0 to $\pm180°$ and quaternions with scalar part always positive. The navigation software may also be incorporating these features. However, it is wiser for control engineer to incorporate the above features in control software before using the angles or quaternions to ensure that there is no possibility of positive feedback situation in control behaviour. Hence, the control designer needs to use additional care on the control errors or error quaternions also.

This is illustrated by a situation where the vehicle desired orientation is, say, 175° and the vehicle actual orientation is, say, −175°. The angle through which vehicle needs to be rotated is only 10° to align the vehicle actual axis to desired axis.

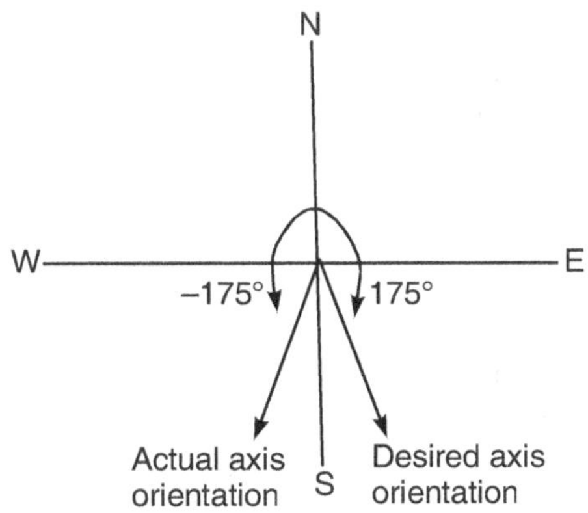

Fig. 3.6

However, the error computed will give

$$\phi_d - \phi_a = 175 - (-175) = 350°$$

Similarly, the error quaternion is given by

$$q_e = \begin{bmatrix} \cos\left(\dfrac{-175}{2}\right) \\ \sin\left(\dfrac{-175}{2}\right) \\ 0 \\ 0 \end{bmatrix}^{-1} \begin{bmatrix} \cos\left(\dfrac{175}{2}\right) \\ \sin\left(\dfrac{175}{2}\right) \\ 0 \\ 0 \end{bmatrix} = \begin{bmatrix} \cos\left(\dfrac{175}{2}\right) \\ \sin\left(\dfrac{175}{2}\right) \\ 0 \\ 0 \end{bmatrix} \begin{bmatrix} \cos\left(\dfrac{175}{2}\right) \\ \sin\left(\dfrac{175}{2}\right) \\ 0 \\ 0 \end{bmatrix} = \begin{bmatrix} \cos\left(\dfrac{350}{2}\right) \\ \sin\left(\dfrac{350}{2}\right) \\ 0 \\ 0 \end{bmatrix} = \begin{bmatrix} -0.996 \\ 0.0872 \\ 0 \\ 0 \end{bmatrix}$$

$$\dots (3.61)$$

Thus, if a protection is not given on the angular error or error quaternion, the control system will attempt to align the body axes frame with the desired orientation by a longer route *i.e.*, through 350° instead of going through the shorter route of 10° as seen in Fig. 3.6. Hence, the control engineer must provide protection on error angle or error quaternion also as given by Eqs. 3.57 and 3.59 to avoid control going through a longer route and in the process developing large angular rates.

3.2.3   Roll Manoeuvre to Launch in Any Azimuth

For launch vehicles, the launch azimuth direction is known well in advance and the vehicle axes are properly aligned when the vehicle is on the launch pad or a minor roll correction is done after the vehicle takes off and clears the launch tower. Tactical missiles are positioned on the launchers which can move about the vertical axis as well as horizontal axis fast enough to fix the launch azimuth and elevation before launching the missile. Long range or medium range missiles or missiles launched from ships etc. are positioned on the launch platform in a particular way. However, the plane of the launch point and target will be decided only after the target is determined. In such cases, it is better to provide a software provision for aligning the missile axes system instead of wasting valuable time in realigning the launch platform to align the missile axes.

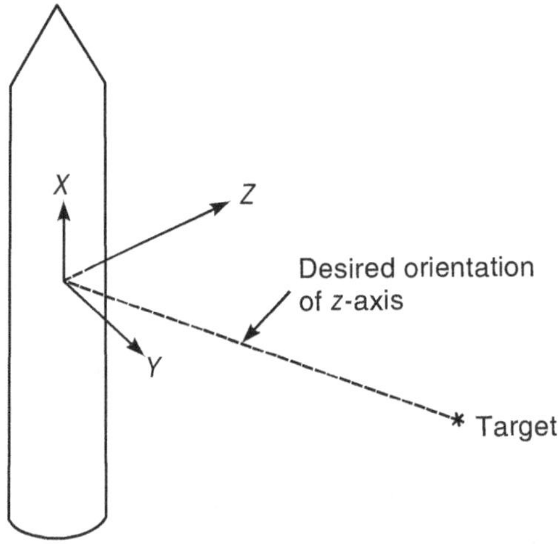

Fig. 3.7

Two possibilities exist:

(1) The roll manoeuvre can be carried out during the vertical rise of the missile after it rises above the nearby obstacles or the launch tower and align the z-axis to the launch direction and then follow the pitching profile. This is possible if there is adequate time available for vertical phase and also if there is adequate roll control torque available to execute the manoeuvre. However, if the roll control torque is provided only by aerodynamic

control surfaces, it will not be wise to have a long vertical rise trajectory till adequate dynamic pressure is developed to execute the roll manoeuvre. In such case, the second possibility should be followed.

(2) vehicle pitching can be done in the required azimuth plane even though the vehicle axes are not properly aligned. The axes alignment can be subsequently carried out once the dynamic pressure builds upto a sufficient value.

In both the cases described above, the desired pitch program or desired quaternion history Q_D is updated to account for the roll angle correction and the attitude errors are resolved through the instantaneous roll correction angle so that the vehicle pitches in the required plane. In Equations 3.18 to 3.20, the initial value of roll angle command $\phi_c(0)$ is initially made same as actual missile roll angle or $\phi_c = \phi(0)$ and then it is slowly brought to zero depending on the time duration over which one intends to correct the roll manoeuvre.

In quaternion approach also, the desired quaternion is modified for a period over which one intends to correct the roll.

Pitch program is obtained for shaping the trajectory assuming that the motion takes place in x-z plane and missile z axis is pointing in the direction of target. However, when the target is determined, the plane of launch point and target may not be passing through the x-z plane of the missile. Let the missile x-z plane be at an angle ϕ w.r.t. launch point-target plane. Then missile needs to be rolled through $(-\phi)$ so that missile x-z plane coincides with desired flight plane. The rotation axis Y_R for the desired pitching is at an angle ϕ w.r.t. missile y axis. Then the desired quaternion w.r.t. missile axis is given by[6,7].

$$\begin{bmatrix} \cos\dfrac{\theta}{2} \\ \cos\alpha\sin\dfrac{\theta}{2} \\ \cos\beta\sin\dfrac{\theta}{2} \\ \cos\gamma\sin\dfrac{\theta}{2} \end{bmatrix} = \begin{bmatrix} \cos\dfrac{\theta}{2} \\ 0 \\ \cos\phi\sin\dfrac{\theta}{2} \\ \cos(90+\phi)\sin\dfrac{\theta}{2} \end{bmatrix} = \begin{bmatrix} \cos\dfrac{\theta}{2} \\ 0 \\ \cos\phi\sin\dfrac{\theta}{2} \\ -\sin\phi\sin\dfrac{\theta}{2} \end{bmatrix} \qquad (3.62)$$

The error quaternion $Q_M^{-1}Q_D$, gives the errors for the control system as discussed in earlier sections. The angle ϕ can then be progressively decreased to zero so that missile x-z plane matches with the trajectory plane.

3.2.4 Quaternion Interpolation

We have seen earlier that multiplication of two quaternions is similar to addition of two angles. A natural question arises about interpolation of quaternions. For example, if a pitch program is stored in computer memory as

t	θ	Q_{D0}	Q_{D1}	Q_{D2}	Q_{D3}
5	90	0.707	0	0.707	0
10	80	0.766	0	0.643	0
15	70	0.819	0	0.574	0

Then, the value of θ at 6 sec. is obtained by interpolation as

$$\theta = 90° + \left(\frac{80-90}{10-5}\right)(t-5) = 88° \ \text{(for } t = 6) \qquad \text{... (3.63)}$$

If a small incremental quaternion corresponding to $-2°$ angle is defined as q, then

$$Q_5 qqqqq = Q_{10}$$

$$Q_6 = Q_5 q, \ Q_7 = Q_6 q \ \text{etc.} \qquad \text{... (3.64)}$$

From the equation 3.64, it is not clear how to obtain the incremental quaternion q for interpolation. In actual practice, the interpolation will be required at much finer time interval. Hence, it is necessary to be clear about the process of interpolation with quaternions.

From the definition of quaternion, the difference quaternion $Q_5^{-1}Q_{10}$ is given by

$$Q_e = Q_5^{-1}Q_{10} = \begin{bmatrix} \cos\dfrac{\theta}{2} \\[2mm] \cos\alpha\sin\dfrac{\theta}{2} \\[2mm] \cos\beta\sin\dfrac{\theta}{2} \\[2mm] \cos\gamma\sin\dfrac{\theta}{2} \end{bmatrix} \qquad \text{... (3.65)}$$

where θ = difference in pitch angle at 10 sec. and 5 sec.

If the angle θ is very small, we can use the approximation $\cos\dfrac{\theta}{2} = 1$ and $\sin\dfrac{\theta}{2} = \dfrac{\theta}{2}$ ($\cos\alpha$, $\cos\beta$ and $\cos\gamma$ are assumed to remain nearly constant over the interval). Thus,

$$Q_e = \begin{bmatrix} 1 \\[2mm] \dfrac{\theta}{2}\cos\alpha \\[2mm] \dfrac{\theta}{2}\cos\beta \\[2mm] \dfrac{\theta}{2}\cos\gamma \end{bmatrix} \qquad \text{... (3.66)}$$

Since θ is now appearing linearly we can use $\theta = \theta_1 + \theta_2 + \theta_3 + \theta_4 + \theta_5 = 5\Delta\theta$

This makes it clear, that if the desired quaternion is stored with an interval such that the incremental angle between two time instants is small and approximation $\cos\theta = 1$ and $\sin\theta = \theta$ is valid, we can use linear interpolation of quaternion components similar to the interpolation of angles. Thus,

$$Q_{D(t)} = Q_{Dt_1} + \left(\frac{Q_{Dt_2} - Q_{Dt_1}}{t_2 - t_1} \right)(t - t_1) \qquad \ldots (3.67)$$

3.2.5 Pitch Program Tracking Errors

The errors in following the pitch program are determined by several factors, among them some important parameters are:

– Rate of change of pitch attitude angle.

– Thrust misalignment disturbance torque.

– Controller gains and integrator in control loop if it is present.

– Wind profiles.

– Vehicle static margin.

Following approximate analysis is given to get a feel of the dependence of the attitude errors on various parameters.

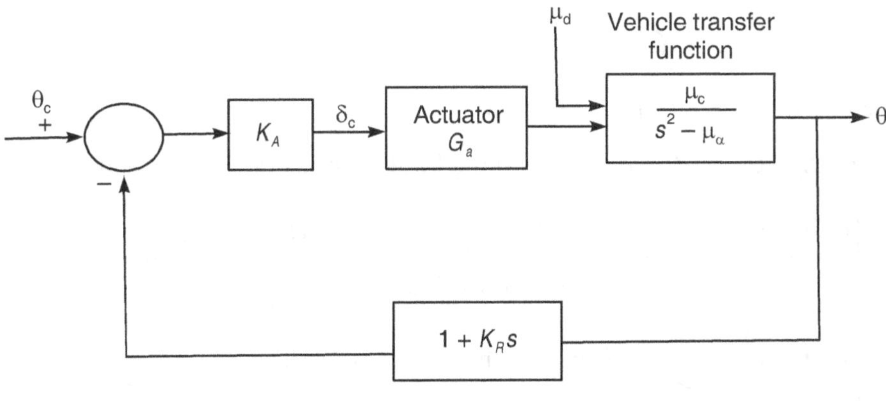

Fig. 3.8

The simplified vehicle dynamic equation (see chapter 4 for detailed derivation) and actuator transfer function are given as follows:

$$\theta = \mu_\alpha \alpha + \mu_\delta \delta + \mu_d \qquad \qquad \text{... (3.68)}$$

where $\quad \mu_\alpha$ = angular acceleration per unit angle of attack

$\qquad \qquad \mu_\delta$ = angular acceleration per unit control deflection

$\qquad \qquad \mu_d$ = angular acceleration due to thrust misalignment

As will be discussed in chapter 4, the control system stability is studied using time slice approach for a perturbed vehicle state assuming the vehicle parameters remain constant over a short period of time. In equation (3.68), the vehicle initially is assumed to be aligned along the velocity vector. Then, a small perturbation θ is given to vehicle attitude, since initial θ and α are zero, the perturbation angle $\theta \approx \alpha$. The launch vehicles generally follow a pitch attitude program which is given in terms of different pitch rates for different time durations.

Let $\qquad \qquad \qquad \theta_c = \theta_{c_0} + \theta_c t \qquad \qquad \text{... (3.69)}$

The actuator is expected to give a unity steady state gain. Hence,

$$\delta_c = \delta = K_A(\theta_c - \theta - K_R\theta)$$

$$= K_A(\theta_{c_0} + \theta_c t - \theta - K_R\theta) \qquad \qquad \text{... (3.70)}$$

$$\therefore \qquad \theta = \mu_\alpha\theta + \mu_c K_A(\theta_{c_0} + \theta_c t - \theta - K_R\theta) + \mu_d \qquad \qquad \text{... (3.71)}$$

Let the error $y(t)$ be defined as

$$y(t) = \theta_c(t) - \theta(t)$$

$$y(t) = \theta_c(t) - \theta(t) = \theta_c - \theta$$

$$y(t) = 0 - \theta \ \text{ since } \ \theta_c = \text{constant} \qquad \qquad \text{... (3.72)}$$

Substituting from eq. 3.72 in eq. 3.71, taking Laplace transform and simplifying, one gets

$$y(s) = \frac{s^2\left[(s + \mu_c K_A K_R)y(0) + y(0)\right] + (s\mu_c K_A K_R - \mu_\alpha)\,\theta_c - s\mu_d - \mu_\alpha s\theta_c(0)}{s^2\left[s^2 + \mu_c K_A K_R s + (\mu_c K_A - \mu_\alpha)\right]} \quad \text{... (3.73a)}$$

with $\theta_{c_0} = 0, \ \theta(0) = 0, \ \theta(0) = 0$ which gives $y(0) = 0, \ y(0) = \theta_c$

$$y(s) = \frac{\theta_c(s^2 + \mu_c K_A K_R s - \mu_\alpha) - \mu_d s}{s^2[s^2 + \mu_c K_A K_R s + (\mu_c K_A - \mu_\alpha)]} \qquad \qquad \text{... (3.73b)}$$

This shows that in steady state $y(t)$ will be a ramp with a slope

$$y(t) = \frac{-\theta_c \mu_\alpha}{\mu_c K_A - \mu_\alpha} \qquad \text{... (3.74)}$$
$$\scriptstyle t \to \infty$$

This gives a slope difference between θ_c and θ depending on μ_α and the error $(\theta_c - \theta)$ will grow indefinitely. If we assume $\mu_\alpha = 0$, we get steady state tracking error:

$$y_{ss} = y(t) = K_R \theta_c - \frac{\mu_d}{\mu_c K_A} \qquad \text{... (3.75)}$$
$$\scriptstyle y \to \infty$$

Eq 3.75 gives a very good representative value for tracking error for a properly designed pitch program which follows nearly gravity turn trajectory during high dynamic pressure region but may not give good results for nominal trajectories having large angle of attacks during high dynamic pressure region. The Fig. 3.9(b) gives $(\theta_c - \theta)$ history for a pitch program given in Fig. 3.9(a) for the three cases of thrust misalignment *i.e.*, (1) zero, (2) pitch up and (3) pitch down cases.

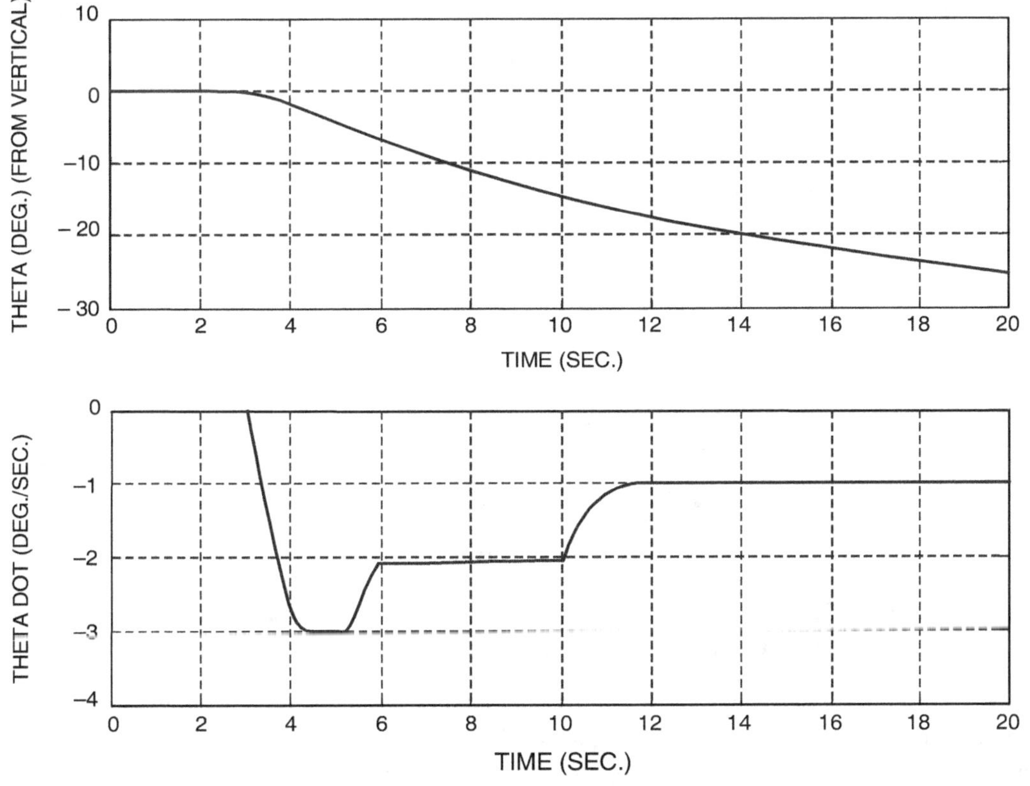

Fig. 3.9 (a) and (b)

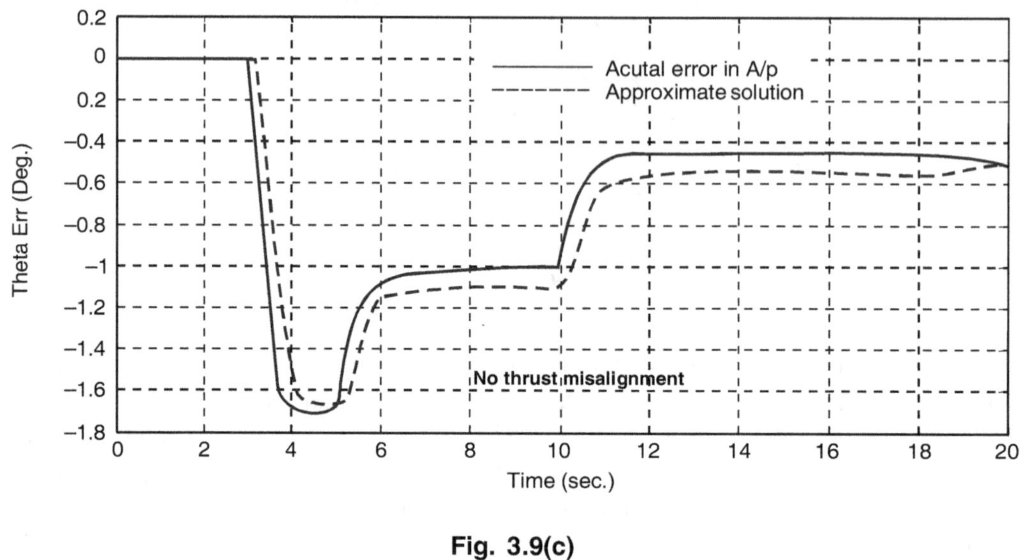

Fig. 3.9(c)

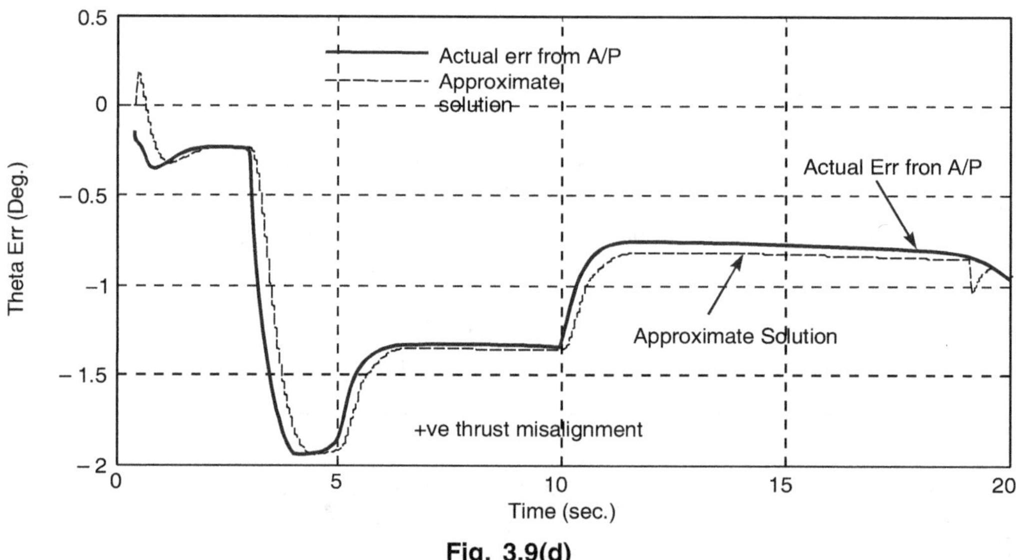

Fig. 3.9(d)

It also gives the variation of error estimate using Eq. 3.75. It is seen that the equation gives a very good estimate of the attitude error profile.

The error has two components. The first component is $K_R\theta_c$ and is a known quantity and can be compensated by giving a feed forward demand. The other component depends on the thrust misalignment for which the direction and magnitude cannot be anticipated in advance. The steady state errors can however be corrected by providing proportional plus integral controller as shown in Fig. 3.10.

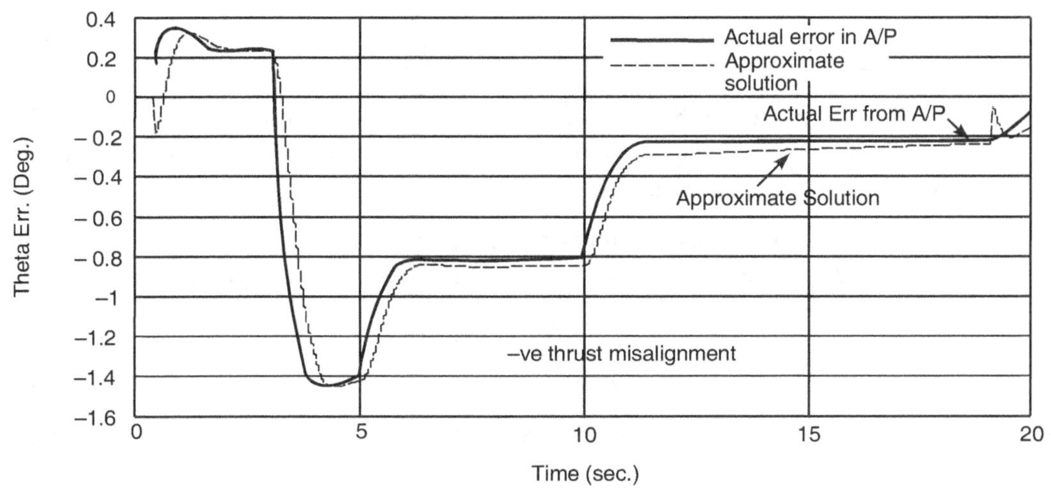

Fig. 3.9(e)

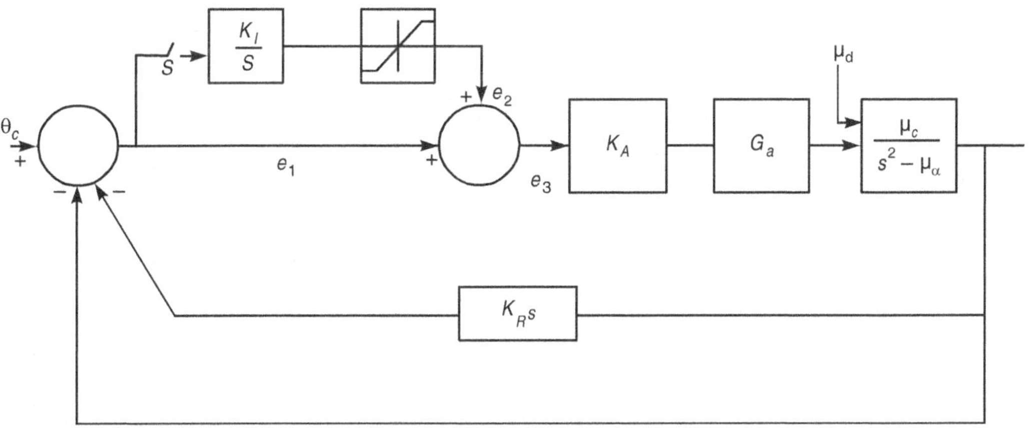

Fig. 3.10

To understand the functioning of the integrator, consider the switch S to be open initially. The system will settle down with an approximate steady state error given by Eq. 3.75 and the control deflection δ will generate control torque to exactly balance the disturbance torque. Therefore,

$$\mu_c \delta = -\mu_d \qquad (\quad \mu_\alpha \text{ is assumed } \approx 0) \qquad \qquad \text{... (3.76)}$$

or

$$\delta = -\frac{\mu_d}{\mu_c} = K_A(\theta_c - \theta - K_R \dot{\theta}) \qquad \qquad \text{... (3.77)}$$

$$\therefore \qquad \theta_c - \theta - K_R\theta = -\frac{\mu_d}{\mu_c K_A} \qquad \qquad \dots (3.78)$$

If switch were open, this would be the values of e_1 and e_3 signals during steady state. If we close the switch now, e_1 becomes the input to the integrator and it gets integrated to a value e_2. The torque balance requires that $e_3 = e_1 + e_2$.

Hence, as the value e_2 increases, e_1 must decrease to maintain e_3 constant. The process continues till e_1 becomes zero and $e_2 = e_3$. Thus, with integrator configured as in Fig. 3.10

$$e_1 = \theta_c - \theta - K_R\theta = 0$$

or
$$\theta_c - \theta = K_R\theta \approx K_R\theta_c \qquad \qquad \dots (3.79)$$

Since θ_c and K_R are known quantities, it is possible to put a feed forward term $K_R\theta_c$ in which case

$$e_1 = \theta_c - \theta - K_R\theta + K_R\theta_c = 0$$

or
$$\theta_c - \theta = 0 \qquad \qquad (\quad \theta \approx \theta_c \text{ in steady state}) \qquad \dots (3.80)$$

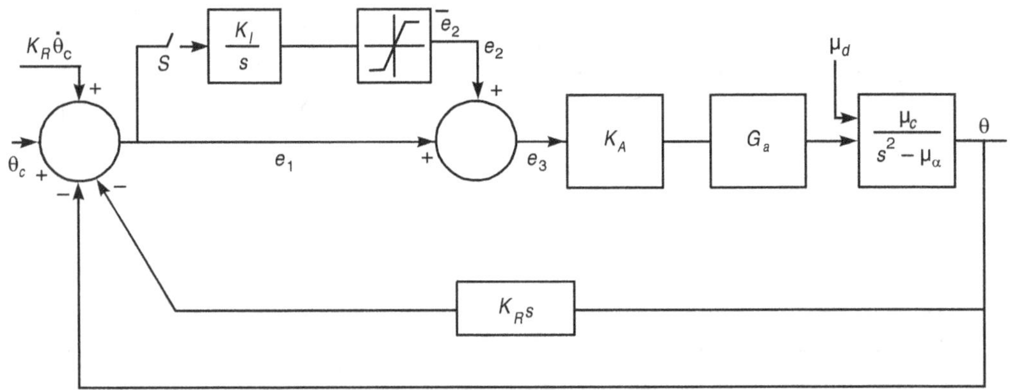

Fig. 3.11

The same result can be achieved by configuring the integrator as shown in Fig. 3.12.

$$e_3 = e_1 + e_2 - K_R\theta = -\frac{\mu_d}{\mu_c K_A} \qquad \qquad \dots (3.81)$$

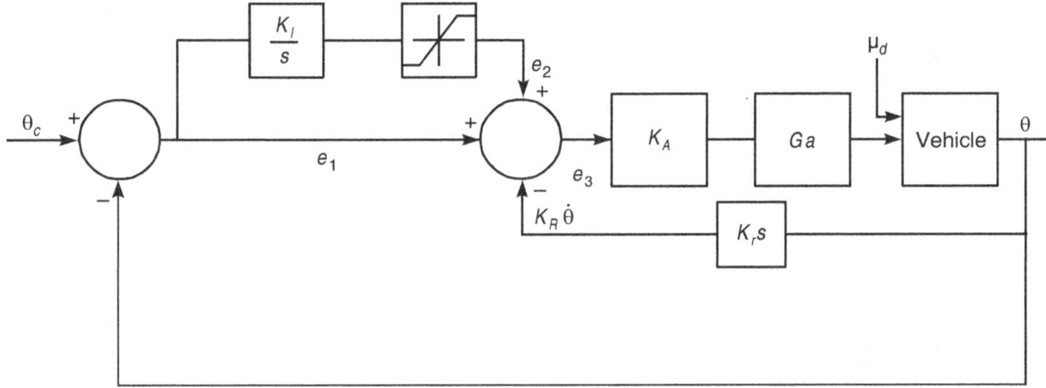

Fig. 3.12

Before switch is closed,

$$e_2 = 0$$

$$e_1 = \theta_c - \theta \qquad \qquad \text{... (3.82)}$$

After the switch is closed e_1 and e_2 will change and settle to a value

$$e_2 = \theta_c - \theta \qquad \qquad \text{... (3.83)}$$

$$e_1 = 0$$

Thus, reducing the tracking errors to nearly zero.

Fig.3.10 and 3.11 and 3.12 show a limiter at the output of integrator. Its value is computed based on the steady state value which one intends to correct with some margin. If one does not use limiter, there is a possibility that during a transient scenario the integrator value can rise to a large value. This will permit e_1 going negative and will create an oscillatory response till it finally settles down. The oscillations can be minimized by putting limiter value just above the intended steady state correction value.

3.2.6 Effect of Attitude Errors on Mission

Control system performance is determined by the accuracy of the pitch program tracking error. We have seen in the previous section that the attitude error depends on

- Pitch rate demand

- Control system gains and

- Disturbance torques due to thrust misalignment and winds.

We have also discussed how this tracking error can be minimised by using proportional plus integral control law. Provision of integrator has an implication of reducing phase margin of the control system. Hence, it is necessary to know the effect of tracking error on the mission profile so that the specifications on tracking errors can be judiciously fixed.

Previous section has shown that the tracking error is high when the demanded pitch rate is high and low for most of the remaining flight. For the sake of worst case study, trajectory has been worked out assuming that the pitch angle is different from the desired angle by a constant angle of 2°. Fig. 1.32 from Chapter I (reproduced below) shows the dispersion in trajectory from the nominal.

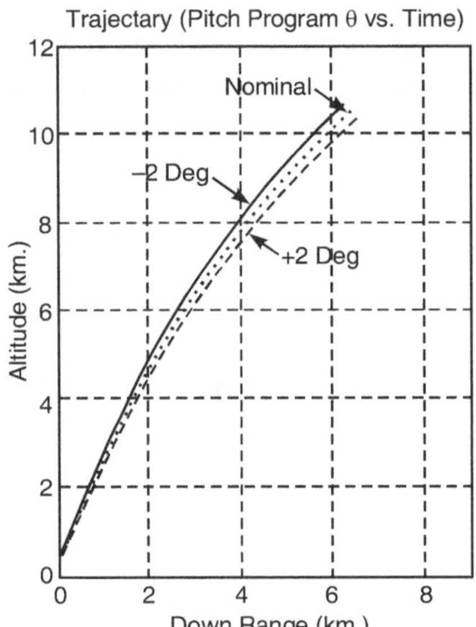

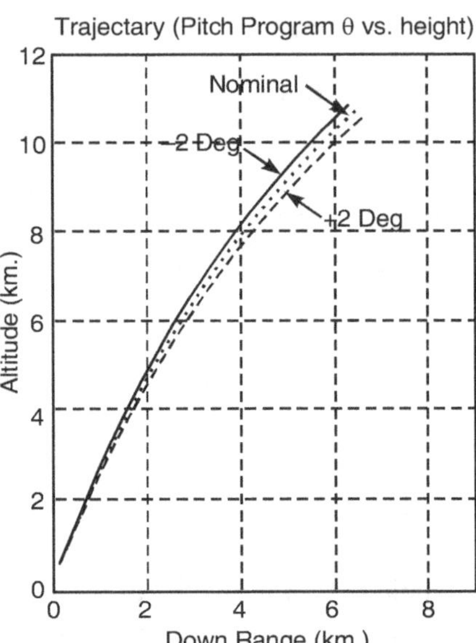

It is seen that the trajectory dispersion is not significant even for a constant attitude difference of 2°. Hence, the specifications on tracking error of the control system need not be very tight and stability margins must be given primary consideration over the tracking errors.

3.3   CONTROL LAWS

3.3.1   Attitude Control Law

The launch vehicles, in general, use attitude control system with predetermined attitude (pitch and yaw) history during atmospheric flight and closed loop guidance after clearing the sensible atmosphere. Fig. 3.8 gives the basic block diagram for the attitude control system and the control law can be written as

$$\delta_c = K_A(\theta_c - \theta - K_R \dot{\theta}) \qquad \qquad ...(3.84)$$

3.3.2   Load Relief and Drift Minimization Control Laws

In a desire to make launch vehicles more efficient, there is a natural tendency to reduce the structure factor of the vehicle which is defined as inert weight to total weight of the vehicle where

$$\text{Total weight} = \text{payload weight} + \text{weight of propellants} + \text{weight of structure} \quad \ldots (3.85)$$

Hence, any reduction in external load acting on the vehicle is a desirable feature for the total system. However, as will be seen in subsequent pages, the reduction in aerodynamic loads in flight is only at the expense of trajectory dispersion. Thus, even though the vehicle may appear to be more efficient at the outset, it may not be really so because the closed loop guidance scheme has to cater for larger initial dispersion zone. It may require additional fuel (thus more weight on higher stages) to be stored on board as there would be more losses due to larger manoeuvres. Hence, a trade off study needs to be carried out.

A flight vehicle is required to withstand several types of loads such as, thrust, inertial, aerodynamic, control and handling loads, in case, it is required to undergo transportation and erection to launching position. As discussed in Chapter I, the pitch program is designed to follow a gravity turn trajectory *i.e.*, the vehicle attitude is programmed to follow the flight path angle giving near zero angle of attack during high dynamic pressure region. However, it will experience angle of attack due to winds. It is possible that the maximum of thrust/inertial load may not be occurring at the same time as maximum of aerodynamic loads.

By modifying control laws, only the aerodynamic loads and associated control loads can be managed. However, since it is associated with trajectory dispersion, it is necessary to fix a level upto which aerodynamic loads should be limited by control system[12]. This limit should be such that the total load on the vehicle with so limited aerodynamic load is not higher than total load at the peak of other loads (Ref. 12). Any further reduction in aerodynamic load will only result in a needless trajectory dispersion.

One of the method of reducing aerodynamic load or reducing trajectory dispersion due to winds is to design wind biased pitch program. However, that means generating number of pitch programs suitable for different winds and using it appropriately. Other method of reducing the load is through modification of control law.

3.3.2.1   *Aerodynamically Stable Vehicle*

A few cases are described below to understand the mechanism of lateral load reduction. Only a yaw control system is discussed. However, the results are valid for pitch plane also. The vehicle is assumed to have the attitude control law with desired yaw angle $\psi_c = 0$ and the control torque will tend to reduce the attitude error to zero.

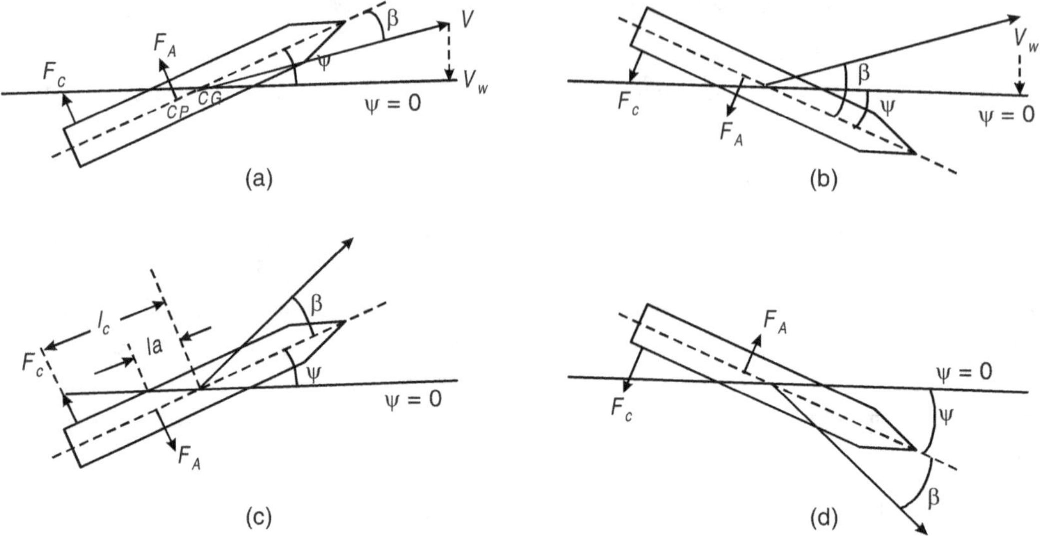

Fig. 3.13

Since, the vehicle is assumed to be aerodynamically stable, the aerodynamic torque will tend to reduce the side slip angle β to zero. Consider the instantaneous vehicle orientation as shown in Fig. 3.13(a) and (b) due to some reason. These are unstable scenarios since both the control torque and aerodynamic torques act in the same direction and vehicle will rotate towards orientation shown in Fig. 3.13(c). In this scenario, a condition will be reached when the control torque will balance the aerodynamic torque in the steady state.

$$F_c l_c = F_A l_a \qquad \qquad \text{... (3.86)}$$

and total normal force acting on the vehicle will be

$$F_N = F_A - F_C = F_A\left(\frac{l_c - l_a}{l_c}\right) = -C_{y\beta}\beta QS\left(\frac{l_c - l_a}{l_c}\right) \qquad \text{... (3.87)}$$

since the locations of control force and aerodynamic force are different, the vehicle will experience a bending moment.

$$M_b = K_1\alpha + K_2\delta \qquad \qquad \text{... (3.88)}$$

The objective for load relief control is to reduce this load on the vehicle.

Equation 3.87 indicates that the load can be reduced by reducing the steady state β. This can be done by reducing the control force F_C and allowing ψ to increase (See Fig 3.13(c)). This is possible by only reduction of control gain. A separate gain K_ψ needs to be provided so that the gain on angle can be varied without affecting the rate feedback gain and the control law can be written as

$$\delta = K_A(K_\psi(\psi_c - \psi) - K_R\dot\psi) \qquad \text{... (3.89a)}$$

where $\dot\psi$ = yaw body angular rate

or
$$\delta = -K_A(K_\psi\psi + K_R\dot\psi) \qquad \text{... (3.89b)}$$

If we reduce K_ψ to zero, the ψ angle will build till vehicle gets aligned with velocity vector and $\beta = 0$, thus, reducing total lateral load on the vehicle to negligible value. However, in the process, the orientation is away from the desired $\psi_c = 0$ and the vehicle thrust will be acting in the direction of resultant ψ (or along the relative velocity) and this will lead to dispersion in the trajectory. Since, it is not necessary to reduce the lateral load to zero, one needs to make a trade off between the allowable load and trajectory dispersions.

It is thus seen that for an aerodynamically stable vehicle, the lateral load can be reduced by reducing the stiffness of the attitude control system (or reducing the gain on attitude error) and this is possible only at the expense of trajectory dispersion.

3.3.2.2 Aerodynamically Unstable Vehicle

Fig. 3.14 shows various situations for aerodynamically unstable configuration.

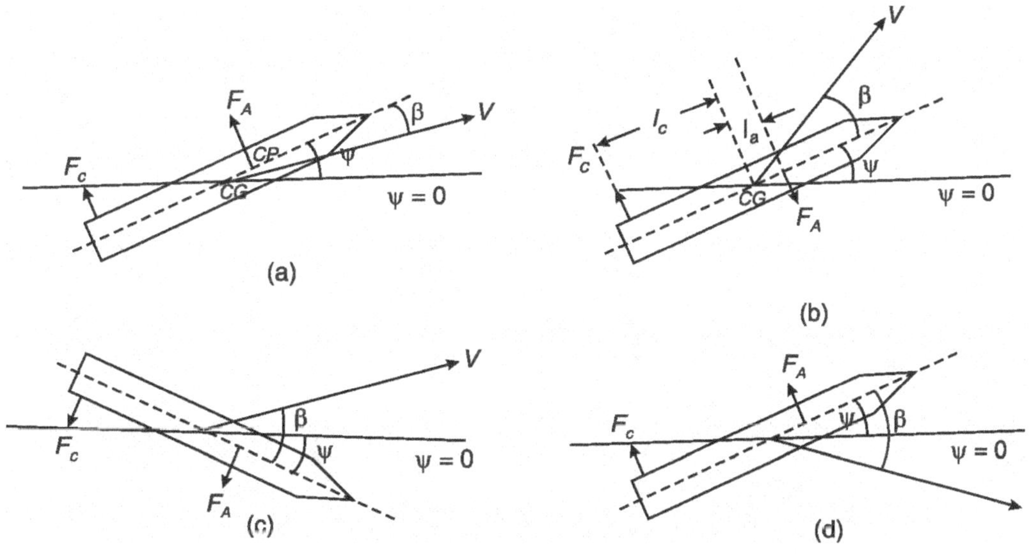

Fig. 3.14

It is easy to see that the scenarios in Fig. 3.14(*a*) and (*b*) are unstable whereas (c) and (d) give the stable situations. The total load on the vehicle is given by:

$$F_N = F_A + F_C = F_A \left(\frac{l_c + l_a}{l_c} \right) = C_{y\beta} \beta QS \left(\frac{l_c + l_a}{l_c} \right) \qquad \text{... (3.90)}$$

where l_a and l_c are magnitudes of moment arms.

The loads on the vehicle can be reduced by reducing β (Eq. 3.90). For aerodynamically unstable vehicle, the β can be reduced by increasing the control stiffness which means increasing the control gain (See Fig. 3.15) to generate more control force at a lower value of ψ. One can imagine very high control stiffness to reduce ψ to a very low value. However, the load cannot be reduced to zero for the unstable vehicle by using only ψ and $\dot\psi$ in the control law. Moreover, the stiffness will have an upper bound limitation from the point of view of stability margins as will be seen in subsequent chapter.

A general practice is to use either side slip angle (or angle of attack in pitch plane) or lateral acceleration for a load relief or drift control law[10,11,12].

The dynamic equations and the control law in yaw plane are given as follows: (See Fig. 3.15).

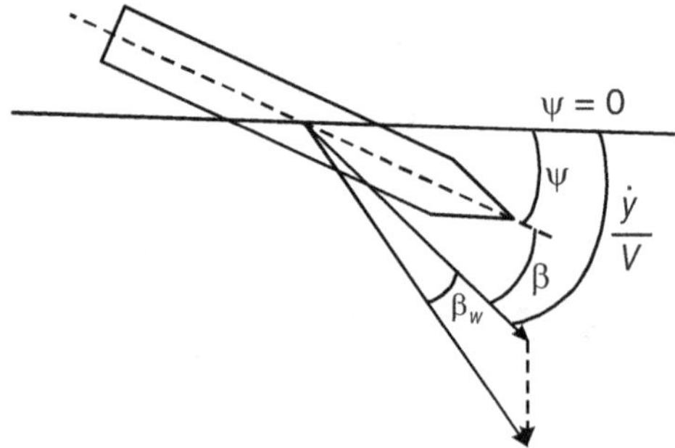

Fig. 3.15

$$\dot y = \frac{T - D}{m} \psi - \frac{L_\beta}{m} \beta - \frac{T_c}{m} \delta \qquad \text{... (3.91)}$$

$$\ddot\psi = -\mu_\beta \beta + \mu_c \delta \qquad \text{... (3.92)}$$

$$\beta = -\psi + \frac{\dot y}{v} + \beta_w \qquad \text{... (3.93)}$$

and the control law is given by

$$\delta = -K_A (K_\psi \psi + K_R \dot\psi - K_\beta \beta) \qquad \text{... (3.94)}$$

For pitch plane the control law would be

$$\delta = K_A [K_\theta (\theta_c - \theta) - K_R \dot\theta - K_\alpha \alpha)] \qquad \text{... (3.95)}$$

The transfer functions for various variables are given as

$$\frac{\left(\frac{y}{v}\right)}{\beta_w} = -\frac{a_2 s^2 + a_1 s + a_0}{s^3 + B_2 s^2 + B_1 s + B_0} \qquad \qquad \dots (3.96)$$

$$\frac{\psi}{\beta_w} = \frac{(\mu_c K_A K_\beta - \mu_\beta)s}{s^3 + B_2 s^2 + B_1 s + B_0} \qquad \qquad \dots (3.97)$$

$$\frac{\beta}{\beta_w} = \frac{s(s^2 + \mu_c K_A K_R s + \mu_c K_A K_\psi)}{s^3 + B_2 s^2 + B_1 s + B_0} \qquad \qquad \dots (3.98)$$

$$\frac{\delta}{\beta_w} = \frac{K_A s \left[K_\beta s^2 + K_R \mu_\beta s + K_\psi K_\beta \right]}{s^3 + B_2 s^2 + B_1 s + B_0} \qquad \qquad \dots (3.99)$$

where

$$B_2 = \mu_c K_A K_R + \frac{L_\beta + T_c K_A K_\beta}{mv} \qquad \qquad \dots (3.100)$$

$$B_1 = \mu_c K_A (K_\psi + K_\beta) - \mu_\beta + \frac{T_c K_A K_R}{mv}\left(\mu_\beta + \frac{\mu_c L_\beta}{T_c} \right) \qquad \qquad \dots (3.101)$$

$$B_0 = \frac{T_c K_A K_\psi}{mv}\left(\mu_\beta + \frac{\mu_c L_\beta}{T_c} \right) - \left(\frac{T - D}{mv} \right)(\mu_c K_A K_\beta - \mu_\beta) \qquad \qquad \dots (3.102)$$

$$a_2 = \frac{L_\beta + T_c K_A K_\beta}{mv} \qquad \qquad \dots (3.103)$$

$$a_1 = K_A K_R T_c \left(\mu_\beta + \frac{L_\beta \mu_c}{T_c} \right) \qquad \qquad \dots (3.104)$$

$$a_0 = T_c K_A (K_\psi + K_\beta)\left(\mu_\beta + \frac{\mu_c L_\beta}{T_c} \right) - [T - D + T_c K_A (K_\psi + K_\beta) + L_\beta] \qquad \qquad \dots (3.105)$$

The characteristic equation has two dominant complex poles pair corresponding to rotational motion and one real pole corresponding to drift velocity. The steady state drift can be made to stabilize at wind velocity, remain constant or diverging slowly by selecting gains so as to make the constant term B_0 positive, zero or negative. It may be mentioned that negative value of B_0 gives one pole positive, thus, making system unstable. However, the drift pole is normally of very small magnitude, it will grow very slowly and in practice the control law will be changed soon after the zone of high dynamic pressure is passed.

Ref. 10 calls the condition $B_0 = 0$ as the drift minimum control law. It also discusses the use of lateral acceleration sensor in place of angle of attack sensor and states that use of accelerometer will sometimes be preferable as it provides two variables (*i.e.*, accelerometer gain and its location) to adjust thus giving designer more freedom.

3.3.3 Lateral Acceleration (Latax) Control Law

It has been seen in earlier section that the control law purely based on ψ and $\dot{\psi}$ is inadequate to manage vehicle load and drift from the nominal trajectory. Also, it is seen that the load relief is associated with drift from the nominal trajectory. Hence, it is necessary to fix a level of maximum or allowable loads during high dynamic pressure region.

The stresses at any section of the vehicle during flight are due to

   (*i*) Thrust

  (*ii*) Inertial load due to forward acceleration

 (*iii*) Lateral loads and bending moments.

The maximum of inertial load will be occurring at the peak of forward thrust which may not be occurring at the same time as maximum dynamic pressure region. Hence, one needs to find the stress levels due to combined loading at various flight instants such as high dynamic pressure region, peak thrust region etc. and decide the allowable level of maximum loads during high dynamic pressure region (Ref. 12). The load relief control law can then be iteratively designed till the trajectory simulation gives load profile within the satisfactory bound.

A more positive approach to the load control is provided by lateral acceleration (latax) control. Fig. 3.16 gives a block diagram for yaw plane latax control.

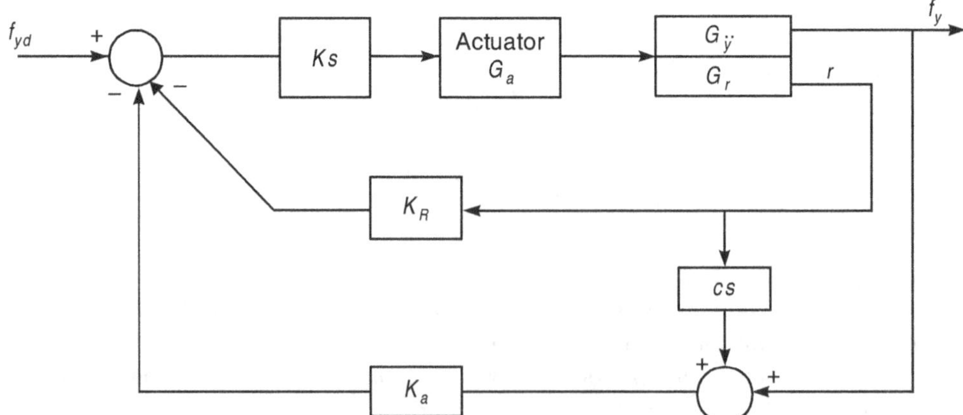

Fig. 3.16 Block Diagram of Latax Control System

$$\left(f_{yd} = \text{Desired Lateral Acceleration} = \frac{\text{Allowable Lateral Force}}{\text{Mass}} \right)$$

$$f_y = v + ur = y_v v + y_r r + y_\delta \delta \qquad \ldots (3.106)$$

$$r = n_v v + n_r r + n_\delta \delta \qquad \ldots (3.107)$$

where

$$y_v = -\frac{C_{y\beta} QS}{mu} \qquad \ldots (3.108)$$

$$y_\delta = -\frac{K_{DS}}{m}, \quad K_{DS} \text{ is side force per unit control deflection} \ldots (3.109)$$

$$n_v = \frac{C_{y\beta} QS(C_p - C_g)}{I_{zz} u} \qquad \ldots (3.110)$$

$$n_\delta = \frac{K_{DS} l_c}{I_{zz}}; \quad \text{where } l_c \text{ is control moment arm} \qquad \ldots (3.111)$$

The transfer functions are given by

$$G_y = \frac{y_\delta s^2 - y_\delta n_r s - u(n_\delta y_v - n_v y_\delta)}{s^2 - (y_v + n_r)s + y_v n_r + n_v u - n_v y_r} \qquad \ldots (3.112)$$

$$G_r = \frac{n_\delta \left[s - y_v + \dfrac{n_v y_\delta}{n_\delta} \right]}{s^2 - (y_v + n_r)s + y_v n_r + n_v u - n_v y_r} \qquad \ldots (3.113)$$

$$G_a = \text{Actuator transfer function}$$

K_S, K_R, and K_a are the control gains. The detailed design of latax autopilot is given in subsequent chapter.

TRANSFER FROM ATTITUDE CONTROL TO LATAX CONTROL AND VICE VERSA

Fig. 3.17 gives a block diagram for a possible scheme for transfer from attitude control to latax control.

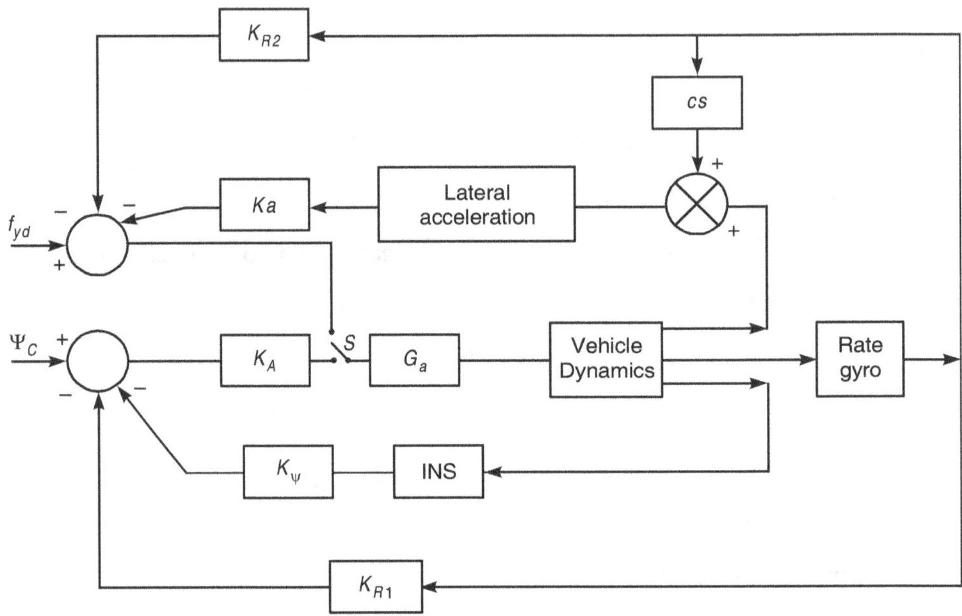

Fig. 3.17

During attitude control, the lateral acceleration can be monitored. If the level exceeds the allowable limit one can transfer from attitude control law to lateral acceleration control law by changing the position of switch 's'.

The transfer from attitude control to latax control is thus clear. The reverse is, however, not quite clear since the control loop will attempt to maintain demanded acceleration level by changing the side slip angle or orientation of the vehicle even after the region of high dynamic pressure during which load relief is required, has passed. Hence, it is necessary to estimate the dynamic pressure using the velocity information from INS and air density from atmosphere model and revert back to attitude control when the dynamic pressure falls below the preset level.

The lateral acceleration control law is very commonly used in all types of missiles. However, it is not used in launch vehicles in the form discussed above. Ref. (10, 12) discuss the use of lateral acceleration in load relief or drift control law, however, the command signal is the desired attitude. Ref. 12 also indicates the need to reduce the attitude angle feedback gain K_ψ to zero (See Fig. 18 for schematic gains schedule) since the maintaining of attitude and relieving load will be working against each other and the vehicle needs to be allowed to change the orientation to relieve the load. Ref. 12 also suggests a deadzone in the sensed acceleration channel so that the attitude control law works as long as the lateral acceleration is within the dead zone and load relief control law comes into operation when the acceleration level becomes more than the dead zone with only an excess acceleration being used for load relieving.

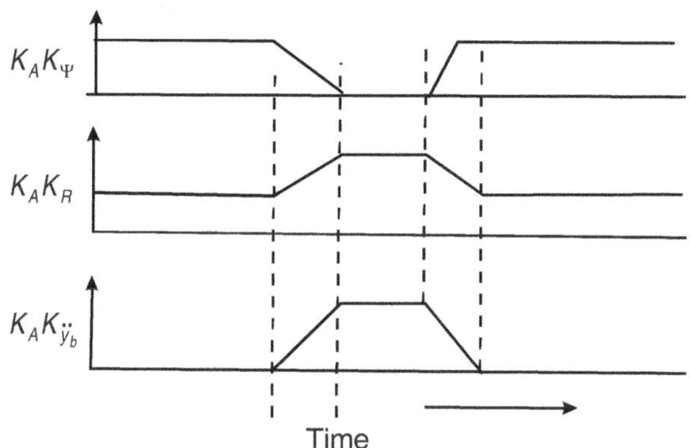

Fig. 3.18 Schematic Gain Schedule

The author of Ref.12 needed several iterations to get a satisfactory control law.

The lateral acceleration law as discussed in this section gives a more definitive control law compared to the load relief law of Ref.12. As discussed in subsequent chapters the lateral acceleration control loop can be designed to give a unity steady state gain and the desired damping characteristics and hence no iterations are required to get a satisfactory control law. Moreover, the same control law can be used to bring down the trajectory dispersion after the critical region of high dynamic pressure is passed by making the desired lateral acceleration proportionate to the trajectory dispersion (perpendicular distance of the vehicle from the nominal trajectory) as is done in case of some missiles.

3.4   CONTROL POWER PLANT SIZING

By 'Control power plant sizing', we mean planning adequate control for various phases of trajectory till the mission is accomplished. It consists of launch (take off) phase, thrusting phase during high dynamic pressure region, coasting phase during atmospheric phase, strap-on separation, stage separation, thrusting phase and coasting phase out of atmosphere, thrust tail off region etc. for launch vehicles. In addition to above, for surface to air or air to air missiles one needs to plan adequate control for the end game phase *i.e.*, near target interception since inadequate control response can result in a big miss distance meaning non accomplishment of the mission.

3.4.1 Aerodynamic Static Margin

3.4.1.1 *Trim Condition Requirement*

Fig. 3.19 shows a general configuration for a vehicle.

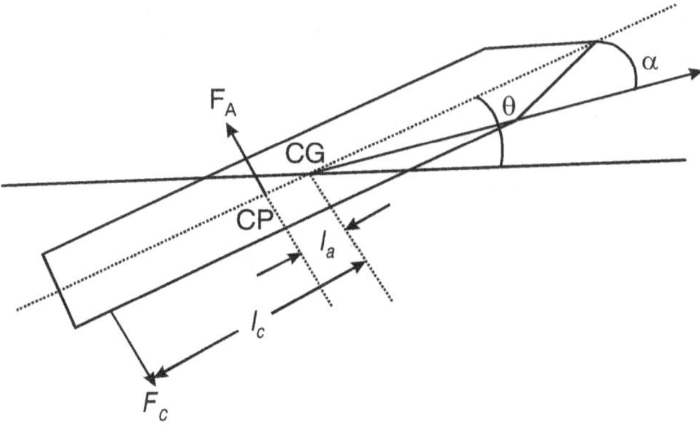

Fig. 3.19

The aerodynamic force is given by

$$F_A = -C_{N\alpha}\alpha\, QS \qquad\qquad \dots (3.114)$$

and the resultant aerodynamic force acts at the aerodynamic centre of pressure. The relative position of this centre of pressure with reference to centre of gravity indicates the aerodynamic stability of the vehicle. It can be seen from Fig. 3.19, that if the centre of pressure is at CP *i.e.*, on the tail side from CG, the aerodynamic moment tends to reduce the angle of attack and the vehicle is said to be aerodynamically stable (or statically stable). If the centre of pressure is on the nose side from CG, the aerodynamic moment will tend to increase the angle of attack. The vehicle is said to be aerodynamically unstable.

If the centre of pressure occurs at the centre of gravity, the vehicle is said to be neutrally stable as there will be no aerodynamic moment acting on the vehicle. If we define aerodynamic static margin as

$$l_a = C_p - C_g \qquad\qquad \dots (3.115)$$

where C_p and C_g are distances measured from nose, then, following table gives the vehicle stability.

Static margin	Aerodynamic Stability
Positive	Stable
Negative	Unstable
Zero	Neutrally stable

Consider, now a control force acting on the vehicle at a distance l_c from the CG of the vehicle. The net torque acting on the vehicle for zero angular acceleration is

$$F_c l_c + F_A l_a = 0 \qquad \qquad \text{... (3.116)}$$

Or
$$F_c l_c = -F_A l_a$$

The resultant lateral force acting on the vehicle is given by

$$F = F_C + F_A = F_C - F_C \frac{l_c}{l_a} = F_C\left(\frac{l_a - l_c}{l_a}\right)$$

$$= -F_C\left(\frac{l_c - l_a}{l_a}\right) = -F_A\left(\frac{l_c - l_a}{l_c}\right) \qquad \text{... (3.117)}$$

$(l_c - l_a)$ gives the distance between the points of application of control forces and aerodynamic forces. It is easy to see that the net lateral force acting on the vehicle in the balanced condition or "Trim condition", is less in case of statically stable vehicle than in case of statically unstable vehicle.

Hence, the design of launch vehicles and long range missiles, which are designed to carry less lateral loads during high dynamic pressure region, should aim towards better static stability and tactical missiles which need high manoeuvrability during high dynamic pressure region should aim for slightly negative stability margins so that the vehicle gives better lateral acceleration for the same control force limits.

For a typical vehicle, the centre of gravity moves towards nose as the propellant gets burnt during powered phase. Thus, the control moment arm l_c increases. Similarly, the centre of aerodynamic pressure also moves towards nose as a function of mach no. (*i.e.*, the ratio of missile relative velocity and sound velocity) and thus $(l_c - l_a)$ changes. However, both the parameters are not likely to move by the same factor. Hence, in some cases, the vehicle is statically stable to start with and becomes unstable towards the stage burn out whereas in other cases, the vehicle is statically unstable to start with and becomes stable towards the stage burn out. In some other cases, the stability character remains same (either stable or unstable) but the level changes. The centre of pressure is also a function of vehicle angle of attack in addition to mach number and needs to be considered when significant angle of attack exists. As the aerodynamic force depends on angle of attack which in turn depends on trajectory pitch program as well as the atmospheric winds, it is essential to take winds into account while planning the maximum control force levels for launch vehicle and long range missiles.

Thus
$$F_{cl} = \frac{l_a}{l_c} C_{N\alpha} Q S(\alpha + \alpha_w) \qquad \text{... (3.118)}$$

where α_w = angle of attack due to winds.

The permissible wind velocities as a function of altitude upto which the vehicle can be launched may be decided beyond which either launch may be suspended or some load relief scheme may be used.

The tactical missiles are mainly designed with a lateral acceleration control loop. Hence, the net control force acting on the missile gets automatically controlled. However, it will involve trajectory dispersion due to wind and needs to be assessed. Fortunately, the winds will be affecting not only the missile but also the target and hence, the relative effect may be considerably lower. The normal force increases with angle of attack upto some value, however, beyond a certain value of angle of attack it again decreases. The angle of attack above which this condition occurs is said to be "stall" angle of attack and the missile should never go near to this value to avoid stalling problem. An angle of attack $\alpha = 20°$ may be considered as safe maximum value though it may differ for different vehicles. In order to be able to operate the missile at its maximum manoeuvring capability the control force Fc_2 can then be calculated from

$$Fc_2 = \frac{C_{N\alpha}(C_p - C_g)QS\alpha_{allowed}}{l_c} \qquad \text{... (3.119)}$$

Surface to surface missiles will be required to execute manoevures only in a fixed plane. Hence, the significant angle of attack can occur only in that plane. Hence, if the control is provided by aerodynamic control surfaces, one can use 'X' position of control surfaces in normal flight instead of '+' position to reduce the control surface size or the maximum control deflection requirement.

Pitch plane control force in '+' position $= C_{N\delta}\delta\, QS$ $\qquad$ ... (3.120)

Pitch plane control force in '×' position $= \sqrt{2}C_{N\delta}\delta\, QS$ $\qquad$ ... (3.121)

Where $C_{N\delta}$ is the aerodynamic coefficient defined for a pair of (diametrically opposite) control surfaces. Thus, control deflection requirement in 'X' position is $\sqrt{2}$ times less than in '+' position.

However, surface to air missile may be required to execute manoeuvre in any plane and hence only '+' configuration needs to be considered for deciding the control force levels.

Thus, for a given lateral acceleration (latax) requirement and the mass of the vehicle and maximum allowed α and δ requirements, the aerodynamic moment coefficients C_{m_α} and C_{m_δ} can be worked out using Equations 3.117 to 3.121.

3.4.1.2   *Aerodynamic Misalignments*

The vehicle configuration will always be having some protrusions on the surface such as antennas, cable ducts for signal cables or electric power cables or wing misalignments, strap-on misalignments etc. One must estimate the aerodynamic disturbance torque on vehicle due to them and provide for an additional control force or control deflection requirement to counter disturbance torque due to misalignments. This type of disturbance torque will be particularly more relevant for roll control system. The disturbance torque due to these effects may be represented as

$$\text{Disturbance torque} \; = \; C_L QSd \qquad \qquad \text{... (3.122)}$$

Where C_L is the aerodynamic rolling moment coefficient for a given misalignment.

3.4.1.3 Dynamic Stability Requirement

Consider the block diagram (Fig. 3.8) for an attitude control system with $\mu_d = 0$ and a first order actuator transfer function (for simplicity) with

$$G_a = \frac{K_c}{s + K_c} \qquad \qquad \text{... (3.123)}$$

and

$$\mu_\alpha = -\frac{C_{N\alpha} QSd(C_p - C_g)}{I_y} \qquad \qquad \text{... (3.124)}$$

Then, the characteristic equation is given by the numerator of

$$1 + GH = 1 + \frac{K_A K_c \mu_c (1 + K_R s)}{(s + K_c)(s^2 - \mu_a)} \qquad \qquad \text{... (3.125)}$$

or

$$s^3 + K_c s^2 + (K_A K_R K_c \mu_c - \mu_\alpha)s + K_A K_c \mu_c - K_c \mu_\alpha = 0 \qquad \text{... (3.126)}$$

For the closed loop system to be stable

$$K_A K_c \mu_c - K_c \mu_\alpha > 0 \qquad \qquad \text{... (3.127)}$$

and

$$K_A K_R K_c \mu_c - \mu_\alpha > 0 \qquad \qquad \text{... (3.128)}$$

Both the conditions are satisfied if

$$\mu_c K_A > \mu_\alpha \qquad \qquad \text{... (3.129)}$$

and

$$K_R > \frac{1}{K_c} \qquad \qquad \text{... (3.130)}$$

If we consider a small perturbation angle $\alpha \approx \theta$, the condition (3.129) says that the restoring control torque developed must be greater than the aerodynamic torque and the condition (3.130) says that the lead provided by rate gyro feedback must be greater than the lag due to actuator. If $\mu_\alpha < 0$ (*i.e.*, the vehicle is statically stable), the condition (3.129) is satisfied for any positive value of gain K_A. Therefore, theoretically, the gain margin is infinity since the phase lag never crosses $180°$.

In actual practice, the vehicle dynamics becomes a third order system when lateral acceleration equation is considered. The actuator will have higher order dynamics. Sensor has its own dynamics and natural frequency. The digital control systems will have additional lags due to sample and hold mechanism and computational delays and the total phase lag of the loop gain (GH) will rapidly increase with frequency and cross 180°. This will put an upper limit on the forward gain K_A at which the system will become dynamically unstable. This automatically restricts the usable maximum value of gain and autopilot bandwidth (which gives the speed of response) for satisfactory performance of the control system. If one desires to increase the bandwidth of the autopilot, he has to correspondingly increase the specifications on the sensor and actuator bandwidth and reduce the phase lags due to sample and hold and computational delay so that the 180° cross over occurs at a higher frequency which enables the designer to aim for higher bandwidth for the autopilot.

We now illustrate the role played by static margin of the vehicle on the design of autopilot. In Equation 3.125 (Fig. 3.8). we have assumed a very simplified vehicle transfer function by ignoring completely the aerodynamic damping. Due to this, the phase lag provided by the

vehicle transfer function $\left(\dfrac{\mu_c}{s^2 - \mu_\alpha}\right)$ is zero for $\omega < \sqrt{-\mu_\alpha}$ for statically stable vehicle and

180° for $\omega > \sqrt{-\mu_\alpha}$. If we consider the aerodynamic damping term, the denominator of the vehicle transfer function is given by

$$D = s^2 + a_1 s - \mu_\alpha \qquad \text{... (3.131)}$$

The phase lag provided by the vehicle transfer function is given by

$$\phi = \tan^{-1}\left[\frac{-a_1\dfrac{\omega}{\mu_\alpha}}{1+\left(\dfrac{\omega^2}{\mu_\alpha}\right)}\right] \qquad \text{... (3.132)}$$

The damping ratio due to aerodynamics is much less than 0.1%. Table 3.1. gives the variation of phase lag for 0.1% aerodynamic damping assuming $a_1 = 0.4$ and

(*i*) Stable vehicle with $\mu_\alpha = -4.0$ and

(*ii*) Unstable vehicle with $\mu_\alpha = 4.0$ and $a_1 = 0.4$.

It can be seen from Table I that the phase lag for stable vehicle approaches 90° as ω approaches $\sqrt{-\mu_\alpha}$ and then rapidly increases to 180° for $\omega > \sqrt{-\mu_\alpha}$ since the coefficient a_1 related to aerodynamic damping is very small.

For unstable vehicle, the phase lag is nearly 180° for all frequencies. Therefore, the phase provided by vehicle transfer function is more or less same for both the stable and unstable vehicle for frequencies $\omega >> \sqrt{-\mu_\alpha}$.

TABLE 3.1 : Phase Lag due to Vehicle Transfer Function

$$a_1 = 0.4, \ \mu_\alpha = \pm 4$$

(aerodynamic damping ratio = 0.1)

ω	$\left|\dfrac{\omega}{\mu_\alpha}\right|$	$\dfrac{\omega^2}{\left|\mu_\alpha\right|}$	$\left|a_1 \dfrac{\omega}{\mu_\alpha}\right|$	*Phase lag (Deg.)*	
				Stable vehicle $\mu_\alpha = -4$	*Unstable vehicle* $\mu_\alpha = 4$
1.0	0.25	0.25	0.1	7.6	175.4
1.5	0.375	0.5625	0.15	18.92	174.5
1.75	0.4375	0.7656	0.175	36.74	174.3
2.0	0.5	1.0	0.2	90.0	174.3
2.25	0.5625	1.2656	0.225	130.7	174.3
2.5	0.625	1.5625	0.25	156.0	174.4
3.0	0.75	2.25	0.3	166.5	174.7
4.0	1.0	4.0	0.4	172.4	175.4
5.0	1.25	6.25	0.5	174.55	176.0
6.0	1.5	9.0	0.6	175.7	176.0

Thus, it may be seen that the frequency at which positive gain margin (or 180° phase crossover) occurs will be mainly decided by the lags due to sensor, actuator, computational delay, zero order hold etc. with lag due to vehicle remaining nearly constant. Fig. 3.20 shows the plots of gain and phase vs. frequency for an actual design of an autopilot for a missile which has been successfully flight tested several times. Two cases are plotted.

    (1)   Static margin $= -0.8D$ with $\mu_\alpha = 18.40$

    (2)   Static margin $= -0.4D$ with $\mu_\alpha = 9.7$

From Fig. 3.20, it can be seen that the phase variation of the loop gain *GH* with frequency is almost identical at all frequencies except for the fact that the 180° cross over frequency marginally differs. This is also clear from Table I since unstable vehicle gives a lag of nearly 180°. For different values of μ_α, the lag may differ by 1 or 2 degrees and accordingly 180° may be crossed at slightly different frequencies. The gain variation is, however, significantly different for different μ_α or static margin cases. From the vehicle transfer function for unstable vehicle, it is quite clear that the magnitude of the vehicle transfer function at low frequencies is mainly decided by μ_α. Thus, the negative stability margin for the autopilot becomes better if the stability margin is less negative.

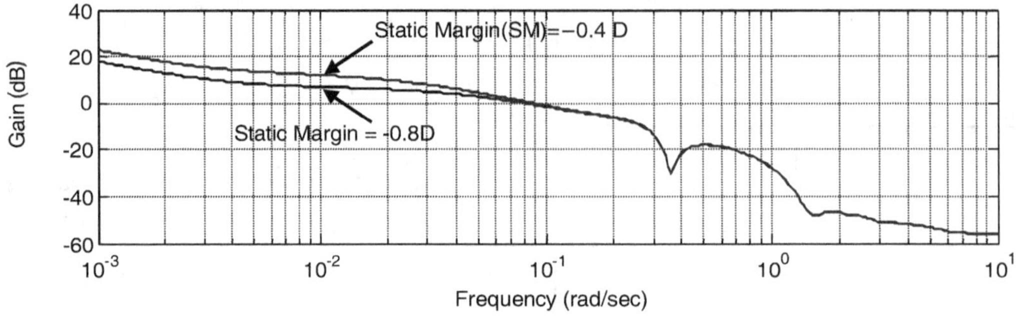

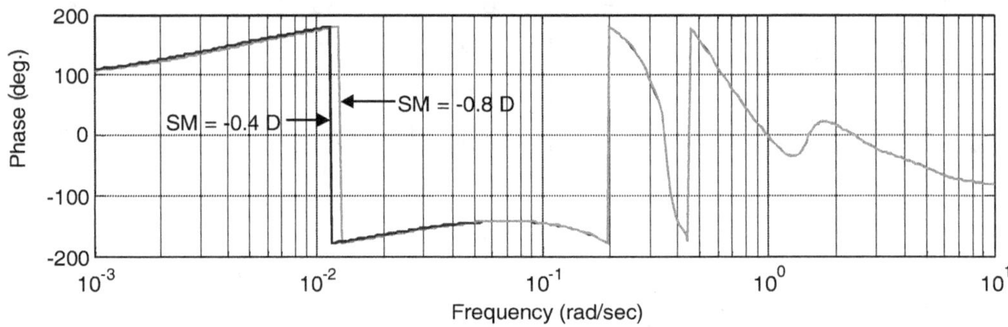

Fig. 3.20 Gain and phase vs. frequency for two cases of Static margin

Recalling Eq. 3.87

$$F_N = -F_c\left(\frac{l_c - l_a}{l_a}\right) = -F_A\left(\frac{l_c - l_a}{l_c}\right)$$

$$= C_{N\delta}QS\left(\frac{l_c - l_a}{l_a}\right)\delta_{max} \qquad \qquad \ldots (3.133a)$$

$$= -C_{N\alpha}\alpha(1 - \frac{l_a}{l_c})QS \qquad \qquad \ldots (3.133b)$$

We get following considerations for deciding the static margin for the vehicle:

(*i*) For a given maximum allowable angle of attack (α_{max}), the achievable latax increases as the static margin becomes more negative.

(*ii*) For a given maximum allowable control deflection δ_{max}, the achievable latax increases as the static margin becomes less negative.

(*iii*) The negative gain margin of the autopilot decreases as the static margin becomes more negative and it may become difficult to provide adequate negative stability margins for the vehicle beyond certain value of negative static margin. This puts an upper limit for acceptable negative static margin.

For good controllability of the vehicle, one must aim for a comfortable negative static margin so that the prediction uncertainties on aerodynamic coefficients do not compromise the stability margins of the autopilot.

It will be interesting to study the variation of static margin for a stable vehicle and its effect on autopilot design. We have already seen that, for frequencies $\omega \gg \sqrt{-\mu_\alpha}$, the gain and phase of the vehicle transfer function is more or less same. The gain at frequencies $\omega \ll \sqrt{-\mu_\alpha}$ is given by $\dfrac{-\mu_c}{\mu_\alpha}$ and at $\omega = \sqrt{-\mu_\alpha}$, it is decided only by the aerodynamic damping, since $s^2 - \mu_\alpha = 0$ *i.e.*, gain peak is given by

$$G_{peak} = \frac{\mu_c}{a_1\sqrt{-\mu_\alpha}} \qquad \qquad \ldots (3.134)$$

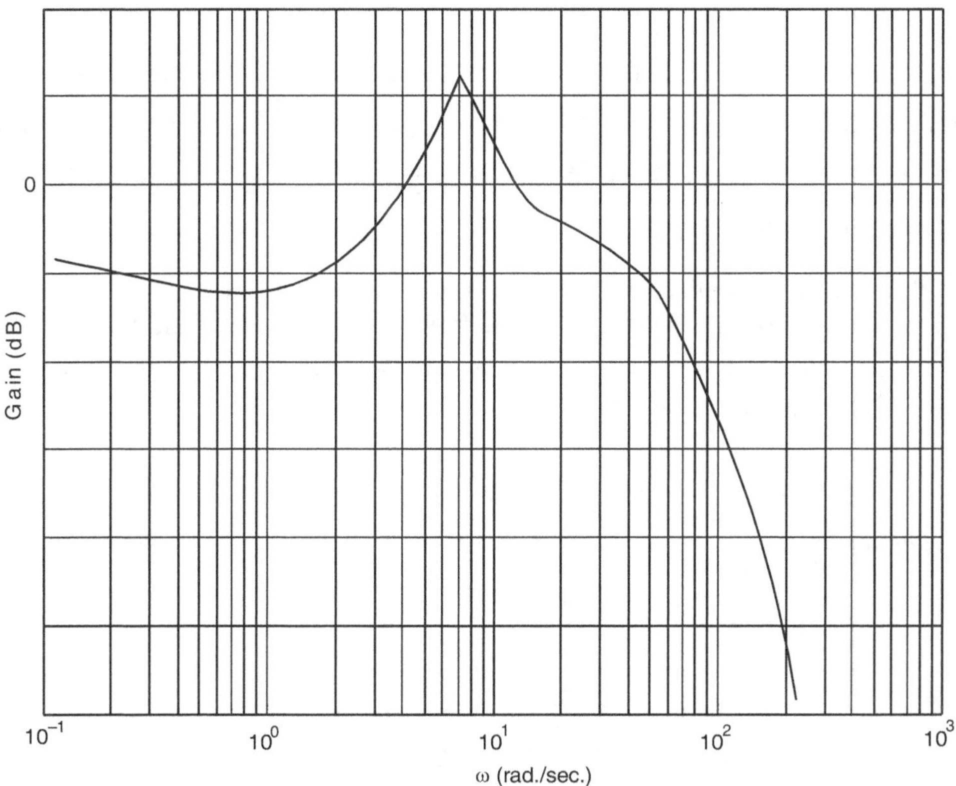

Fig. 3.21

The gain margin point, as seen earlier is decided by several other elements in the loop. However, if positive static margin and hence $(-\mu_\alpha)$ increases, the gain peak will move nearer to the required gain cross over frequency (or phase margin point). It may happen that one has to provide forward gain $K_A \ll 1$ to adjust the loop gain cross over (*i.e. GH = 1.0*) at the required frequency. This further attenuates the loop gain at $\omega \approx 0$. It is known that the sensitivity of autopilot errors to various disturbances, misalignment etc. is determined by steady state value of the loop gain as $\omega \to 0$. Higher the steady state loop gain, less the sensitivity to disturbance. Thus we see that, as the positive static margin or $(-\mu_\alpha)$ increases, the system becomes more sensitive to the disturbances. This problem can be partly overcome by providing proportional plus integral control or giving a higher forward gain and then adjusting the gain cross over frequency and the phase margin by providing lag and lead filters.

It has been found that this problem becomes particularly important for long range or medium range missiles during re-entry phase. During this phase, the dynamic pressures are very high and hence even for small static margin, the μ_α becomes very high. Forward gain K_A becomes too low. Hence, even though stability margins are quite satisfactory and digital simulation shows quite satisfactory response, the hardware-in-loop simulation and flight performance show oscillations in the vehicle body rates.

With low value of $-\mu_\alpha$, not only the hump in loop gain shifts away from the required gain cross over frequency but also increases the steady state loop gain $\dfrac{-\mu_c}{\mu_\alpha}$, requires more forward gain K_A which further enhances the steady state loop gain. Thus, the control system becomes much less sensitive to disturbances.

It is thus clear that neither high positive static margin nor high negative static margin is good from the point of view of satisfactory control performance. A good guideline will be to have stability margin such that $\sqrt{-\mu_\alpha}$ or $\sqrt{\mu_\alpha}$ is less than the desired gain cross over frequency by a factor of about 3 (See Table 3.1).

3.4.1.4 *L/D Ratio for the Vehicle*

The aerodynamic load distribution for a typical vehicle without wings is shown in Fig. 3.22.

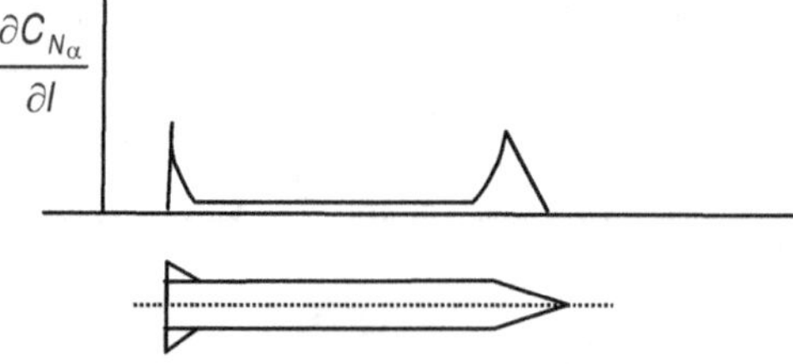

Fig. 3.22 Aerodynamic Load distribution

If the configuration has wings, an additional peak load will be present at the wing location. Assuming the control force acts near the base, the total bending moment acting on the vehicle can be represented as

$$BM = K_1\alpha + K_2\delta \qquad\qquad \dots (3.135)$$

The elementary structural mechanics gives a relation

$$\frac{M}{I} = \frac{E}{R} \qquad \text{or} \qquad \frac{1}{R} = \frac{M}{EI} \qquad\qquad \dots (3.136)$$

where

M = Bending moment at a section

E = Young's modulus

I = area inertia of cross section

R = radius of curvature of the bent section.

I is proportional to d^2 where d is the diameter of cylindrical shell.

It is very easy to see that for given α and δ values, the bending moment will increase as the vehicle length increases and the inertia will increase as the square of diameter of the vehicle cross section under consideration. Thus, the curvature $\dfrac{1}{R}$ increases as the vehicle length increases and diameter reduces. It is customary to indicate the vehicle slenderness in terms of the ratio $\left(\dfrac{L}{D}\right)$ of the vehicle. Higher the $\dfrac{L}{D}$ of the vehicle, more it is flexible.

Table 3.2 gives $\dfrac{L}{D}$ ratios for some launch vehicles and missiles to indicate the range of values in practice.

TABLE 3.2: L/D Ratio For Typical Vehicles

Type of vehicle	L/D Ratio	Country
I. LAUNCH VEHICLES		
SLV-3	22.8	India
ASLV	23.8	India
PSLV	15.8	India
GSLV	10.6	INDIA
Ariane	–	Europe
Saturn	–	US
Titan	–	US
II. MISSILES		
A. Silo Based Long Range Missiles		
Minuteman II	10.11	US
Minuteman III	9.84	US
Peace Keeper	9.3	US
CSS-2	10.7	China

Table Contd.

Table Contd.

	CSS-3	12.4	
	CSS-4	9.73	
	SS-13	11.8	
	SS-18	12.2	
	Stilleto	10.8	
B.	Surface To Surface Short Range Missiles		
	SS-300	14.3	Brazil
	SS-600	11.0	Brazil
	CSS-6	10	China
	CSS-8	16.6	China
	PRITHVI	9	India
	AGNI	15	India
C.	Submarine Launched Long Range Missiles		
	S-3	9.2	France
	M-4	5.7	France
	M-5	5.3	France
	SS-N-18	8.67	Russia
	Polaris	6.77	Russia
	Trident 5	6.36	US
	Poseidon	4.91	US
D.	Surface To Air Missiles		
	Crotale	18.1	China
	Mistral	19.67	France
	Aster-30	13.33	France
	Sea Sprint	13.68	Russia
	Strella-3	20.4	Russia
	Javelin	18.3	UK
	Red Eye (Stinger)	21.7	US
	Akash	16.57	India
	Trishul	15	India
E.	Air To Surface Missiles		
	YJ-2	14.72	China
	Exocet	13.4	France
	Trigat	10	France
	Popeye-2	7.5	Israel
	AS-11	13.5	Russia
	AS-17 (Krypton)	13	Russia
	AGM-158	7.75	US
	AGM-88	16.4	US
F.	Air To Air Missiles		
	Side Winder	22.8	USA
	Stinger	21.7	USA
	Mistral	20.67	France
	Mica	18.34	France
	Atoll	22.34	Russia
	Sky Flash	18	UK

When the vehicle flexes under aerodynamic load, the local angle of attack and the load distribution increases on nose side and reduces on tail side, resulting in shifting of centre of pressure for aerodynamic load. This increases the control torque requirement for stabilizing the vehicle. Hence, one needs to keep in mind the additional control force requirement due to vehicle flexibility. An estimate of 20% increase in control force requirement for SLV-3 was made in Ref. 14.

One may define the factors such as

$$\xi_{C_{N\alpha}} = \frac{C_{N\alpha\ flex}}{C_{N\alpha\ rigid}} \qquad\qquad \dots (3.137)$$

$$\xi_{Cp} = \frac{C_{p\ flex}}{C_{p\ rigid}} \qquad\qquad \dots (3.138)$$

These factors will be function of dynamic pressure and may be used for design or evaluation of design at the relevant operating conditions.

3.4.1.5   Nonlinear Aerodynamics

The normal force C_N and moment coefficient C_M of aerodynamic parameters are not only a function of mach number but also of angle of attack.

Thus
$$C_N = A_1\alpha + A_2\alpha^2 \qquad\qquad \dots (3.139)$$

$$C_M = B_1\alpha + B_2\alpha^2 \qquad\qquad \dots (3.140)$$

and A_1, A_2 and B_1, B_2 are then given as a function of mach number.

The launch vehicles mostly operate with very low angles of attack. But the surface to air missile operate at high angles of attack with large lateral acceleration manoeuvres. Hence, the control system design may be made at the normal expected values of angle of attack and the design verification can be carried out with extensive simulation of trajectories incorporating not only the non-linear aerodynamics but also the prediction and measurement uncertainties on the aerodynamic coefficients.

3.4.2   Propulsion Disturbances

3.4.2.1   Thrust Misalignment

We have discussed in chapter 1, the scheme for evaluating the disturbance force and torque due to propulsion including the case of number of strap-on boosters. We must provide an adequate control torque to balance these forces. The case is very simple when a single booster is present since one can assume that the misalignment can be present either in pitch plane or in yaw plane. Then

$$F_{c3} l_c = T\, \delta_m l_m \qquad \qquad ...(3.141)$$

and adequate control needs to be provided in both the planes.

The case becomes a little uncertain when number of strapons are present. One may consider that the misalignment for all the strapons and the core vehicle may be additive either for pitch plane, yaw plane or about the roll axis of the vehicle. In such cases, it is desirable to use a thrust vector control system with adequate vectoring angle so that even the worst case combination of misalignment can be handled though in actual case the net disturbance may be much less.

3.4.2.2   *Unequal Thrust of Strap-ons—Design of Nozzle Cant Angle*

The strapon boosters are designed to have same thrust to minimize the disturbance levels due to unequal thrust. In actual practice, the strapons located on diametrically opposite sides of core vehicle will have unequal thrust within the design tolerance band. The disturbance torque due to unequal thrust needs to be balanced by control torque. The instantaneous unequal thrust becomes particularly high when the slope of thrust vs. time curve is high which occurs during ignition phase and tail off region. Since, the ignition phase lasts for a very short duration and also vehicle is not released from launch platform till adequate thrust builds up, it is not critical from the point of view of control system. However, tail off region lasts for substantial time and even the tail off may start at different times giving a situation that one of the oppositely mounted booster has high thrust but other has a very low thrust, thus, giving a large disturbance torque on the vehicle.

To avoid this situation, the solid propellant strapons are provided with a cant angle to the nozzle in such a way that the thrust lines of the strapons pass very nearly through the *CG* of the entire vehicle at strapon burn out time.

Following example illustrates the design of nozzle cant angle for a case of two strapon boosters.

The distance between vehicle *CG* at strap-on burn out and nozzle throat = *L*

Distance between the core and strap-on axes = *D*

Then the cant angle,

$$\text{Tan } \delta = \delta = (D/L)*57.3 \text{ deg.} \qquad \qquad ...(3.142)$$

The cant angle reduces the forward thrust by a factor of cos δ. Hence, the cant angle calculated as above needs to be readjusted as a compromise between the loss of forward thrust on one hand and controllability on the other hand. Further, the vehicle *CG* may change from the estimated value during design phase to the value when vehicle is fully realized. Hence, an additional control force required to take care of *CG* variation needs to be found out.

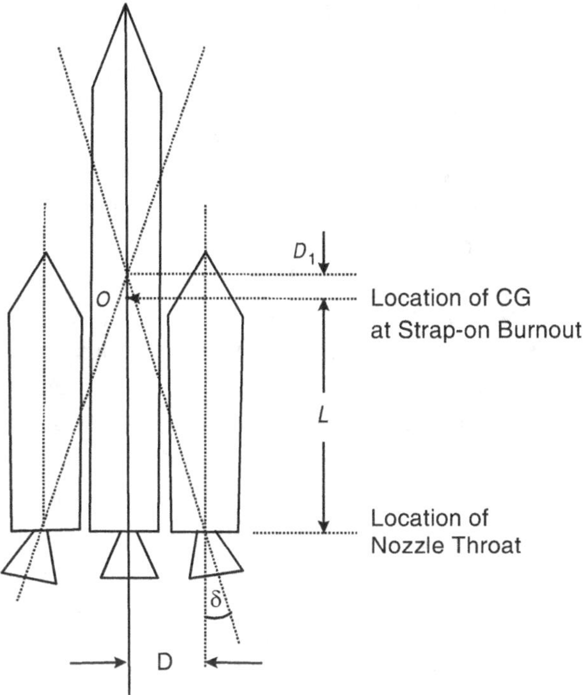

Fig. 3.23

Let D_1 be the distance of actual CG at burn out from the point of intersection ('O') of thrust lines of strapons

T_D = the difference of thrusts of two strap-ons

L_c = be the distance between point 'O' and control force location. Then the control force required to balance the unequal thrust is given by

$$F_{c4}\, Lc = T_D\, D_1 \sin \delta \qquad \qquad \text{... (3.143)}$$

This control force level will be significantly less than the value required without cant angle provision *i.e.*, $D.T_D$.

One can also estimate the control force required in case the two strapon cant angles are different due to manufacturing tolerances or misalignments. Thus

$$\Delta F'_{C4} \cdot L_c = T_D D_1 \cos \delta * \Delta \delta$$

$$= T_D D_1 \cos \delta \cdot (\delta_2 - \delta_1) \qquad \qquad \text{... (3.144)}$$

where ΔF_{C4} is incremental change in control force due to difference in cant angles of two nozzles.

3.4.2.3   Controllability of the Vehicle During Tail-off Region of Strap-ons

Vehicles such as ASLV having two strapons may have initially only strap-ons burning phase and the core will be ignited at appropriate time and strapons separated from the vehicle. The control will be provided by thrust vectoring on the strapon. The thrust vector control using secondary injection control (See Chapter I) or flexible nozzles will not be very effective when the thrust or the chamber pressure falls below certain level. Further, the configuration is likely to have a high rolling moment coefficient (C_L), and the strapon burnout time may be occurring in high dynamic pressure region. It is essential that the core vehicle be ignited based on the measurement of chamber pressures of the strapons instead of making it a purely time based event and the control be transferred from strapons to core vehicle before the strapon control becomes ineffective. In such case, a provision for adequate roll control needs to be made to cater for the aerodynamic and thrust misalignment disturbances till the strapons are discarded.

3.4.2.4   Lateral CG Shift of the Vehicle

The main thrust will give a disturbance torque on the vehicle when there is a *CG* (centre of gravity) offset from the thrust line. This will require an additional control force requirement in both pitch plane and yaw plane depending on the *CG* offset (Y_{cg} or Z_{cg}) in the plane.

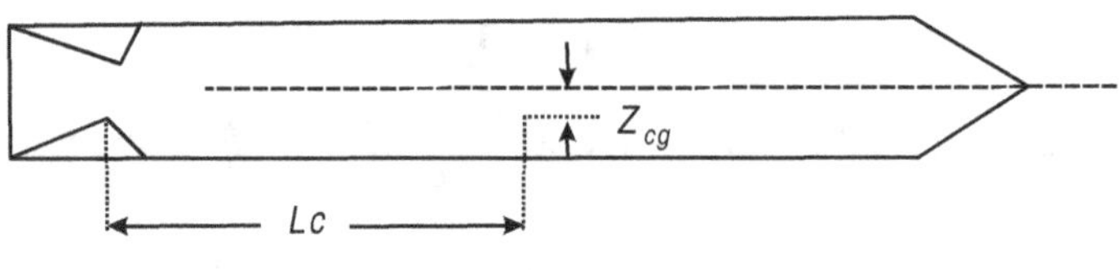

Fig. 3.24

The control force required is given by

$$F_{cp5}\, L_c = T\, Z_{cg} \qquad\qquad \text{... (3.145)}$$

$$F_{cy5}\, L_c = T\, Y_{cg} \qquad\qquad \text{... (3.146)}$$

Eq. 3.145, 3.146 assumes control by thrust vectoring. In that case, the thrust vectoring angle required to compensate the disturbance is given by

$$\delta_p = \frac{Z_{cg}}{L_c} \qquad\qquad \text{... (3.147)}$$

$$\delta_y = \frac{Y_{cg}}{L_c} \qquad\qquad \text{... (3.148)}$$

If the control is provided by some other systems such as aerodynamic control (during atmospheric flight) or reaction control system in case the flight is out of atmosphere, appropriate control moment needs to be used.

3.4.2.5 Disturbance Moment about Roll Axis

Let us consider the disturbance moment about the roll axis.

Case I : **Thrust Vector Control**

Since we have assumed main thrust without any misalignment, it will have no contribution to roll moment. The net roll moment on the vehicle due to pitch and yaw control forces is given by (using Eq. 3.145 and 3.146)

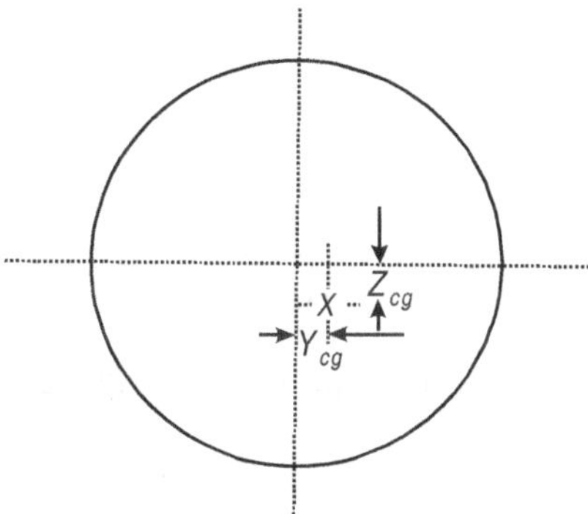

Fig. 3.25

$$M_{dx} = -F_{cz} \cdot Y_{cg} + F_{cy} \cdot Z_{cg} \qquad \qquad \text{... (3.149)}$$

$$= -T \cdot \frac{Z_{cg}}{L_c} \cdot Y_{cg} + T \frac{Y_{cg}}{L_c} \cdot Z_{cg}$$

i.e., $\qquad\qquad\qquad M_{dx} = 0 \qquad\qquad\qquad\qquad\qquad\qquad$ ... (3.150)

Even though a thrust misalignment is considered it will be counteracted by a pitch and yaw control force and net roll disturbance moment will become zero.

Let us consider a thrust vector control system with a resultant deflection such that the thrust components are T_x, T_y and T_z. Let X_{cg}, Y_{cg} and Z_{cg} be the coordinates of *CG*.

Then the position vector of the point of application of thrust is given by

$$\overline{R} = (X_G - X_{cg})i + (Y_G - Y_{cg})j + (Z_G - Z_{cg})k \qquad \text{... (3.151)}$$

and $\qquad\qquad \overline{T} = T_x i + T_y j + T_z k \qquad\qquad\qquad\qquad$ ... (3.151)

Then, the net torque about the vehicle centre of gravity is given by:

$$\bar{M} = \bar{R} \times \bar{T}$$

$$= \{T_z(Y_G - Y_{cg}) - T_y(Z_G - Z_{cg})\}i + \{T_x(Z_G - Z_{cg}) - T_z(X_G - X_{cg})\}j$$

$$+ \{T_y(X_G - X_{cg}) - T_x(Y_G - Y_{cg})\}k \qquad \qquad ...(3.153)$$

since the vehicle is controlled in pitch and yaw by thrust vectoring, the net torque about pitch and yaw axis is zero. Hence

$$T_x(Z_G - Z_{cg}) - T_z(X_G - X_{cg}) = 0$$

or
$$T_z = T_x\left(\frac{Z_G - Z_{cg}}{X_G - X_{cg}}\right) \qquad \qquad ... (3.154)$$

and
$$T_y(X_G - X_{cg}) - T_x(Y_G - Y_{cg}) = 0$$

or
$$T_y = T_x\left(\frac{Y_G - Y_{cg}}{X_G - X_{cg}}\right) \qquad \qquad ... (3.155)$$

Using Eq. 3.154 and 3.155, we can now find the net torque about x-axis, *i.e.*,

$$M_x = T_x\left(\frac{Z_G - Z_{cg}}{X_G - X_{cg}}\right)(Y_G - Y_{cg}) - T_x\left(\frac{Y_G - Y_{cg}}{X_G - X_{cg}}\right)(Z_G - Z_{cg}) = 0 \qquad ... (3.156)$$

We have thus seen that there is no disturbance torque about roll axis due to main thrust and control forces due to thrust vectoring even in presence of *CG* offsets when pitch and yaw is properly controlled in trim condition. There will however be a small torque due to gravity when the *CG* offset occurs out of vertical plane.

Case 2: **Aerodynamic Control**

Let the resultant control force act at $(X_c\, i + Y_c\, j + Z_c\, k)$ and the control force be given by $(o.i + F_{cy}\, j + F_{cz}\, k)$. Using similar analysis and using trim condition for pitch and yaw giving net torque about pitch and yaw axes zero, we get roll moment torque as

$$M_x = T_z\left[\frac{(X_G - X_C)Y_{cg} + (Y_C - Y_G)X_{cg} + Y_G X_C - X_G Y_C}{(X_C - X_{cg})}\right]$$

$$-T_Y\left[\frac{(X_G - X_C)Z_{cg} + (Z_C - Z_G)X_{cg} + Z_G X_C - X_G Z_C}{(X_C - X_{cg})}\right] \qquad ... (3.157)$$

position coordinates X_c and X_G only are having large values, all others are small quantities. If it is a tail control vehicle, the control location is near to the nozzle throat location and the roll moment will be of smaller magnitude. However, if it is canard control, the position X_C and X_G are of opposite sign. Control force and disturbance force in trim condition will have same sign and the net roll disturbance moment will be addition of that due to thrust misalignment as well as control force.

Case 3: **Reaction Control System (RCS)**

This case is applicable for out of atmosphere trajectory. The effect of thrust misalignment will be continuously present. But the effect of control force will be present only when the RCS motors fire. Further, the pitch and yaw motors may not fire simultaneously and the location of control force also may be either on tail side or nose side. The net result over one cycle of oscillation will be

$$M_x = T_y Z_{cg} - T_z Y_{cg}, \text{ for RCS off duration}$$

$$= (T_Y + F_{cy}) Z_{cg} - (T_z + F_{cz}) Y_{cg} \text{ for RCS on duration} \tag{3.158}$$

or $$M_x = T_Y Z_{cg} - T_z Y_{cg}, \text{ for oscillation period } Tp$$

and control torque $M_c = F_{cy} Z_{cg} - F_{cz} Y_{cg}$ for RCS on duration during each oscillations

$$\text{... (3.159)}$$

in which case, one of the F_{cy} or F_{cz} may be zero, or full value or a full value for only part of the time.

The disturbance moment can thus be a pulsating torque about x-axis and magnitude also will depend on control location on tail side or nose side since the force can be either subtractive or additive respectively.

Similarly, the effect of control motor mounting misalignment with vehicle radial line need to be considered for roll disturbance moment. (See Fig. 3.26).

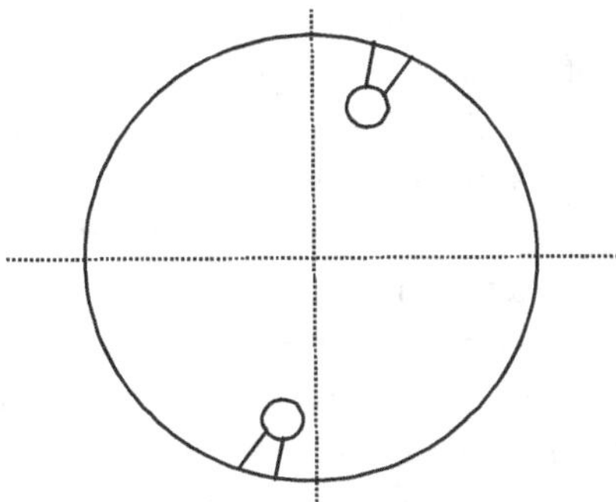

Fig. 3.26 RCS Motors mounting misalignments

Case 4: **Manoeuvring Vehicle**

Above discussion is mostly applicable for launch vehicles except for the case of canard control. Missiles can have high 'g' manoeuvres. The manoeuvre will have similar effect as gravity and should be calculated based on the maximum latax capability of the vehicle in pitch, yaw and both together and adequate control deflection needs to be provided for roll control.

3.4.3   Control during Thrust Tail-off Region

If the control is provided by thrust vectoring, there is effective control during the burning period and part of the region of tail-off when thrust is significantly high. However, thrust vector control (TVC) becomes ineffective below certain value of the thrust. For secondary injection control, it will become ineffective below certain chamber pressure. Studies carried out on the SITVC for SLV-3 and ASLV indicated that control becomes ineffective (or unreliable) for chamber pressure below 8Ksc. For flexible nozzle or gimballed engine TVC, tail-wag-dog (TWD) effect due to nozzle or engine inertia becomes pronounced. One may prolong control for sometime by reducing the autopilot frequency as the TWD frequency due

to falling thrust $\left(\omega = \sqrt{\dfrac{T\,L_C}{I_R + M_R L_R L_C}} \right)$ starts decreasing. Since, the stage cannot be

separated till thrust dies down below the retro motors capability, it is desirable to ensure control using either aerodynamic control during atmosphere or by switching on reaction control system at appropriate time.

3.4.4   Disturbance due to Stage Separation

The separation of stage may not be perfectly smooth. The sudden disturbance at the time of separation is impulsive and usually described by a net result as attitude error and angular rate.

When the separation occurs out of atmosphere, the disturbance in vehicle state is of much less concern since the control of subsequent stage brings down the errors and rate in a short time. When the separation occurs within high dynamic pressure region, it needs to be studied properly, particularly when the vehicle after stage separation has unstable aerodynamic configuration. The likely wind profiles also need to be taken into account at the time of separation. It is desirable that the angle of attack of the vehicle before separation is brought down to near zero value. The present day onboard computers have enabled to implement such schemes during flight and has been successfully implemented in case of a surface to surface missile. The angle of attack and sideslip angles are computed onboard the vehicle during flight using velocity vector and direction cosine matrix (DCM). Flight path angle in pitch and yaw is calculated using relative velocity vector and vehicle attitude is calculated using DCM matrix.

Then
$$\alpha = \theta - \gamma_p$$

And
$$\beta = \gamma_y - \psi \qquad \qquad \text{... (3.160)}$$

Additional commands are then given to the attitude control system so as to align the vehicle attitude to the velocity vector well before the separation command is issued. This brings down the level of angle of attack to be catered for by the control system immediately after separation. Further, it will enable smoother and collisionless separation of the two bodies.

A sensitivity study on ASLV during initial phase at the time of strapon separation (Ref. 13) showed that the control force requirement for core vehicle shoots up to 1200 kg. for core without fins from $300 - 400$ kg. for core with fins thus indicating sensitivity of unstable configuration to disturbances and data variation.

Hence, it is necessary to carry out the capturability study of the subsequent stage control system with appropriate perturbations in data and wind profiles.

3.4.5 Control System for Out-of-Atmosphere Trajectory

3.4.5.1 *Burning Phase*

Control for this phase is possible using either thrust vector control system or reaction control system. The main considerations for fixing the maximum control deflection angle are:

- Capturability of separation disturbance of earlier stage

- *CG* offset

- Alignment errors.

The main consideration for RCS control force levels in addition to above factors are:

- Thrust misalignment

- Control impulse level.

It will be shown in chapter on RCS system that the total control impulse for RCS system is fixed for a given disturbance. However, if the thrust misalignment disturbance for a particular plane happens to be negligible or zero, the control force provided would be unnecessarily high and the system will be hunting giving rise to high impulse consumption. It is found that a control force level of 1.2 to 1.5 times the maximum disturbance force level gives good performance of the control system during burning phase.

3.4.5.2 *Coast Phase*

Launch vehicle trajectories, generally, have long duration of coast phases to gain the altitude and only RCS control force is possible for these phases. The only significant disturbances acting on the vehicle during this phase of flight are:

(1) Residual thrust misalignments in the initial part of coast phase

(2) Disturbances due to control motor mounting misalignment whenever the RCS motors fire.

The disturbances are generally small. A control force level which can give an angular acceleration of 5 to $10°/\text{sec.}^2$ appears adequate for this phase.

There is no stringent requirement on accuracy for control system for major portion of coast phase. Hence, sufficiently large control deadzones can be used for most of the coast phase duration to minimize control impulse consumption. However, the deadzone must be narrowed down to a small value well before the separation of the stage or the start of next stage. The time should be adequate for the maximum initial error due to large dead zone to settle down and have few steady state limit cycle oscillations to ensure good accuracy before the stage separation is initiated.

3.4.6   Control Impulse and Actuator Force Requirement

We have considered so far most of the factors that need to be considered to determine the control force to be provided which also means the maximum deflection required for thrust in case of thrust vector control system and control surfaces for aerodynamic control system. As one can see, all the disturbances need not be additive. But the contribution of some of the factors is not very large.

For launch vehicle during atmosphere, the main considerations will be thrust misalignments, aerodynamic disturbances in presence of likely winds, unequal thrusts of strapons and roll moment coefficients (C_L). Out of atmosphere, the considerations will be thrust misalignment and CG offsets during powered phase and long duration coast phases. For medium to long range surface to surface missiles, considerations will be similar to launch vehicles except for re-entry phase.

For surface to air missiles, the maximum latax capability will be the most dominant factor to determine the control deflection required.

3.4.6.1 Control Impulse Requirement

The trajectory simulation is the most convenient method for determining the total impulse required. However, a rough estimate for atmosphere phase and quite accurate estimate for out of atmosphere phase can be done analytically.

Let $M_d(t)$ be disturbance torque as a function of time assuming a thrust misalignment δ_m to occur in pitch plane.

Then
$$M_d(t) = T(t) \cdot \delta_m \ (L_G\text{-}L_{CG})$$
... (3.161)

Balancing control force $F_c(t) = \left(\dfrac{M_d(t)}{(L_c - L_{CG})} \right)$ ... (3.162)

Then control impulse $\int F_c(t)\, dt = $ Area under the $F_c(t)$ vs. time curve ... (3.163)

Actual thrust misalignment can occur in any plane. Hence, though the magnitude of $M_d(t)$ is same, the pitch and yaw control forces will be given by

$$F_{cz}(t) = \frac{M_d(t)}{(L_C - L_{CG})} \cdot \cos\eta = F_c(t)\cos\eta$$

$$F_{cy}(t) = \frac{M_d(t)}{(L_C - L_{CG})} \cdot \sin\eta = F_c(t)\sin\eta$$

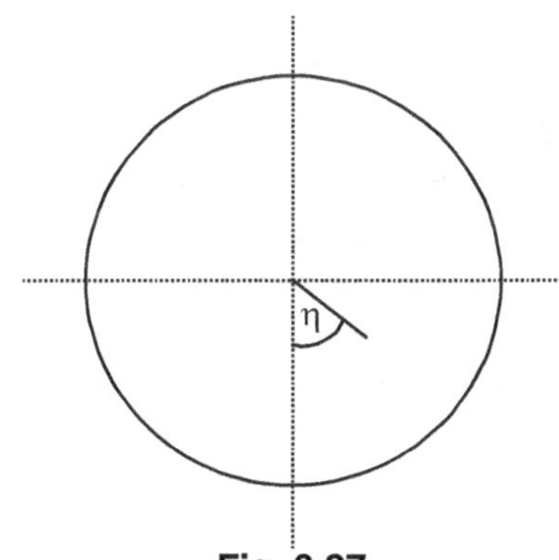

Fig. 3.27

where η indicates the orientation of the misalignment.

The total impulse will then be

$$\text{Impulse} = \int F_c(t)\,(\cos\eta + \sin\eta)\, dt$$

It can be easily shown that this becomes maximum when $\eta = 45°$. Then the control impulse required to take care of thrust misalignment.

$$I_{TM} = \sqrt{2} \int F_c(t)\, dt \qquad \text{... (3.164)}$$

This will be the major component of impulse for out of atmosphere stage. Though equivalent control force for nominal trajectory α and wind angle of attack α_w can be obtained as above, the impulse estimate by trajectory simulation will give more realistic value. For control impulse estimation, it is safer to use percentile profile winds (95 percentile, for example) instead of synthetic wind profile.

The additional requirement to take care of unequal thrust can similarly be computed.

The computation of control impulse requirement for Reaction Control System during coast phase will be discussed in chapter on RCS control systems.

3.4.6.2 *Actuator Force Requirement*

Following main factors need to be considered to decide the actuator force requirement:

1. Inertial force and torque acting on the nozzle due to vehicle accelerations (forward, lateral and angular).

    The detailed expressions for the inertial forces acting on the engines and torque to be supplied by actuator about the gimballing point is given in chapter II, section 2.3.5.

2. The boot friction forces in case of flexible nozzle.

3. Maximum angular accelerations for nozzle/engine deflection required to provide the specified band width for the actuation system.

An amplitude equal to 10% of maximum deflection capability is usually considered as a linear zone of operation for launch vehicle actuation system and the maximum angular rate and accelerations are calculated accordingly for estimating the power and force requirement for actuator. Since, there is no requirement on fast manoeuvres for launch vehicles, the actuator deflection rates are limited to about 10 to 12°/sec for lower stages controlled by large inertia actuation system such as flexible nozzle.

The surface to air missiles have high latax capability and use aerodynamic controls. The actuator force capability will be decided by

(*i*) the hinge moment acting on the shafts of control surfaces due to aerodynamic centre of pressure not lying on the hinge line and

(*ii*) the torque required to provide adequate angular acceleration to control surfaces. The linear zone of operation is considered as 20 to 25% of maximum deflection angle. The hinge moment is given by

$$M_h = (C_{h\delta}\delta + C_{h\alpha}\alpha)\, QSd \qquad\qquad \text{... (3.165)}$$

The power requirement for actuator is then given by

$$P = M_h \delta \qquad\qquad \text{... (3.166)}$$

Estimation of these values as a worst case condition are likely to be high and it may pose practical problems in realization of actuators within the weight and volume constraints imposed by the missile configuration. It would be better to obtain the maximum values by simulating number of typical trajectories covering the zone of operation. Ref. 15 illustrates determination of actuator δ, hinge moment and power requirement for a typical interceptor missile.

3.4.7 Autopilot Bandwidth

The autopilot bandwidth varies depending on the mission. For launch vehicles, there are no rapid manoeuvres. Hence, low bandwidth is used. The main factor which restricts the bandwidth is the lowest bending mode frequency. The autopilot bandwidth is generally 4 to 5 times less than the first bending mode. If it is too close to the bending mode frequency, it will present more difficulties to provide adequate phase margin since any compensation or filtering used to stabilize the first bending mode will eat away some of the phase margin at the autopilot frequency. Larger the launch vehicle, lower is the first mode frequency and so lower is the autopilot bandwidth. The range of bandwidth for representative launch vehicles is between 2 r/s for large vehicles to 5 to 6 r/s for smaller launch vehicles.

The missiles on the other hand are designed for high lateral acceleration manoeuvres. The first bending mode frequencies are also correspondingly higher. Hence, the bandwidth is decided mainly from the guidance requirement instead of from the consideration of separation from the first bending mode frequency. The autopilot bandwidth is 4 to 5 times higher than the guidance bandwidth. Thus, the representative bandwidths are 1 Hz for surface to surface missiles to $3 - 5$ Hz for surface to air missiles.

3.4.8 Actuator Bandwidth

We have already seen that the bandwidths of actuator, sensor etc. become the limiting factors for achieving higher bandwidth for the autopilot. Hence, the bandwidths of these elements must be sufficiently high so that they cause only a small lag at autopilot cross over frequencies. Higher bandwidth requires higher power and hence trade off is essential. In general, the actuator bandwidth is chosen 5 to 6 times the required band width of the autopilot.

Following simple analysis demonstrates the relative benefit on autopilot accuracy if we increase the actuator bandwidth. (Ref. 8)

Consider the attitude control system given by the block diagram of Fig. 3.8 and first order actuator transfer function given by Eq. 3.98. The characteristic Equation 3.104 can be represented by one real pole and one complex pair of poles as follows:

$$s^3 + K_c s^2 + (K_A K_c K_R \mu_c - \mu_\alpha)s + (K_A K_c \mu_c - \mu_\alpha) = (s + p)(s^2 + 2\zeta_c \omega_c s + \omega_c^2)$$

$$\dots (3.167)$$

Equating coefficients of like powers of s on both sides and simplifying one gets

$$\mu_c K_A = \omega_c^2 \left(1 - \frac{2\zeta_c \omega_c}{K_c}\right) + \frac{\mu_\alpha}{K_c}$$

and
$$K_R = \frac{2\zeta_c \omega_c^2 + \mu_\alpha}{\mu_c K_A K_c}$$

Assuming $\mu_\alpha \approx 0$, we get a simplified equation

$$\mu_c K_A = \omega_c^2 \left(1 - \frac{2\zeta_c \omega_c}{K_c} \right) \qquad\qquad \text{... (3.168)}$$

and

$$K_R = \frac{2\zeta_c}{\omega_c} + \frac{1}{K_c - 2\zeta_c \omega_c} \qquad\qquad \text{... (3.169)}$$

Let the desired poles of the autopilot be given by ζ_c and ω_c. Then assuming $\zeta_c = 0.7$, variation of $\mu_c K_A$ and K_R is obtained as a function of K_C (*i.e.*, actuator bandwidth). Fig. 3.28 shows the variation of these gains for two values of

(1) $\omega_c = 5$ r/s and

(2) $\omega_c = 10$ r/s.

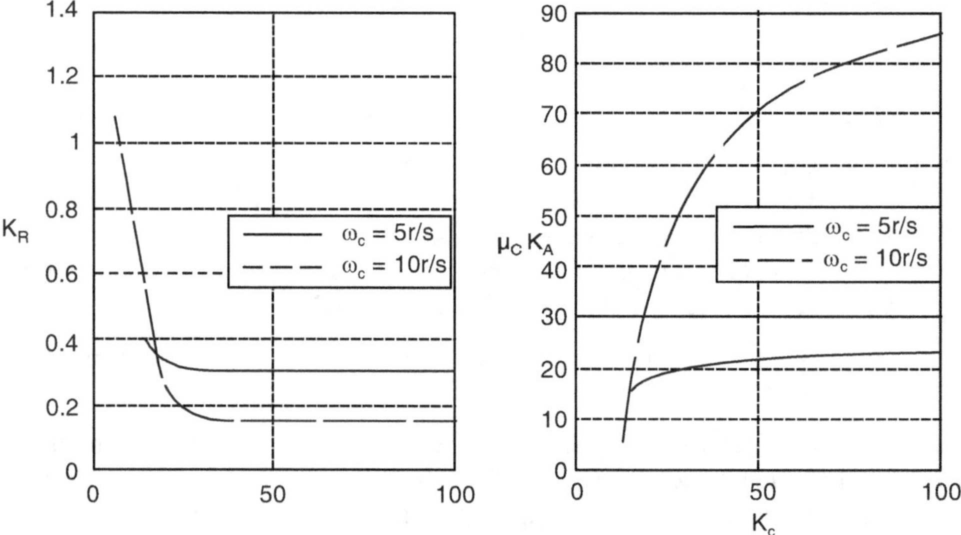

Fig. 3.28 Variation of Control Gains With Actuator Bandwidth

It may be seen that the gains $\mu_c K_A$ and K_R vary rapidly initially and then reach almost constant values. $\mu_c K_A$ increases with K_c and K_R decreases with K_c. From Eq. 3.75, it is seen that, for $\mu_\alpha \approx 0$, the autopilot tracking error is approximately given by

$$(\theta_c - \theta) = K_R \, \theta_c - \frac{\mu_d}{\mu_c K_A}$$

From Fig. 3.28, it is seen that the contribution of thrust misalignment decreases as $\mu_c K_A$ increases (*i.e.*, as K_c increases) and contribution due to constant pitch rate also decreases as

K_R decreases (*i.e.*, as K_C increases). The misalignment component can be positive or negative depending on the direction of misalignment. It can thus be seen that increasing the actuator bandwidth enables one to select gains which help in reducing the tracking error of the autopilot for the same autopilot natural frequency ω_c. From figure it is seen that the gain curve reaches nearly flat portion when K_c is nearly 6 times (or $K_c = 30$) the autopilot frequency ($\omega_c = 5$ r/s). Fig. 3.28 also shows gain curves for another value of $\omega_c = 10$ r/s and it is seen that the curve reaches nearly flat portion for $K_c = 70$ to 80 r/s. Thus, it can be seen that the relative benefit on autopilot accuracy due to increasing actuator bandwidth becomes only marginal if we increase actuator bandwidth beyond 6 times. Since, autopilot bandwidth cannot be increased for launch vehicles beyond a certain value to avoid closeness from the first bending mode and increasing actuator bandwidth involves increase in power, weight and volume requirement, selection of actuator bandwidth equal to 5 to 6 times the autopilot bandwidth appears a very reasonable choice. The analysis also demonstrates the need of increasing the actuator bandwidth, if one intends to increase the autopilot bandwidth to improve the guidance accuracy.

3.5 DESIGN OF BACK-UP CONTROL SCHEMES

3.5.1 Requirement for Back-up Control

Chapter I has described different control effectors or devices for generating control forces and moments. Those are categorized as

(*i*) Aerodynamic control systems (ADC)

(*ii*) Thrust vector control systems (TVC) and

(*iii*) Reaction control systems (RCS).

Each one of these is best suited for a particular phase of flight trajectory.

The aerodynamic control depends on aerodynamic pressure for generating control force. Therefore, this control system is not suitable when aerodynamic pressure is very small which occurs at the initial take off phase till sufficient velocity is built up and again at higher altitudes when air density becomes very low. Further, for a given mach number and dynamic pressure, the control force generated varies linearly with control surface deflection upto a certain value and becomes nonlinear at higher deflections and starts decreasing at further higher deflections (aerodynamic stall effect). Hence, maximum control deflection generally available is of the order of $20°$ to $25°$.

The thrust vector control systems generate control force by changing the direction of exhaust gases. For liquid propulsion systems, this is done by mounting the engines on gimbal structures. For solid propulsion systems, this is done by either making nozzle joint flexible or by using jet vanes for deflecting the gases or by injecting a reactive fluid at an angle in the exhaust gases. This control is available only during burning or powered phase of flight and not available for coasting phase of the trajectory. Higher the thrust deflection angle, more will be the loss of forward component. Mechanical engineering considerations in providing higher gimballing angle or providing flexible joint will also limit the maximum thrust vectoring angle and hence the maximum thrust deflection angles are generally limited to 3° to 5°.

The reaction control systems have fixed force levels and are used mostly out of sensible atmosphere or when the dynamic pressure is very low.

Due to above features of different control systems, generally thrust vector control is used in powered phase, aerodynamic control is used in high dynamic pressure region and in coast phase and reaction control is used when aerodynamic pressure becomes very low. However, following considerations make it necessary to switch over from one type of control to another type to further optimize the configuration:

1. Secondary injection thrust vector control systems (SITVC) requires storage of injectant onboard the vehicle. Hence, it would be desirable to minimize the requirement of injectant by switching over from SITVC to aerodynamic control when the dynamic pressure becomes adequately high.

2. In case of jet vane control system, the vanes are exposed to high temperature exhaust gases and they wear out progressively leading to performance degradation. Hence, it is desirable to change over to aerodynamic control as soon as the aerodynamic pressure becomes adequate and jettison the jet vanes which otherwise would cause a loss in forward thrust.

3. If the main thrust is not adequately high, the maximum thrust vectoring angle also may not be adequate to meet the aerodynamic disturbances due to aerodynamic misalignments, atmospheric winds and trajectory angle of attack requirements. This will become particularly more important if the vehicle is aerodynamically unstable. Therefore, it becomes desirable to switch over from TVC to ADC when dynamic pressure becomes adequately high.

A natural question would arise as to why not keep the system which has been switched off in a back up mode so that the mission requirements would still be met even if the primary functioning system reaches its saturation level due to some unforeseen disturbances or due to temporary malfunctions. It has been observed that the engine gimballing system leads to thrust misalignment disturbances even when its actuators are given zero commands due to unsymmetric deflection of thrust frame under thrusting load and due to actuator biases. This thrust misalignment torque puts additional burden on aerodynamic control system which tends to saturate when dynamic pressure starts falling. Hence, it will be desirable to use the TVC in back-up mode instead of leaving it idle and causing disturbing moment on the vehicle.

3.5.2 Aerodynamic Control with SITVC as a Back-up

The aerodynamic control and SITVC actuation systems are relatively low inertia systems. The required actuator bandwidths are also nearby each other. In such cases, the logic becomes quite simple.

Let　　　　F_{adc}　C_N $QS.$　K_D　　　　fin deflection　　　... (3.170)

$$F_{sitvc} = K_{DS}x \qquad\qquad x - \text{pintle valve deflection} \qquad ... (3.171)$$

$$K_A(\ _c \qquad K_R\ )\ K_A e \qquad\qquad ... (3.172)$$

Let　be higher than the saturation value　$_{sat}$ such that

$$_E \qquad _{sat} = \text{excess fin deflection} \qquad\qquad ... (3.174)$$

We must give such a signal to SITVC that it would generate same control torque as ADC would generate for　$_E$, *i.e.*,

$$K_{DS}l_{sitvc}.x = K_D l_{adc}.\delta_E$$

or　$x = \dfrac{K_D l_{adc}}{K_{DS}l_{sitvc}}.\delta_E = \dfrac{1}{K_{FE1}}.\delta_E$; where $\mathrm{K_{FE1}}$ is fin equivalence coefficient　　　... (3.175)

Thus, $x = 0$ for　$_E$　0　or　when　$_{sat}$

Or $x = 0$ if $|e| < \dfrac{\delta_{sat}}{K_A}.$ This is equivalent to putting a dead zone

$$d_z \quad \frac{_{sat}}{K_A} \qquad\qquad ... (3.176)$$

in the SITVC channel as shown in the Fig. 3.29.

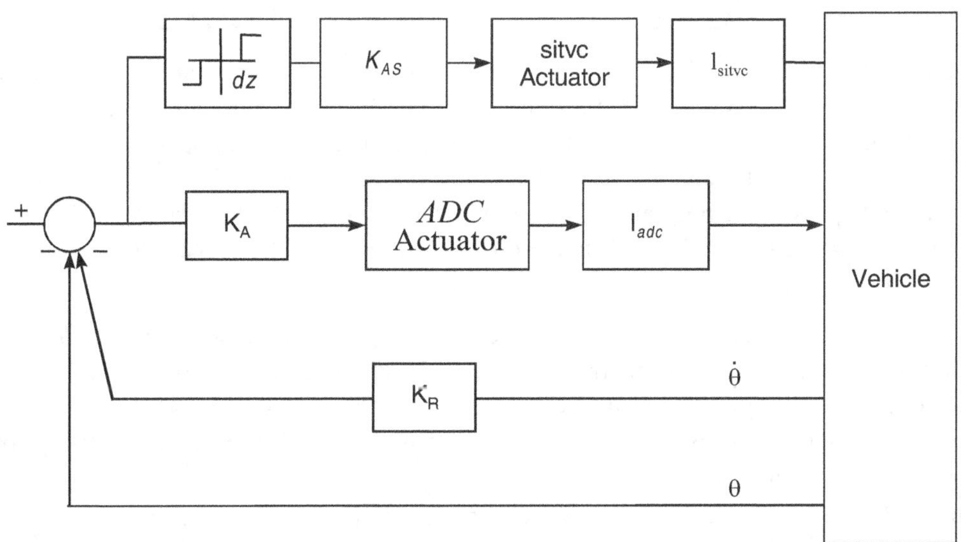

Fig. 3.29 Block Diagram for Autopilot With *ADC* as Primary and SITVC as Backup Control Systems

The control torque vs. error function curve is as shown in Fig. 3.30.

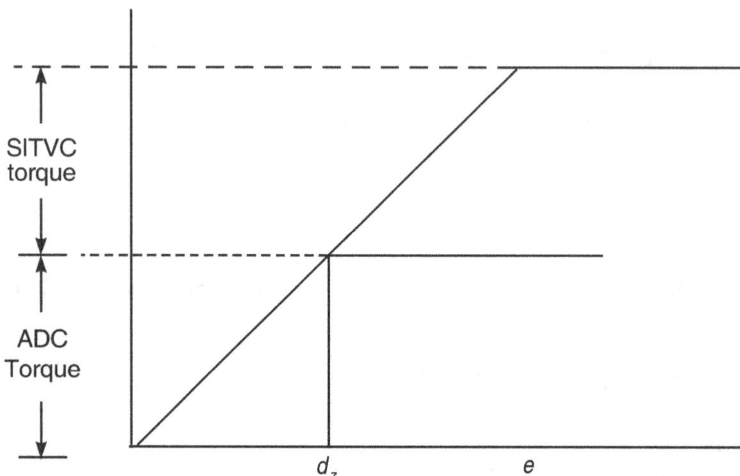

Fig. 3.30 Control Torque vs. Error Function for ADC Primary and SITVC Backup Control

For stability analysis one can assume that only one channel exists. The other channel either increases (SITVC) the saturation level or fills up (ADC) the deadzone without affecting the proportionality.

This scheme has worked quite satisfactorily in several flight trials .

3.5.3 Aerodynamic Control with Gimballed Engine Thrust Vector Control System as a Back-up[17] (Dynamic Trimming of Disturbances)

The ADC actuation system is a relatively low inertia actuation system and gimballed engine thrust vector control system is a high inertia system. The tail-wags-dog effect becomes an important issue when TVC is activated and needs to be adequately compensated for. Hence, it would be preferable to devise a scheme which will not excite TWD effect but at the same time share adequately the burden of providing control torque to the vehicle with ADC, thus, avoiding ADC going to saturation. Such a scheme is possible and has been successfully flight tested several times.

Fig. 3.31. shows a typical control deflection command profile when ADC is used alone. The profile consists of a slowly varying mean deflection command superimposed by a fluctuating signal. The slowly varying mean deflection basically balances the constant or very slowly varying disturbance torque and fluctuating signal provides stability requirement and guidance command requirement. The slow variation occurs mainly due to dynamic pressure and $C_{N\delta}$ variation. If TVC is used to provide control torque to balance the steady disturbances, ADC would need to meet only the stability requirement and relatively fast guidance requirement and thus will not reach saturation.

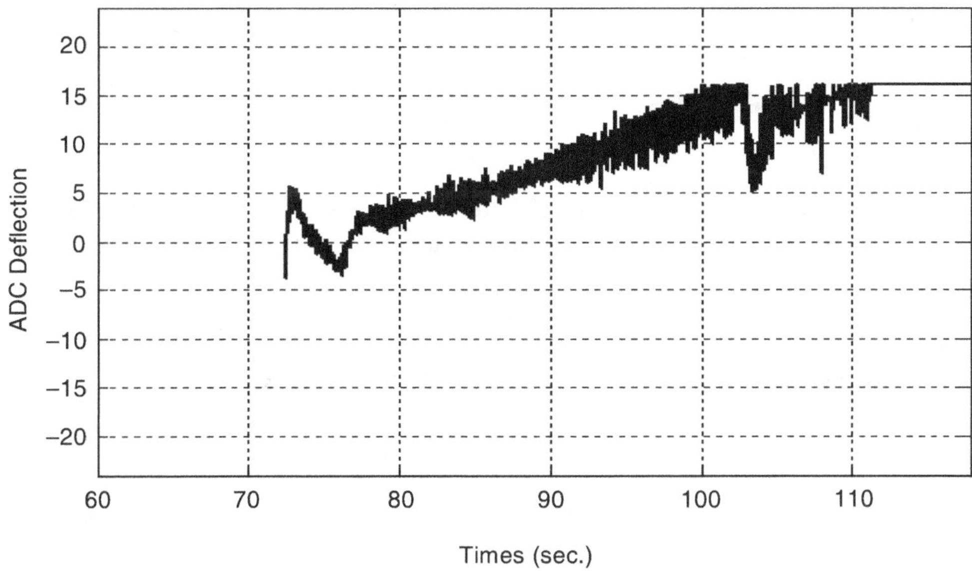

Fig. 3.31 ADC Deflection Command Profile

3.5.3.1 *Extraction Of Mean ADC Deflection Command*

The mean of ADC deflection command can be computed by passing it through a low pass filter so that the amplitude of fluctuating signal is significantly attenuated, say, by a factor of 50 to 100. In that case, the filtered output would look like a mean curve slightly shifted due to filter lag. This mean curve can be used to compute an equivalent TVC command value.

Fig. 3.32 Filtered ADC Deflection Command Profile

For the typical case under consideration, a digital filter corresponding to S-domain filter $\left(\dfrac{1}{1+2s}\right)$ was used to provide an attenuation, of $\dfrac{1}{100}$. For any other case, the filter should be chosen based on the observed frequency of δ oscillations.

Fig. 3.32 shows the filtered output of ADC deflection command given in Fig. 3.31 and is treated as the required mean of ADC deflection command variation.

3.5.3.2 Computation of Equivalent TVC Deflection Command

Let f_0 be the filtered value of ADC deflection command at a given flight time.

Then
$$\delta_{TVC} = \frac{C_{N_\delta} QSl_{adc}}{T_E l_{TVC}} f_0 = \frac{1}{K_{FE2}} f_0 \qquad \qquad \text{... (3.177)}$$

Where $\qquad\qquad\qquad T_E$ – thrust

l_{adc}, l_{TVC} – control moment arms for ADC and TVC

K_{FE2} - fin equivalence coefficient

If this command is given to TVC, the disturbance torque requirement will be met by TVC and the mean value of ADC deflection will decrease to nearly zero. If the compensation is not perfect or if the aerodynamic disturbances change slowly due to variation in dynamic pressure, the mean value will change slightly. Hence, it is desirable to compute the mean value continuously and compute the incremental equivalent TVC deflection command periodically and add to the previously computed value. This will ensure that the mean ADC deflection command remains zero and TVC command gets minor updatements periodically. The update interval should be chosen adequately large so that TVC actuators remain mostly stationary and two systems *i.e.*, TVC and ADC are totally decoupled.

Fig. 3.33 shows the flowchart for the computation scheme and Fig. 3.34 shows the schematic block diagram. In the flow chart T_{set} is the time at which TVC backup is invoked, T_{update} is the time interval at which TVC Command is updated, T_{cycle} is variable time during update interval with an initial value equal to T_{update}. The designer may fix the value for T_{set} based on different considerations or logic such as the time at which the filtered mean becomes greater than a fixed percentage of the full scale deflection δ_{sat}.

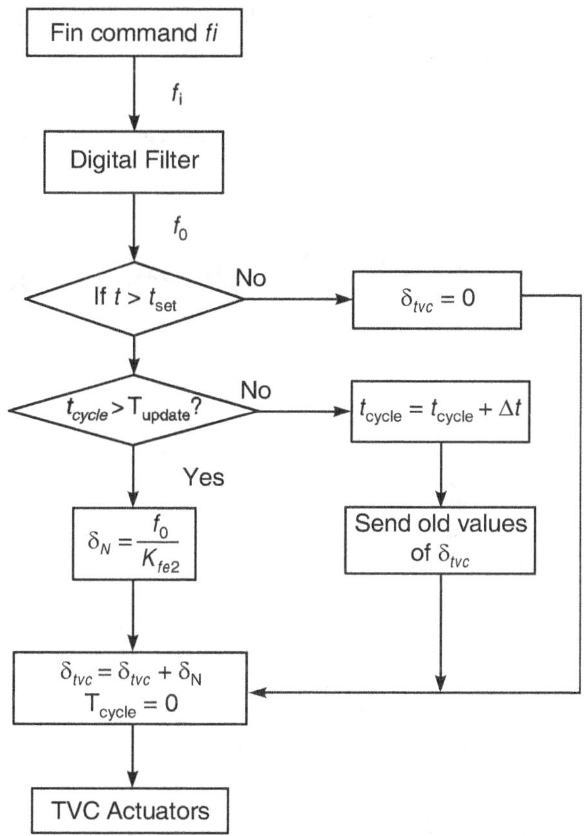

Fig. 3.33 Flow Chart for *TVC* Command Computation

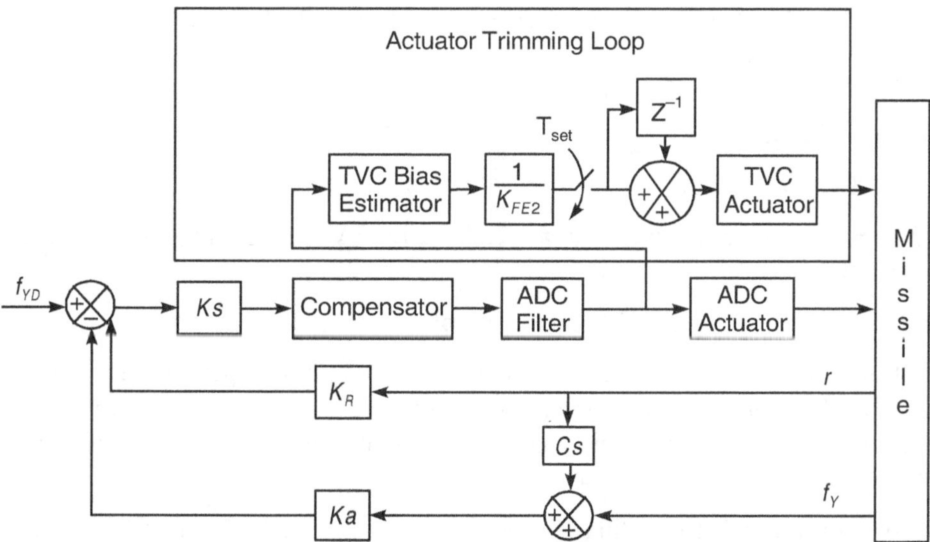

Fig. 3.34 Block Diagram for *ADC* Primary And Gimballed Engine *TVC* as a back-up

Fig 3.35 gives the ADC and TVC deflections during six-degrees-of freedom trajectory simulation of a typical missile using this scheme.

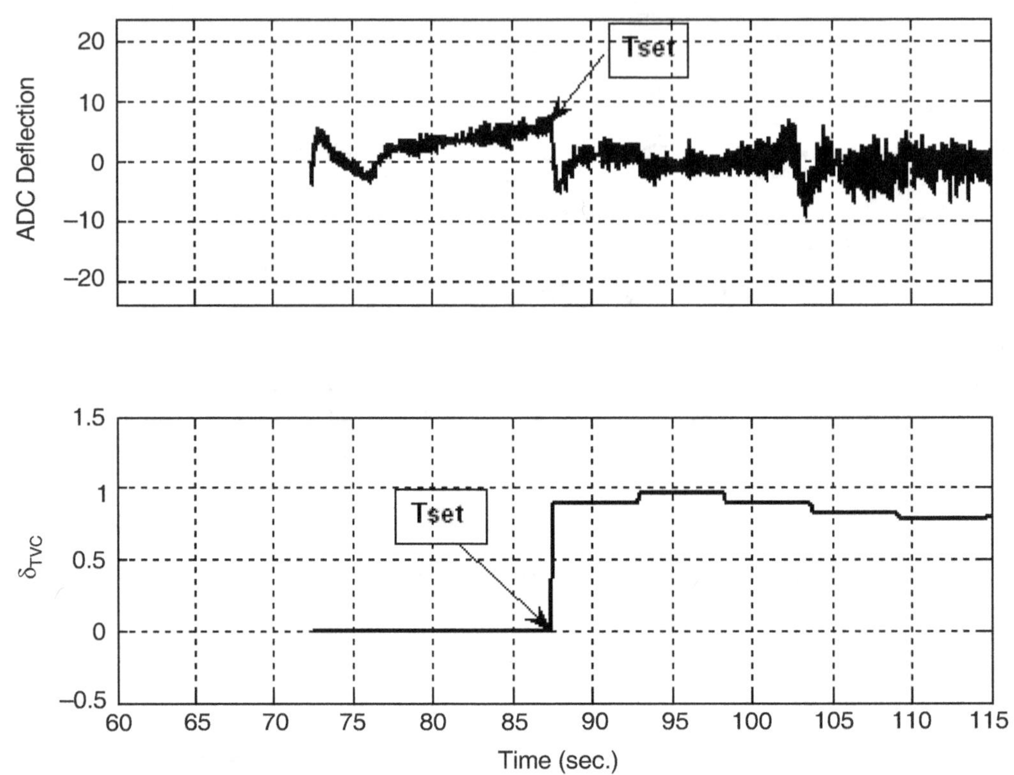

Fig. 3.35 Actuator Deflection Profiles with Backup Control

(For the illustrative example T_{update} = 5 sec)

This scheme will thus

(*i*) Avoid TWD excitation

(*ii*) Will require minimal power requirement to drive TVC actuators (since they are mostly stationary) and

(*iii*) Relieve the burden on ADC to meet control requirements.

As the steady and slowly varying disturbances are totally compensated by the TVC, the performance of the primary system *i.e.*, ADC will look like as if there are no disturbances acting on the missile. Therefore, this scheme may also be called as the Dynamic Trimming of Steady Disturbances.

In principle, it is possible to use this scheme also to the case of ADC with SITVC as back-up discussed in earlier section 3.5.2. However, it is not desirable to use it in practice because it requires continuous consumption of SITVC injectant and is not essential to use it in this way to avoid any other problems such as TWD excitation. However, if ADC and SITVC actuators are using a common power source such as hydraulic pump motor package and the capacity of the pump motor package is not adequate to meet the hydraulic flow requirement of 8 actuators (four for ADC and four for SITVC) it would be possible to use this scheme as a compromise solution, that is,. use the available limited power source at the expense of additional SITVC injectant.

There is one more scenario where this scheme can be used for ADC and SITVC. If there is more than adequate injectant stored on board, one may like to use it as much as possible to get rid of unnecessary weight on board the vehicle to gain more velocity. Since the ADC deflection in this case is lower than the case of without dynamic trimming, the drag coefficient component related to control deflection will also be smaller thus enabling further gain in velocity. However, these benefits are comparatively very small or incremental in nature and "Adequate control capability" should be the primary consideration.

3.6 CONCLUDING REMARKS

In this chapter, we have covered most of the general aspects of autopilot design such as error resolution in body coordinate frame (using both Euler angles as well as quaternions), different control laws followed in launch vehicles and missiles and various factors which need to be considered for determining the control power plant size. It mainly included aerodynamic static margin, L/D ratio of the vehicle, aerodynamic misalignment, non-linear aerodynamics, thrust misalignments, unequal thrust, design of nozzle cant angle and lateral *CG* offsets. We have also discussed the controllability of the vehicle during various segments of trajectory such as strapon separation, thrust tail off region of strapons, capture phase after stage separation, α correction to reduce angle of attack at separation, powered phase during atmosphere and out of atmosphere and coast phase out of atmosphere. Subsequently, we considered the estimation of control impulse, autopilot bandwidth, actuator force capability and actuator bandwidth.

In the end, we discussed about the best options for actuator systems for different phases of trajectory and provision of back-up control schemes to utilize the available resources on board the vehicle optimally, in case there arises a need for higher control force due to some unforeseen circumstances.

We now proceed with the actual loop design for the autopilot in the next chapter and then design of reaction control system in subsequent chapter.

REFERENCES

1. **N.V. Kadam, B. Saseendran:** Resolution of attitude errors in vehicle body axes frame for a general case of pitch, yaw and roll manoeuvres, VSSC:TR:026:82, May 1982.

2. **William, E. Wagner, Arno, C. Serold:** Formulation on statistical trajectory estimation programs, NASA CR 1482, Jan. 1970.

3. **E.T. Whitaker:** 'Analytical dynamics' Dover, New York, 194.

4. **B.P. Ickes:** A new method for performing digital control system attitude computations using quaternions. AIAA Journal, Vol. 8, No. 1 January 1970, pp 13–17.

5. **N.V. Kadam:** Use of Euler angles and quaternions for control error generation DRDL.6160.1082.513 Apr. 2000.

6. **Anonymous:** Agni note on updates

7. **Naresh Kumar, K.K. Mangrulkar:** Updates on Interface and Guidance algorithms, DRDL/DOWS/SDG/K-15/12 Aug. 2004.

8. **N.V. Kadam:** On the most desirable bandwidth for actuators for applications in launch vehicle control systems, VSSC/CGD/CSA/05/83, Oct 1983. Also National Systems Conference, IIT, Bombay, 1985.

9. **N.V. Kadam:** Conceptual design of load relief control system for ASLV missions, Oct. 1988.

10. **R.F. Hoelkar:** Theory of artificial stabilization of missiles and space vehicles with exposition of four control principles NASA TN - D 555 June 1961.

11. **A.L Greensite:** Control theory; Vol. II - Analysis and design of space flight control systems, Spartan Books, 1970 (Chapter V).

12. **Robert, R. Harris:** Analysis and Design of Space vehicle flight control systems, Vol. XIV, Load Relief NASA CR - 833 Aug. 1967.

13. **A.D. Murthy, E.M. George, N.V. Kadam:** Studies on fin requirement for ASLV, VSSC/CGD/TM/01/81, Apr. 1981.

14. **K.L Handoo, N. Somrajan:** Response of SLV-3 treated as an Elastic body to inflight winds and Control forces, VSSC-TR-39-133-79.

15. **Pawan Kishore Tiwari, Abhijit Bhattacharya:** Actuator specifications requirements for AAD missile system, DRDL/PGAD/193/04/20/10, Aug. 2003.

16. **N.V. Kadam:** Design of Flight Control System for the boost stage of a satellite launch vehicle using two control power plants, IEEE India Council Convention on Space Electronics, Bombay 1977.

17. **C.S. Adisesha, N.V. Kadam, V. Srinivas Rao, A.A. Bhagat:** A Novel actuator blending scheme for adaptive missile autopilot, National Systems Conference NSC-2007, Manipal.

CHAPTER 4

Control Systems Design–2: Linear System

4.1 INTRODUCTION

4.1.1 Simplification of Vehicle Dynamics

The launch vehicles and missiles are flying objects in three dimensional space with six degrees of freedom motion when considered as a rigid body. The six degrees of freedom are: Linear motion along X, Y, Z and rotational motion about X, Y and Z. When flexibility and slosh are also considered, one may say that there are many other degrees of freedom. In Chapter II, we have derived generalized equations of motion. If we ignore the displacements of mass elements due to flexibility, slosh etc. we get the standard six degrees of freedom equations of motion. The equations are, however, highly nonlinear and not amenable to easy analysis though it is possible to get the required result through simulation. However, simulation does not give the required insight and understanding of the complex scenario which is very much essential for proper design of the control system. Though there is literature which handles nonlinear state equations and nonlinear control systems design, the most familiar and appealing techniques are those for the linear control systems design. The most commonly followed design approach is to simplify the system equations to such an extent that they can be handled easily and get a first cut design solution and then add complexity by adding higher order dynamics and modify the design in such a way that the higher order dynamics does not affect the basic performance adversely and create unwanted instabilities. This is the most sensible approach instead of getting the most complex model in the beginning itself and start design with that model and get lost in the complexities.

It may sound a little contradiction that we started with the most generalized form of equations, described in detail its ramifications and use for deriving models to include structural flexibility, propellant sloshing and engine inertia effects and now making a case for simplified equations. The generalized equations are in a way a simplistic model which helps in seeing through the effect of various modeling features just by appropriate interpretation of various terms instead of going through a very elaborate derivation by Lagrangian approach followed in many books and reports and will be very handy for developing a trajectory simulation program incorporating these effects. These simple looking generalized equations will now be further simplified in following steps so as to make them amenable to in the design of control system.

1. Suppress or ignore the displacements of elemental masses and reduce the model to rigid body dynamic equations. This also makes the rates of change of inertias equal to zero. (Note that the rate of change of moment of inertia due to burning of the propellant is not to be used as clarified in Chapter II).

2. Use the centre of gravity of the vehicle as the origin of body coordinate system. This reduces the *CG* offsets x_{CG}, y_{CG}, z_{CG} to zero.

3. Most of the missiles and launch vehicles are symmetric about the x-y and x-z planes. This reduces the product inertias I_{xy}, I_{yz} and I_{zx} to zero.

4. In order to enable the use of linear control system design techniques, the non-linearities are ignored in the design process and their effect is subsequently studied by detailed simulation.

   This step of simplification reduces the terms pq, qr. rp and $(p^2 + q^2)$, $(q^2 + r^2)$ and $(r^2 + p^2)$ to zero.

5. Ignore the effect of forward acceleration and gravity on the force and moment terms which are very small compared to the control torque and force terms. This reduces f_{gy}, f_{gz} terms and $(A_x + g\sin\gamma_0)\, M_R L_R \delta_p$, $(A_x + g\sin\gamma_0)\, M_R L_R \delta_y$ terms to zero.

6. The aerodynamic force and torque due to distributed aerodynamic load (Ref Eqs. 2.103 and 2.104) are identified with the total $C_{N\alpha}$, C_p and C_{Nq} and C_{mq} coefficients as follows:

$$\int \frac{\partial C_{N\alpha}}{\partial l}\, \bar{\alpha}\, dl = \int \frac{\partial C_{N\alpha}}{\partial l} \left(\alpha - \frac{(l_{cg} - l)}{u} \right) \theta\, dl$$

$$= C_{N\alpha}\alpha + C_{Nq} \frac{d}{2u} \cdot \theta \qquad\qquad \dots (4.1)$$

$$\int \frac{\partial C_{N\alpha}}{\partial l}\, \bar{\alpha}\, x\, dl = \int \frac{\partial C_{N\alpha}}{\partial l}\, \bar{\alpha}(l_{cg} - l)\, dl$$

$$= \int \frac{\partial C_{N\alpha}}{\partial l} \left[\alpha - \frac{(l_{cg} - l)}{u}\theta \right] (l_{cg} - l)\, dl$$

$$= C_{N\alpha}\left[L_{cg} - L_{cp} \right]\alpha + C_{mq} \cdot \frac{d}{2u}\theta \qquad\qquad \dots (4.2)$$

where L_{cg} and L_{cp} are measured from nose tip.

7. The only coupling still remaining in the model is due to roll rate p appearing in the expressions for A_x, A_y and A_z (Eq. 2.13). We assume that the roll of the missile is well controlled, so that $p \approx 0$. Then the pitch plane, yaw plane and roll motion can be separately studied. The equations 2.103 and 2.104 can then be written down as follows:

PITCH PLANE EQUATIONS OF MOTION

$$w = -\frac{2T_E}{M}\delta_p - \frac{2M_R L_R}{M}\delta - \frac{C_{N\alpha}QS}{M}\alpha + \left(u - \frac{C_{Nq}\frac{d}{2u}QS}{M}\right)\theta \qquad \dots 4.3(a)$$

$$q = -\left(\frac{2T_E L_C + 2M_R L_R(u + g\sin\gamma_0)}{I_{yy}}\right)\delta_p - \frac{2(I_R + M_R L_R L_C)}{I_{yy}}\delta p$$

$$+ \frac{C_{N\alpha}(L_{cg} - L_{cp})QS}{I_{yy}}\alpha + \left(\frac{C_{mq}\frac{d}{2u}QSd}{I_{yy}}\right)\theta \qquad \dots 4.4(a)$$

In terms of notations used in books on missile control systems, these are (using $\alpha = \dfrac{w}{u}$):

$$w = z_w w + (z_q + u)q + z_\delta\,\delta + z_\delta\,\delta_p \qquad \dots 4.3(b)$$

$$q = m_w w + m_q q + m_\delta\delta + m_\delta\,\delta_p \qquad \dots 4.4(b)$$

Actuator equation (using Eq. 2.107)

$$\delta_p = -2\varsigma_a\omega_a\,\delta_p - (\omega_a^2 + \frac{u + g\sin\gamma_0}{I_R})\delta + \omega_a^2\delta_{pc}$$

$$-\frac{M_R L_R}{I_R}w + \frac{M_R L_R u}{I_R}\theta - \frac{M_R L_R(L_R + L_C)}{I_R}\theta \qquad \dots (4.5)$$

where

$$z_w = \frac{-C_{n\alpha}QS}{Mu}, \qquad\qquad Z_q = -\frac{\left(C_{Nq}\frac{d}{2u}QS\right)}{M}$$

$$z_\delta = -\frac{2T_E}{M}, \qquad\qquad z_\delta = \frac{-2M_R L_R}{M}$$

$$m_w = \frac{C_{N\alpha}(L_{cg} - L_{cp})QS}{I_{yy}u}, \qquad m_q = \frac{C_{mq}\frac{d}{2u}QSd}{I_{yy}}$$

$$m_\delta = -\left(\frac{2T_E L_C + 2M_R L_R (u + g\sin\gamma_0)}{I_{yy}}\right);$$

$$m_{\dot\delta} = \frac{-2(I_R + M_R L_R L_C)}{I_{yy}} \qquad \ldots (4.6)$$

YAW PLANE EQUATIONS OF MOTION

$$\dot v = -\frac{2T_E}{M}\delta_y - \frac{2M_R L_R}{M}\dot\delta_y - \frac{C_{y\beta}QS}{M}\beta - \left(u + \frac{C_{yr}\dfrac{d}{2u}QS}{M}\right)\psi \qquad \ldots 4.7(a)$$

$$\dot r = \frac{2T_E L_C + 2M_R L_R\,(u + g\sin\gamma_0)}{I_{zz}}\delta_y + \frac{2(I_R + M_R L_R L_C)}{I_{zz}}\dot\delta_y$$

$$-\frac{C_{y\beta}QSd(L_{cg} - L_{cp})}{I_{zz}}\beta - \frac{C_{mr}\dfrac{d}{2u}QSd}{I_{zz}}\psi \qquad \ldots 4.8(a)$$

Or in terms of missile control literature notations

$$v = y_v v + (y_r - u)r + y_\delta \delta + y_{\dot\delta}\,\dot\delta_y \qquad \ldots 4.7(b)$$

$$r = n_v v + n_r r + n_\delta \delta + n_{\dot\delta}\,\dot\delta_y \qquad \ldots 4.8(b)$$

$$\ddot\delta_y = -2\varsigma_a\omega_a\dot\delta_y - (\omega_a^2 + \frac{u + g\sin\gamma_0)M_R L_R}{I_R}\delta_y + \omega_a^2\delta_{yc}$$

$$-\frac{M_R L_R}{I_R}v - \frac{M_R L_R u}{I_R}\psi + \frac{M_R L_R (L_R + L_C)}{I_R}\psi \qquad \ldots (4.9)$$

where

$$y_v = -\frac{C_{y\beta}QS}{Mu}; \quad y_r = \frac{-C_{yr}\dfrac{d}{2u}QS}{M}; \quad y_\delta = \frac{-2T_E}{M}, \quad y_{\dot\delta} = \frac{-2M_R L_R}{M}$$

$$n_v = -\frac{C_{y\beta}QSd(L_{cg} - L_{cp})}{I_{zz}u}; \qquad n_r = \frac{-C_{mr}\dfrac{d}{2u}QSd}{I_{zz}}$$

$$n_\delta = \frac{2T_E L_C + 2M_R L_R\,(u + g\sin\gamma_0)}{I_{zz}}; \quad n_{\dot\delta} = \frac{2(I_R + M_R L_R L_C)}{I_{yy}}$$

$$\ldots (4.10)$$

Most of the launch vehicles and missiles are symmetric about the roll axis. In such cases, one can use either the pitch plane or yaw plane equations of motion for the design of control system and the same design can be used for both the planes. However, vehicles such as ASLV which has two strap-ons in X-Y plane are not symmetric and their aerodynamic characteristics such as static stability can be quite different in pitch plane and yaw plane. Therefore, one has to use judgement whether the same design of control system is applicable for both pitch plane and yaw plane or a separate design is required for two planes.

For illustration, we use yaw plane equations to get a standard state equation form.

$$
\begin{bmatrix} v \\ \psi \\ \ddot{\psi} \\ \delta_y \\ \delta_y \end{bmatrix} = \begin{bmatrix} y_v & 0 & y_r - u & y_\delta & 0 \\ 0 & 0 & 1.0 & 0 & 0 \\ n_v & 0 & n_r & n_\delta & 0 \\ 0 & 0 & 0 & 0 & 1.0 \\ 0 & 0 & -\dfrac{M_R L_R U}{I_R} & -\bar{\omega}_a^2 & -2\zeta_a \omega_a \end{bmatrix} \begin{bmatrix} v \\ \psi \\ \psi \\ \delta_y \\ \delta_y \end{bmatrix}
$$

$$
+ \begin{bmatrix} 0 & 0 & 0 & 0 & y_\delta \\ 0 & 0 & 0 & 0 & 0 \\ 0 & 0 & 0 & 0 & n_\delta \\ 0 & 0 & 0 & 0 & 0 \\ -\dfrac{M_R L_R}{I_R} & 0 & \dfrac{M_R L_R (L_R + L_C)}{I_R} & 0 & 0 \end{bmatrix} \begin{bmatrix} v \\ \psi \\ \psi \\ \delta_y \\ \delta_y \end{bmatrix} + \begin{bmatrix} 0 \\ 0 \\ 0 \\ 0 \\ \omega_a^2 \end{bmatrix} \delta_{yc} \qquad \text{... (4.11)}
$$

Where
$$
\bar{\omega}_a^2 = \omega_a^2 + \frac{(u + g \sin \gamma_0)\, M_R L_R}{I_R} \qquad \text{... (4.12)}
$$

Defining state vector as

$$
\bar{x} = \begin{bmatrix} v & \psi & \psi & \delta_y & \delta_y \end{bmatrix}^T
$$

Eq. 4.11 can be written as

$$
\dot{\bar{X}} = A\bar{X} + D\dot{\bar{X}} + BU
$$

or

$$
\dot{\bar{X}} = (I - D)^{-1} A\bar{X} + (I - D)^{-1} BU
$$

$$
\dot{\bar{X}} = F\bar{X} + GU \qquad \text{... (4.13)}
$$

EQUATIONS OF MOTION ABOUT ROLL AXIS

$$p = \frac{2T_E L_a}{I_{xx}}\delta_R + \frac{M_R L_R L_a}{I_{xx}}\ddot{\delta}_R - \frac{C_{lp}\dfrac{d}{2u}QSd}{I_{xx}}\,p$$

$$= l_\delta \delta_R + l_{\dot\delta}\,\delta_R - a\,p \qquad\qquad \dots (4.14)$$

$$\ddot{\delta}_R = -2\zeta_a \omega_a \dot{\delta}_R - \overline{\omega}_a^2 \delta_R + \omega_a^2 \delta_{RC} \qquad\qquad .. (4.15)$$

where $\qquad \delta_{y2} = \delta_y + \delta_R$

$$\delta_{y1} = \delta_y - \delta_R$$

$\therefore \qquad \delta_{y2} - \delta_{y1} = 2\delta_R$

and C_{lp} is the aerodynamic roll damping coefficient

$$l_\delta = \frac{2T_E L_a}{I_{xx}} = \mu_{cr}$$

$$l_{\dot\delta} = \frac{M_R L_R L_a}{I_{xx}} = K_{Dr}; \qquad\qquad a = \frac{C_{lp}\dfrac{d}{2v}QSd}{I_{xx}}$$

putting $\ddot{\delta}_R$ from Eq. 4.15 in Eqn. 4.14, one gets

$$p = \dot{\phi} = (\mu_{cr} - K_D\overline{\omega}_a^2)\delta_R - 2\zeta_a\omega_a\,\delta_R + \omega_a^2 K_{Dr}\delta_{R_C} - a\dot{\phi}$$

Then
$$\begin{bmatrix} \dot{\phi} \\ \ddot{\phi} \\ \dot{\delta}_R \\ \ddot{\delta}_R \end{bmatrix} = \begin{bmatrix} 0 & 1 & 0 & 0 \\ 0 & -a & \mu_{cr} - K_D\overline{\omega}_a^2 & -2\zeta_a\omega_a K_{Dr} \\ 0 & 0 & 0 & 1 \\ 0 & 0 & -\overline{\omega}_a^2 & -2\zeta_a\omega_a \end{bmatrix} \begin{bmatrix} \phi \\ \dot{\phi} \\ \delta_R \\ \dot{\delta}_R \end{bmatrix} + \begin{bmatrix} 0 \\ K_{Dr}\omega_a^2 \\ 0 \\ \omega_a^2 \end{bmatrix}\delta_{R_c} \qquad \dots (4.16)$$

In state equation form, with $\overline{X} = \begin{bmatrix} \phi & \dot{\phi} & \delta_R & \dot{\delta}_R \end{bmatrix}^T$ $\qquad\qquad \dots (4.17)$

$$\dot{X} = F\overline{X} + G\overline{U} \qquad\qquad \dots (4.18)$$

Referring to the equations 2.140 and Eq. 2.147 of Greensite's model derived in Section 2.3.9 of Chapter 2, reader is expected to have doubts about the correctness of the term $M_R(L_R + L_C)q$ or $M_R(L_R + L_C)\theta$ term in force equation and $M_R(L_R + L_C)(A_z - g\cos\theta)$ term in torque equation and may feel that either this author's model or Greensite's corrected model must be incorrect w.r.t. this term. Ref. 4.2 investigates this matter in detail and has derived completely one model from the other model. It brings out clearly that the presence of these terms in Greensite's model is due to the following:

1. CG of the reduced vehicle parameter model (Greensite) is different from the CG of total vehicle parameter model (due to this author) by an amount

$$\Delta CG = \frac{2M_R(L_{CO} + L_R)}{M_s} \qquad \text{... (4.19)}$$

    where M_s is the mass of vehicle excluding slosh masses but including mass of engines.

    L_{CO} = distance between gimbal point and CG of reduced vehicle

    ('Reduced vehicle' is the vehicle excluding engines and slosh masses)

2. Moment of inertia to be used in Greensite's model is given by

$$[I_{y0} + 2I_0 + 2M_R(L_R + L_{CO})^2] \qquad \text{... (4.20)}$$

    where I_{yo} = Moment of inertia of reduced vehicle about reduced vehicle *CG*

    I_0 = Inertia of each engine about its own *CG*

3. Moment of inertia to be used in total vehicle parameter model is about total vehicle *CG* and the relation is given as follows:

$$I_{ys} = I_{y0} + 2I_0 + 2M_R(L_R + L_{CS})^2 + M_0(L_{CO} - L_{CS})^2 \qquad \text{... (4.21)}$$

    Where L_{CS} = distance between gimbal point and total vehicle *CG*

    I_{YS} = Inertia of total vehicle (excluding slosh masses)

    L_{CS} = Distance between gimbal point and total vehicle *CG*

    Ref. 2 shows that both the models give identical results if the data is correctly used as described above. This emphasizes the need to compute appropriate data for Greensite's model from the data generally available only for the total vehicle.

4.   Author feels that the use of exact parameters as given above has not been appreciated in general by many users of Greensite's model and the parameters of total vehicle itself may be being used (erroneously) for reduced vehicle parameter model.

5. Author has shown in Ref. 2 that both the models when correctly used give identical results.

4.1.2 Time Slice Approach

The last simplification to be used for the design is the so-called time slice approach.

The various parameters of the vehicle and trajectory continuously change during flight. The vehicle related parameters are mass, moment of inertia and centre of gravity and the trajectory related parameters are vehicle velocity, attitude and flight path angle, density of atmosphere and ambient temperature which determine the speed of sound and the aerodynamic pressure.

Thus, the vehicle dynamics continuously changes during flight and makes it a time varying parameter system. To enable use of classical frequency domain techniques, this system is converted to fixed parameter system by assuming that the system parameters remain constant over a short period of time. Thus, the parameters of the vehicle dynamics are worked out for short segments of trajectory to cover the entire trajectory. A small perturbation motion is then considered about the nominal position and control system parameters or gains are designed to have a satisfactory and well damped response for the perturbed motion. Such a study is repeated at various segments of trajectory and suitable gains for various segments are obtained. This gives a schedule for the control gains to be used during flight. This approach has been used by all the designers of flight control system and is found to be giving quite satisfactory results. Such study in which the parameters are held constant over a short period of time and control performance is studied for a perturbed motion is referred to as 'short period analysis.' Once this study is completed for entire trajectory, the vehicle performance is studied using detailed simulation model where all nonlinearities and disturbances including atmospheric disturbances is studied. Such a study is called a 'Long period' analysis and it includes the effect of autopilot performance parameters on the overall trajectory. The autopilot specifications and parameters are modified if the long period analysis shows that there is a need to do so to meet the mission objectives.

4.2 CONTROL SYSTEM DESIGN STEPS

The various steps in the design of flight control system are therefore summarized as follows:

1. Design of autopilot assuming simplified or rigid body dynamics of the vehicle and obtain the gain schedule, compensators etc. for the entire zone of operation.

2. Incorporate structural flexibility and propellant sloshing effect (if liquid propellant stage is present) in the vehicle dynamics and carry out the stability analysis.

   Design suitable filters if required to have adequate stability margins.

3. Carry out 6DOF trajectory simulation (rigid body) to study the vehicle performance and adequacy of various parameters such as maximum control force available, maximum control deflection δ, hinge moments, total control impulse, vehicle loads, adequacy of control deflections for pitch, yaw, roll etc. and guidance errors.

4. Carry out 6DOF trajectory simulation incorporating vehicle flexibility and slosh dynamics to test adequacy of filters.

5. Carry out detailed validation studies as discussed in the last chapter.

4.3 DESIGN OBJECTIVES

The design objectives for the flight control system are

1. Good stability margins

2. Good speed of response to guidance commands leading to good tracking accuracy

3. Less sensitivity to disturbances.

4.3.1 Stability Margins

Specifications on the stability margins need to take care of the following:

(*a*) Approximations in the analytical models used for vehicle and other control elements inside the loop such as actuators, sensors etc.

(*b*) Uncertainties of the parameter values in the above models

(*c*) The autopilot performance parameters such as damping ratio, overshoot etc.

Experience has shown that following stability margins for nominal case have given good performance during flight trials.

1. Gain margin > 6db

2. Negative gain margin < -6db   (or $> |-6|$ db)

3. Phase margin $> 30°$

4. Stability margin R or $|1 + GH|_{\min} > 0.5$

5. Attenuation for higher modes > 10db.

Though some of the margins given above are well known, a brief explanation of various margins is given below using Nyquist diagram and Bode plots for open loop gain *GH* for the system. Fig. 4.1 shows a typical block diagram of a control system and the closed loop transfer function is given by:

$$M = \frac{C}{R} = \frac{G}{1 + GH} \qquad \qquad \text{... (4.22)}$$

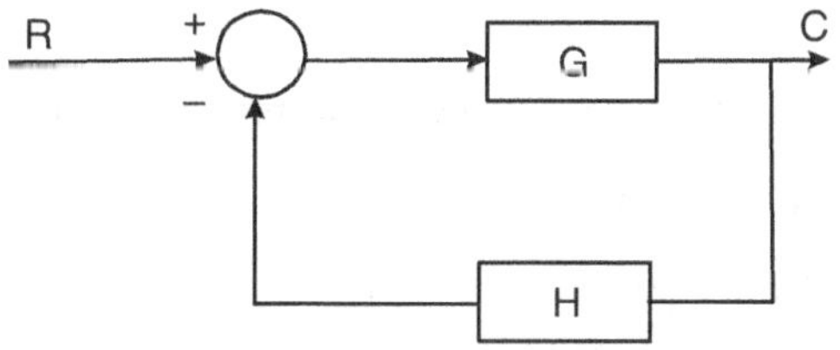

Fig. 4.1

The stability of the system is decided by the roots of the characteristic equation given by the numerator polynomial of $(1+GH)$. The system is stable if all the roots lie in the left half of the s-plane. It is critically stable if atleast one of the roots is on the imaginary axis and unstable if any of the roots is on the right half of the s-plane. Readers can refer to any standard text books (Ref. 5,6) to know how Nyquist diagram can be used to find how many poles of the system are on the right half of s-plane (RHP) to determine the stability of the system.

Fig 4.2 shows a typical Nyquist plot for a conditionally stable system.

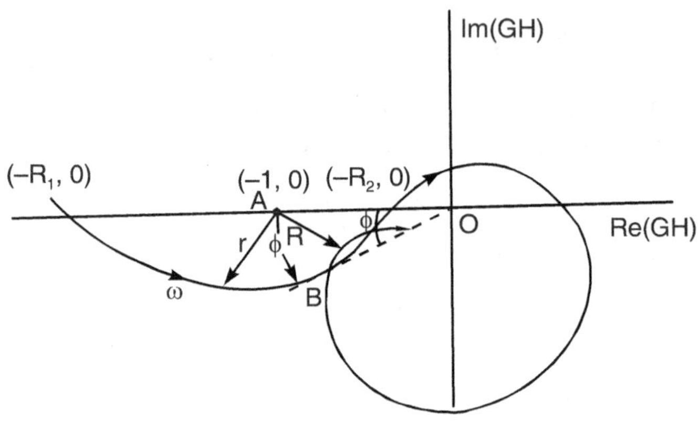

Fig. 4.2

Let r be the complex variable given by

$$r = a + jb = (1 + GH)$$

where

$$a = \text{Real part of } (1 + GH)$$

$$b = \text{Imaginary part of } (1 + GH)$$

The various margins are defined as follows:

1. Gain margin $= 20\log_{10}\left(\dfrac{1}{R_2}\right)$ ... (4.23)

2. Negative gain margin $= 20\log_{10}\left(\dfrac{1}{R_1}\right)$ ... (4.24)

3. Phase margin $=$ Angle ϕ between the negative real axis and

   GH vector given by point of intersection of

   $|GH| = 1$ circle with GH curve.

4. Stability margin $R = \min_{\omega} [r(\omega)]$ ... (4.25)

For a normal system, gain margin and phase margin give a fairly good indication of the stability margins. However, for systems such as launch vehicle, there are number of low damped structural flexibility and propellant slosh modes. The *GH* plot can be something like that shown in Fig. 4.2 at higher frequency. The system may be having enough gain margin as well as enough phase margin but still the *GH* plot at some frequencies could be quite close to $(-1, 0\,)$ point as indicated by the vector $R = \min\limits_{\omega} r\,(\omega)$. Small perturbations in the parameters could drive the plot towards $(-1, 0)$ point and will make the system unstable. Hence for systems having low damped modes, a more general stability margin is defined by the smallest value of $r(\omega)$ over the range of frequencies.

This general definition of stability margin enables us to find a correlation between gain margin and phase margins specifications.

(*i*) At gain margin point,

$$R = 1 - R_2 \qquad\qquad\qquad \text{... (4.26)}$$

$\therefore$ For 6 *db* gain margin $R_2 \approx 0.5$. This is now equivalent to

$$R = 1 - 0.5 = 0.5$$

(*ii*) At phase margin point, consider the isocelles triangle formed by *OAB*.

Then $\qquad\qquad\qquad R = 2\sin\dfrac{\phi}{2}$

or $\qquad\qquad \phi = 2\sin^{-1}\left(\dfrac{R}{2}\right) = 2\sin^{-1}\left(\dfrac{0.5}{2}\right) = 28.955°$

or $\qquad\qquad R = 0.5$ is equivalent to 28.955° phase margin.

This gives us a correlation that a *6dB* gain margin and 29° (or approximately 30°) phase margin are equivalent to a general stability margin of $R = 0.5$

(*iii*) At negative gain margin point

$$20\log_{10}\left(\dfrac{1}{R_1}\right) = -20\log_{10} R_1 = -6\ dB$$

or $\qquad\qquad\qquad R \approx 2.0 \qquad\qquad\qquad \text{... (4.27)}$

Thus, $-6dB$ gain margin is equivalent to $R = 1.0$ instead of 0.5 and $R = 0.5$ at negative gain margin gives only $20\log_{10}\left(\dfrac{1}{1.5}\right) = -3.52\ dB.$

This is the reason why some literature on flight control systems for launch vehicles show an elliptical 'zone of exclusion' instead of circular (see Fig. 4.3) around the critical point (−1,0). The 'zone of exclusion' is the region of *GH* plane in which the *GH* plot should not enter to ensure that the stability margin specifications are not violated.

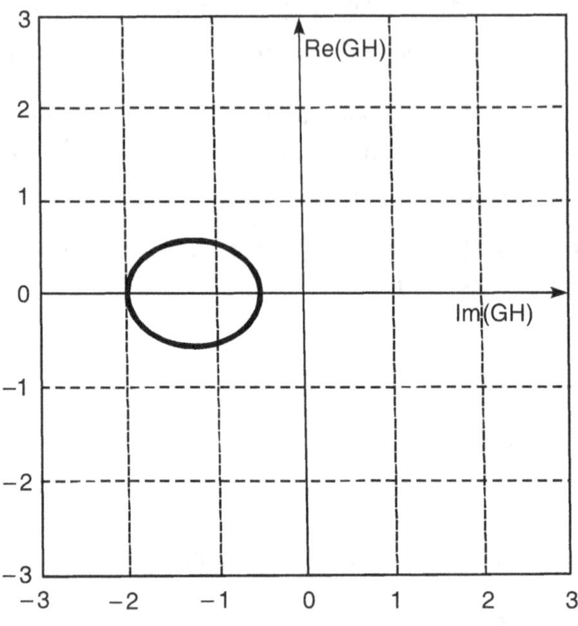

Fig. 4.3 Zone of exclusion

The method for obtaining the equation for 'zone of exclusion' boundary is illustrated below for the following specifications.

Positive gain margin $= 6dB$

Negative gain margin $= -6dB$

Phase margin $= 30°$

We have shown earlier that the ellipse intersects real axis at (−0.5, 0) and (−2.0, 0)

Then the centre of the ellipse is located at (−1.25, 0)

Let the ellipse with centre located at (−1.25, 0) be given by

$$\frac{x^2}{a^2} + \frac{y^2}{b^2} = 1 \qquad \qquad \text{... (4.28)}$$

Then $\qquad\qquad\qquad\qquad a = 1.25 - 0.5 = 0.75 \qquad\qquad\qquad\text{... (4.29)}$

$\qquad R \text{ (at phase margin point)} = 2 \sin 15° = 0.51764$

$\qquad\qquad\qquad y = $ coordinate at phase margin (*PM*) point is given by

$$y^2 = (0.51764)^2 - (1-\cos 30^\circ)^2$$

$$= 0.26795 - 0.017949$$

$$= 0.25$$

$$x^2 \text{ (at } PM \text{ point)} = [0.25 + (1-\cos 30^\circ)]^2$$

$$= 0.38397^2$$

$$= 0.147436$$

$$\therefore \qquad b^2 = \frac{y^2}{\left[1 - \dfrac{x^2}{a^2}\right]} = \frac{0.25}{0.73789} = 0.3388 = (0.582)^2 \qquad \ldots (4.30)$$

Thus the boundary of the zone of exclusion for the above specifications is given by:

$$\frac{x^2}{0.75^2} + \frac{y^2}{0.582^2} = 1 \qquad \ldots (4.31)$$

(It is easy to compute and see that rounding the value of $b = 0.6$ amounts to a phase margin specification of 36°)

Since the boundary is given in GH plane, we must write x and y in terms of real and imaginary parts of the GH.

Let any point on the curve be given by the magnitude K and Phase ϕ of the GH

OR $$GH = K\angle\phi$$

$$= K\cos\phi + j\,K\sin\phi \qquad \ldots (4.32)$$

Then $$x = K\cos\phi + 1.25 \qquad \ldots (4.33)$$

$$y = K\sin\phi \qquad \ldots (4.34)$$

The zone of exclusion is now given by

$$\frac{(K\cos\phi + 1.25)^2}{(0.75)^2} + \frac{(K\sin\phi)^2}{(0.582)^2} = 1 \qquad \ldots (4.35)$$

The definition of the zone of exclusion helps for computerization when some optimization is being carried out such as for optimization of compensator parameters etc. using a computer program.

The equation for the boundary of zone of exclusion for any other specifications can be obtained following the above procedure.

(5) Attenuation for higher modes (or attenuation margin)

In most of the cases, the first bending mode frequency of a launch vehicle is quite low and there is no adequate separation between the first bending mode frequency and control frequency. This makes it difficult to attenuate the lower modes adequately and those are mostly phase stabilized. Higher modes however must be adequately attenuated. This is mainly due to the fact that there will be more uncertainties or errors in prediction of higher mode frequencies and shapes and also phase characteristics of other elements in the control loop such as actuator, sensor etc. also may not be accurately known at higher frequencies and there will be variation between unit to unit.

The attenuation margin gives the maximum peak of *GH* plot beyond certain frequency of interest. Thus an attenuation margin of $20dB$ beyond frequency 'f' means that the maximum or peak of low damped modes in *GH* plot above frequency 'f' is $-20dB$.

4.3.2 Speed of Response

(i) Launch Vehicles

The final aim point for the launch vehicle is well known in advance and there is adequate time for correcting the guidance errors. Hence, the speed of response for autopilot is less critical. The speed of response or autopilot bandwidth is therefore mainly decided from the consideration of structural loads and adequate separation from the first bending mode frequency. The bending moments are also calculated at various sections of the vehicle as a function of time in simulation to assess the vehicle loads during atmospheric flight for various cases of atmospheric disturbances such as winds, wind gusts etc. and the speed of response finalized taking into account structural loads.

(ii) Missiles

For missiles especially surface to air and air to air class, the speed of response is critical. These missiles are designed to execute high lateral acceleration manoeuvres (25g to 30g or even more) therefore, the structural load is not a critical factor for deciding the autopilot bandwidth. The total flight time is very short. Moreover, the targets can undertake escape manoeuvres when the missile attack is detected leaving very little time for missile guidance to react, minimize the guidance errors and intercept the target. The guidance errors take 6 to 8 guidance time constant to settle down. Hence, the guidance time constant should be 6 to 8 times smaller than the time to go for interception from the instant of target acquisition by missile seeker and autopilot time constant should be 4 to 5 times smaller than the guidance time constant to have satisfactory performance of the guidance loop. The achievable speed of response depends on the aerodynamic pressure, airframe characteristics, actuators, sensors etc. and become slower at high altitudes due to low aerodynamic pressure. Thus, the maximum range of target detection capability of the seeker, closing velocity and altitude of interception determine the available time to go and the time constants for guidance and autopilot.

4.3.3 Sensitivity to Disturbances

It is known from any standard text books on feedback control theory that the feedback control gives reduced sensitivity to disturbances and plant parameter variations.

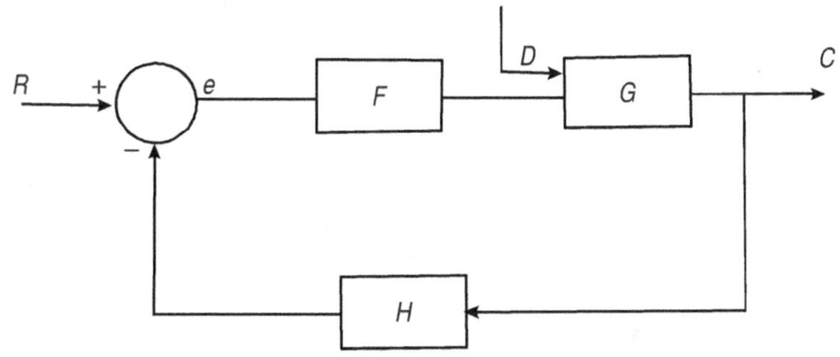

Fig. 4.4

Thus for closed loop system

$$\frac{C}{D} = \frac{G}{1 + GHF} \qquad \qquad \text{... (4.36)}$$

Or
$$\frac{e}{D} = \frac{GH}{1 + GHF} \qquad \qquad \text{... (4.37)}$$

From previous section, we can write that $r = 1 + GHF$. At $|GHF| \gg 1$, we get

$$\frac{e}{D} = \frac{1}{F} \qquad \qquad \text{... (4.38)}$$

The steady state errors due to constant disturbances such as thrust misalignments or other misalignments etc. is thus decided by the forward gain or compensators appearing before the disturbance entry into the loop. Higher the steady state loop gain , lower will be the errors due to disturbances. If F is an integrator $\left(\dfrac{1}{s}\right)$ or proportional plus integral control $\left(\dfrac{K_1 + s}{s}\right)$, one gets $\left(\dfrac{e}{D} = \dfrac{1}{F}\right)$ = proportional to s and the error due to disturbance in steady state (*i.e.* as $s \rightarrow 0$) becomes zero.

If the disturbance is having a frequency content, such as atmospheric turbulence, the errors will depend on the radius $R = (1 + GHF)$.

4.4 AUTOPILOT DESIGN USING RIGID BODY DYNAMICS

4.4.1 A Third Order Attitude Control System

The launch vehicles do not have lifting surfaces since they are not designed for manoeuvrability. In fact many times, there will be a deliberate attempt to reduce the lateral loads on the vehicle. Hence, the lateral acceleration equations in the vehicle dynamics can be ignored (unless when load relief is being studied).

Let us first consider a case where the control is provided either by secondary injection thrust vector control system or aerodynamic control where effect of inertia of control effector can be neglected and the actuator is considered as a first order actuator *i.e.*,

$$\delta + K_c \delta = K_c \delta_c \qquad \qquad \text{... (4.39)}$$

Using $\dfrac{w}{u} = \alpha$ and for a $\alpha \approx \theta$ and $q \cong \theta$ perturbation model, Eq. 4.4 can be written as

$$\theta = m_w \cdot u\theta + m_q\, \theta + m_\delta \delta$$

$$= \mu_\alpha \theta + m_q\, \theta + \mu_c \delta \qquad \qquad \text{... (4.40)}$$

where $\mu_\alpha = m_w u$ and $\mu_c = m_\delta$ are the standard notations used in literature on flight control systems for space vehicles. The vehicle transfer function is then given by

$$\frac{\theta}{\delta} = G = \frac{\mu_c}{s^2 - m_q s - \mu_\alpha} = \frac{\mu_c}{s^2 - \mu_\alpha} \qquad \qquad \text{... (4.41)}$$

and actuator $\qquad \qquad \qquad \qquad \qquad \qquad \qquad \qquad \qquad \qquad \text{... (4.42)}$

$$G_a = \frac{K_c}{s + K_c}$$

and the attitude control system block diagram is as given below:

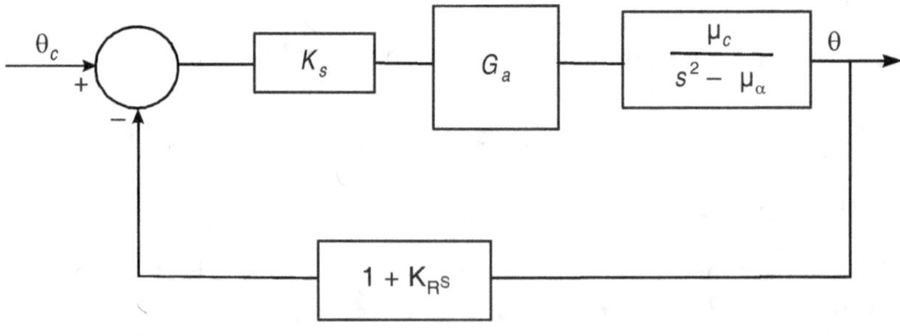

Fig. 4.5

The attitude control systems use inertial sensors (inertial measurement unit) such as gyros for attitude angle and rate gyros for angular rate feed back for providing damping. The aerodynamic damping provided by the m_q term is usually very small and can be neglected in the first cut design.

The closed loop transfer function is given by

$$M = \frac{K_s G_a G}{1 + K_s G_a G(1 + K_R s)}$$

Or

$$M = \frac{\mu_c K_s K_c}{s^3 + K_c s^2 + (\mu_c K_s K_R K_c - \mu_a)s + K_c(\mu_c K_s - \mu_a)} \qquad \ldots (4.43)$$

$$= \frac{K}{(s + p)\,(s^2 + 2\varsigma_c \omega_c s + \omega_c^2)} \qquad \ldots (4.44)$$

where the cubic characteristics polynomial is assumed to give, in general, a real pole p and a complex pair of poles represented by ς_c and ω_c. Equating coefficients of like powers of 's', one gets

$$\mu_c K_s = \frac{\omega_c^2(K_c - 2\varsigma_c \omega_c)}{K_c} + \mu_a \qquad \ldots (4.45)$$

$$K_R = \frac{\omega_c^2 + 2\varsigma_c \omega_c(K_c - 2\varsigma_c \omega_c) + \mu_a}{\mu_c K_s K_c} \qquad \ldots (4.46)$$

and

$$p = K_c - 2\varsigma_c \omega_c \qquad \ldots (4.47)$$

For illustration, let

$$\mu_\alpha = -4$$

$$K_c = 30\ r\,/\,s$$

$$\varsigma_c = 0.7$$

and

$$\omega_c = 5\,r\,/\,s$$

The gain schedule for the entire flight duration can be obtained by taking the appropriate values of μ_α and μ_c at the various flight instants and solving for K_s and K_R.

Fig. 4.6 shows the location of poles for the system with $\mu_\alpha = -4$ and $K_c = 30$ *r/s* for three cases of specified ω_c.

TABLE 4.1

ω_c	$\mu_c K_s$	K_R	Complex pole	Real pole
5	15.167	0.4	$-3.5 \pm j\,3.57$	-23
10	49.333	0.2162	$-7 \pm j\,7.14$	-16
15	63.5	0.2152	$-10.5 \pm j\,10.77$	-9

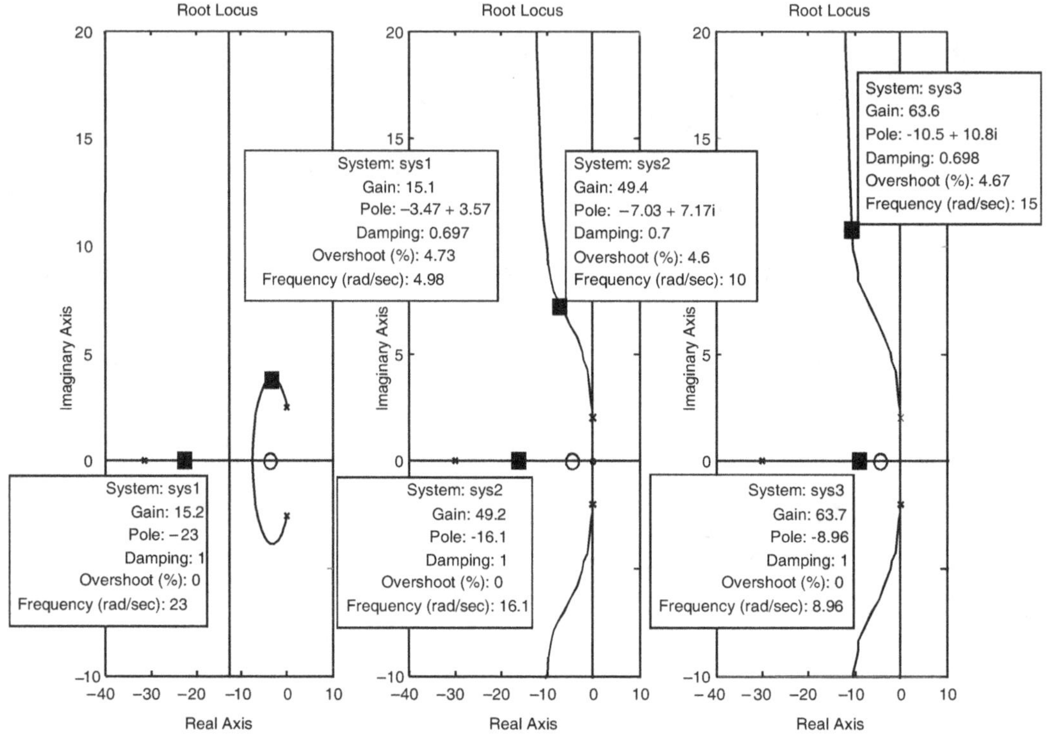

Fig. 4.6

All standard text books give the step response of a second order system for different damping ratios. The effect of the third pole is to reduce the overshoot of that step response. Hence, even though the damping ratio for the complex pole is quite less, the overshoot of the step response can be significantly brought down by the suitable location of the third pole.

The step response of this system (Eq. 4.40) is given by (Ref. 7)

$$c(t) = 1 - \frac{e^{-pt}}{\beta^2 - 2\varsigma_c\beta + 1} + \frac{\beta e^{-\varsigma_c\omega_c t}}{\sqrt{1-\varsigma_c^2}\sqrt{\beta^2 - 2\varsigma_c\beta + 1}} \sin\left(\left(\sqrt{1-\varsigma_c^2}\right)\omega_c t - \psi\right)$$

where $$\beta = \frac{p}{\omega_c} \qquad \qquad \text{... (4.48)}$$

and $$\psi = \tan^{-1}\left(\frac{\sqrt{1-\varsigma_c^2}}{-\varsigma_c}\right) + \tan^{-1}\left(\frac{\sqrt{1-\varsigma_c^2}}{\beta-\varsigma_c}\right) \qquad \qquad \text{... (4.49)}$$

This response is categorized in three types of characteristics as follows (Ref. 7, 8):

1. *Type A:* The response at first overshoot is higher than the response at second or subsequent overshoots and the upper envelope of the response is monotonically decreasing.

2. *Type B:* The response at first peak is smaller than the response at second peak. The upper envelope reaches its maximum value after the first peak.

3. *Type C:* This response is characterised by the condition that the derivative of the response $c(t)$ is never negative during oscillatory rise.

Ref. 8 gives the relation between β and ς_c which separates the three regions according to which, the Type C response is given by the condition that $\beta < \varsigma_c$. This only means that the real pole is nearer to imaginary axis than the complex pole ($p < \varsigma_c \omega_c$). The response of such a system is dominated by the real pole. Following table gives an approximate range of the ratio $\left(\dfrac{p}{\varsigma_c \omega_c}\right)$ for different types of response.

TABLE 4.2

ς_c	Range of $\dfrac{p}{\varsigma_c \omega_c}$ for response type		
	Type A	*Type B*	*Type C*
1.0	No overshoot	1	
0.8	> 1.025	1 to 1.025	< 1.0
0.6	> 1.1	1 to 1.1	< 1.0
0.4	> 1.3	1 to 1.3	< 1.0
0.2	> 2.1	1 to 2.1	< 1.0

Fig. 4.7a shows the regions diagrammatically and Fig. 4.7b shows typical plots of different types.

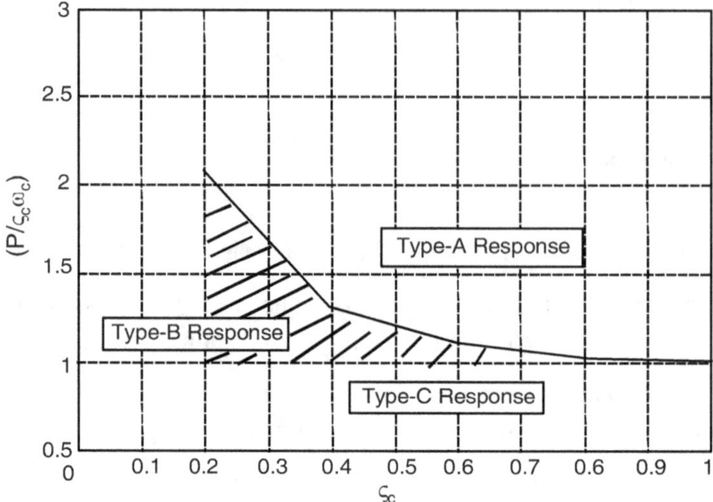

Fig. 4.7(a)

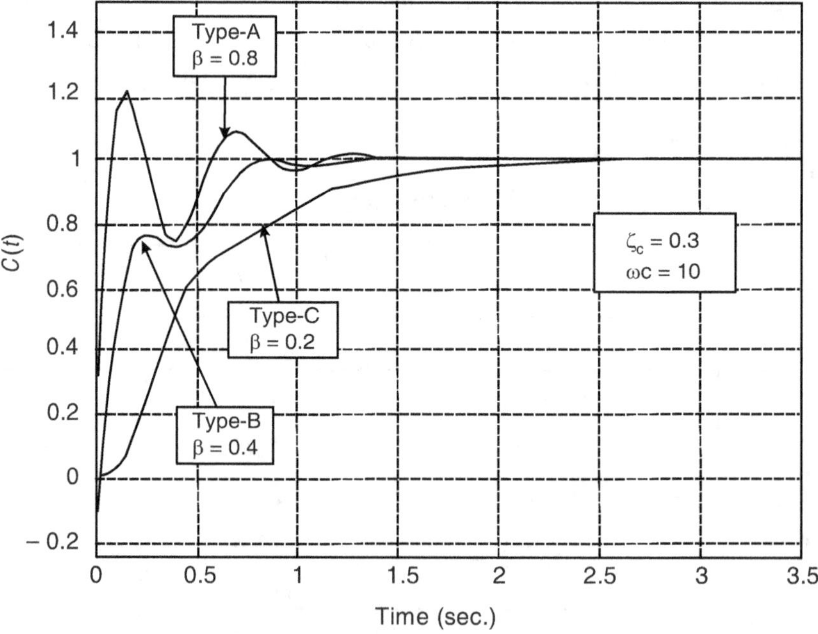

Fig. 4.7(b)

4.4.2 Fourth Order System and Effect of Engine Inertia (Tail-wags-Dog Effect)

When the control torque is generated by gimballing the engine or by deflecting the flexible nozzle, the actuator needs to be considered as a higher order (second or third order) transfer function. The dynamic equations of the system are given by (using Eq. 4.4(b) and 4.6 and $\alpha \approx \theta$)

$$q = \mu_\alpha \theta + \mu_c \delta + K_D\, \delta \qquad \dots (4.50)$$

$$\delta = -2\varsigma_a \omega_a\, \delta - \omega_a^2 \delta + \omega_a^2 \delta_c \qquad \dots (4.51)$$

where
$$\mu_c = m_\delta = \frac{-2T_E L_C}{I_{yy}} \qquad \dots (4.52)$$

$$K_D = m_\delta = \frac{-2(I_R + M_R L_R L_C)}{I_{yy}} \qquad \dots (4.53)$$

$$\mu_\alpha = m_w.u \qquad \dots (4.54)$$

Then
$$G = \frac{\mu_c + K_D s^2}{s^2 - \mu_\alpha} \qquad \dots (4.55)$$

and
$$G_a = \frac{\omega_a^2}{s^2 + 2\varsigma_a \omega_a s + \omega_a^2} \qquad \dots (4.56)$$

Using similar procedure as for third order system in Section 4.4.1, one gets a characteristic equation for the system as follows:

$$s^4 + (2\varsigma_a \omega_a + K_s K_R K_D \omega_a^2)s^3 + (\omega_a^2 - \mu_a + K_s K_D \omega_a^2)s^2$$

$$+(K_s K_R \omega_a^2 \mu_c - 2\varsigma_a \omega_a \mu_\alpha)s + (\mu_c K_s - \mu_\alpha)\omega_a^2 = 0 \quad \dots (4.57)$$

This is considered to be consisting of a product of two quadratics as

$$\left(s^2 + 2\varsigma_c \omega_c s + \omega_c^2\right)\left(s^2 + 2\varsigma\omega s + \omega^2\right) = 0 \qquad \dots (4.58)$$

where ς_c and ω_c define the desired pole locations of the system and ς and ω define the remaining two poles.

Equating coefficients of like powers of s, one gets

$$\left(2\varsigma_a \omega_a + K_s K_R K_D \omega_a^2\right) = \left(2\varsigma_a \omega_c + 2\varsigma\omega\right) \qquad \dots (4.59)$$

$$(\omega_a^2 - \mu_\alpha + K_s K_D \omega_a^2) = \omega^2 + \omega_c^2 + 2\varsigma_c \omega_c \cdot 2\varsigma\omega \qquad \dots (4.60)$$

$$\mu_c K_s K_R \omega_a^2 - 2\varsigma_a \omega_a \mu_\alpha = \omega_c^2 \cdot 2\varsigma\omega + \omega^2 \cdot 2\varsigma_c \omega_c \qquad \dots (4.61)$$

$$\mu_c K_s \omega_a^2 - \omega_a^2 \mu_\alpha = \omega^2 \omega_c^2 \qquad \dots (4.62)$$

These equations are solved for K_s and K_R which give

$$K_s = \frac{1}{\mu_c}\left[\frac{L[1 - K'_D\omega_c^2(1 - 4\varsigma_c^2)] - P[2\varsigma_c\omega_c \cdot \omega_c^2 K'_D]}{[1 - K'_D\omega_c^2][1 - K'_D\omega_c^2(1 - 4\varsigma_c^2)] + 2\varsigma_c\omega_c K'_D(2\varsigma_c\omega_c \cdot \omega_c^2 K'_D)}\right] \quad \ldots(4.63)$$

$$K_R = \frac{2\varsigma_c\omega_c K'_D L + (1 - K'_D\omega_c^2)P}{L[1 - K'_D\varsigma_c^2(1 - 4\varsigma_c^2)] - P[2\varsigma_c\omega_c \cdot \omega_c^2 K'_D]} \quad \ldots (4.64)$$

where $\qquad K'_D = \dfrac{K_D}{\mu_c} \qquad\qquad\qquad\qquad\qquad$ ... (4.65a)

$$M = 2\varsigma_a\omega_a - 2\varsigma_c\omega_c \qquad \ldots (4.65b)$$

$$N = \omega_a^2 - \omega_c^2 - \mu_\alpha - 2\varsigma_c w_c(2\varsigma_a\omega_a - 2\varsigma_c\omega_c) \qquad \ldots (4.65c)$$

$$L = \mu_\alpha + \frac{\omega_c^2}{\omega_a^2}N \qquad \ldots (4.65d)$$

$$p = \frac{2\varsigma_a\omega_a}{\omega_a^2}\mu_\alpha + \frac{\omega_c^2}{\omega_a^2}M + \frac{2\varsigma_c\omega_c}{\omega_a^2}N \qquad \ldots (4.65e)$$

Fig. 4.8 shows the variation of parameters of a typical unstable missile. Table 4.3 gives the computed values of control gains K_s and K_R using equations 4.63 and 4.64. It also gives the gain values if K'_D is assumed to be zero (*i.e.*, engine inertia effect neglected for computing the gains). It is seen that the difference in gains is insignificant.

TABLE 4.3: Gain Schedule with and without K_D

Time (sec.)	$K_D = 0$		With K_D	
	Ks	K_R	Ks	K_R
1.100	−2.246	0.294	−2.266	0.291
6.100	−2.246	0.292	−2.265	0.290
11.100	−2.264	0.288	−2.283	0.286
16.100	−2.309	0.281	−2.327	0.278
21.100	−2.386	0.269	−2.403	0.267
26.100	−2.500	0.255	−2.516	0.253
31.100	−2.653	0.238	−2.668	0.236
36.100	−2.834	0.220	−2.847	0.219

(Contd.)

41.100	−3.037	0.203	−3.047	0.202
46.100	−3.245	0.188	−3.253	0.186
51.100	−3.483	0.172	−3.488	0.171
56.100	−3.745	0.158	−3.748	0.157
61.100	−3.635	0.159	−3.638	0.158
66.100	−3.483	0.161	−3.486	0.160
71.100	−3.262	0.166	−3.266	0.165
76.100	−3.008	0.174	−3.013	0.173
81.100	−2.648	0.189	−2.655	0.188
86.100	−2.332	0.206	−2.340	0.204
91.100	−2.047	0.223	−2.056	0.221
96.100	−1.789	0.240	−1.798	0.239
101.100	−1.570	0.257	−1.579	0.255
106.100	−1.385	0.270	−1.394	0.268

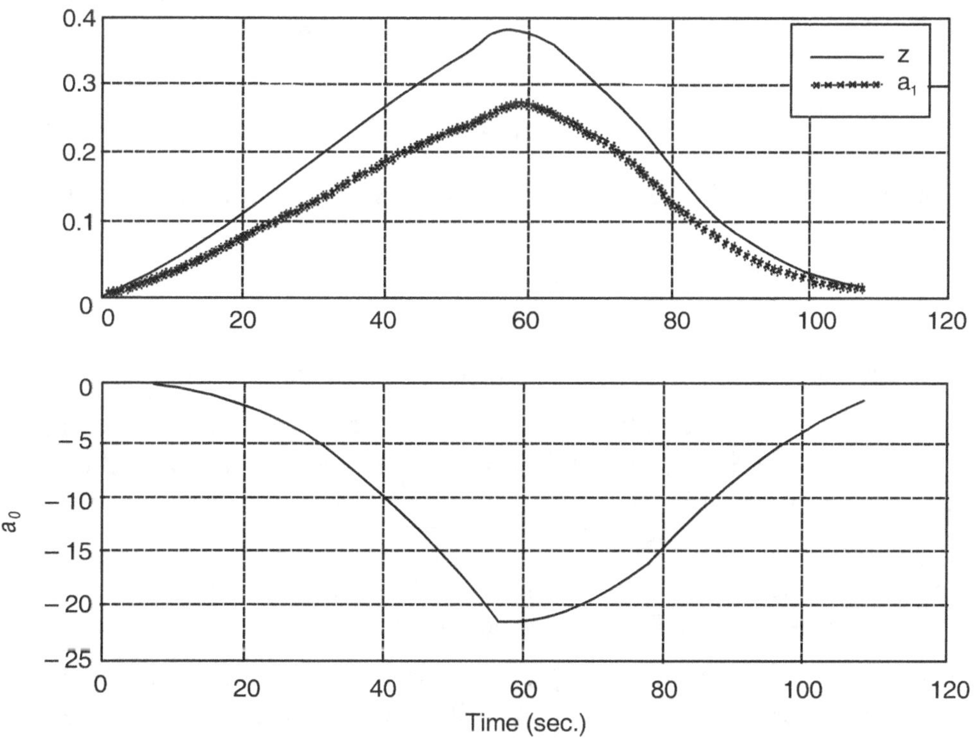

Fig. 4.8(a)

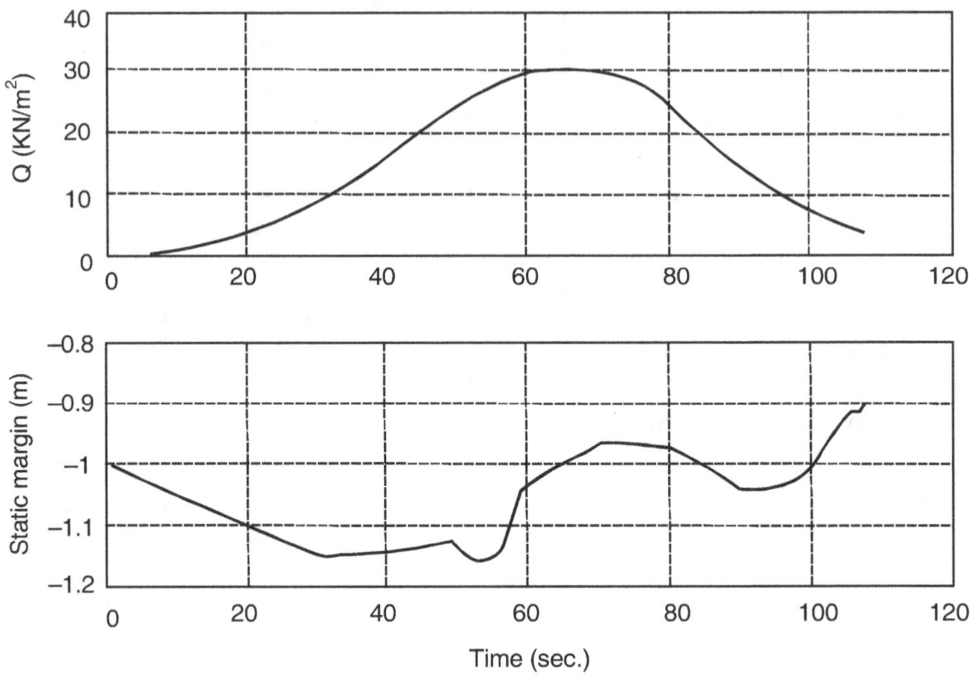

Fig. 4.8(b)

This conclusion holds whenever the Tail-wags-dog (TWD) frequency which is given by

$$\omega_{TWD} = \sqrt{\frac{\mu_c}{K_D}} \quad r/s \qquad \qquad \dots (4.66)$$

is sufficiently far away from the desired dominant pole frequency ω_c. However, it will influence the design when the TWD frequency falls due to decreased value of μ_c which happens when the thrust level starts falling near the burnout and the net torque (*i.e.*, the difference of control torque and inertial torque) acting on the vehicle will reduce and change the direction for frequencies higher than the TWD frequency.

Fig. 4.9 gives a typical bode plot for the system. The gain plot shows a notch occurring at $\omega = 40$ *r/s*. This gives the TWD (tail-wags-dog) frequency.

There is a hump in gain curve beyond this frequency which is due to TWD effect. The designer must provide adequate attenuation for frequencies above TWD frequency since the gain hump has a potential of introducing TWD oscillations if not adequately attenuated.

This is the main effect of engine inertia on the control system.

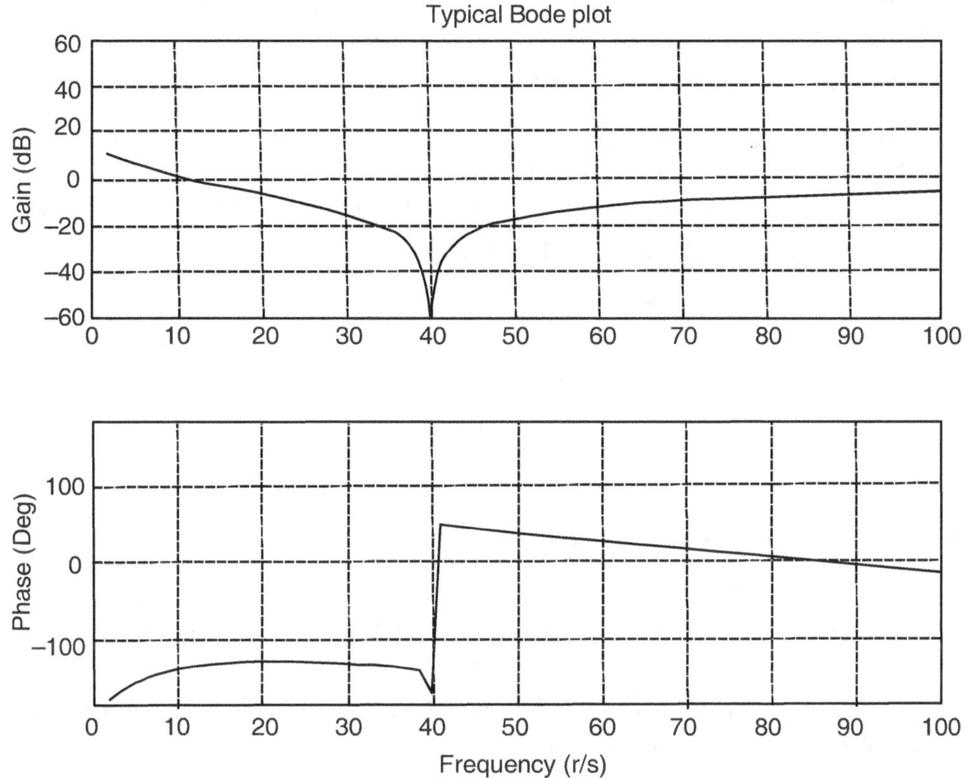

Fig. 4.9 Typical Bode Plot With TWD Effect

The design of compensator for providing adequate stability margins and the required attenuation at higher frequencies will be discussed after incorporating the structural flexibility and propellant sloshing in the model.

In general, the inertia effect has been ignored when the control is through aerodynamic control surfaces since the inertia of the surface about the hinge line is expected to be small value. However, during low dynamic pressure region where forward gain of the control system is high, it is found to excite control-structure oscillations in some missiles when the centre of gravity of the control surface is behind the hinge line.

Equation 4.48 shows that the TWD effect is contributed by the factor K_D where

$$K_D = 2\left(\frac{I_R + M_R L_R L_C}{I_{yy}}\right)$$

For the case of aerodynamic control surface

I_R = Moment of inertia of the control surface about the hinge line

L_R = Distance of *CG* of surface from the hinge line. This is +ve when measured towards trailing edge of the surface

L_C = The distance between hinge line and vehicle *CG*.

In case of control using gimbaled engines or flexible nozzles, the designer has no choice of selecting the sign of L_R. However, in case of control surfaces, the fins can be designed to have its centre of gravity forward of hinge line without affecting aerodynamic characteristics.

Thus, the control surface *CG* can be designed such that

$$I_R + M_R L_R L_C = 0$$

or
$$L_R = -\frac{I_R}{M_R L_C}$$

Thus,

$$\textit{\textbf{The desired CG location of control surface forward of hinge line}} = \frac{I_R}{M_R L_C} \qquad \text{... (4.67)}$$

This will ensure that the control surface inertia does not cause any control-structure oscillations.

4.4.3 Sensitivity of Control Gain Schedule to the Transfer Function Zero and Digital Control System Lags

It has been found in the previous section that the engine inertia does not significantly affect the gains schedule which is designed mainly to have the satisfactory autopilot response at low frequency. Hence, the engine inertia terms are neglected for studying the effect of zero and digital control system lags. Considering both the angular acceleration (Eq. 4.8b) and lateral acceleration (Eq. 4.7b) equations, the vehicle transfer function can be written as

$$G = \frac{\mu_c(s+z)}{s(s^2 + a_1 s + a_0)} \qquad \text{... (4.68)}$$

Where
$$\mu_c = n_\delta$$

$$z = -y_v + \frac{n_v y_\delta}{n_\delta}$$

$$a_1 = -(y_v + n_r)$$

$$a_0 = -\mu_a = y_v n_r + n_v u_0 - n_v y_r \approx n_v u \qquad \text{... (4.69)}$$

The parameters corresponding to dynamic derivative y_r and n_r are assumed zeros. Actuator transfer function is

$$G_a = \frac{\omega_a^2}{s^2 + 2\varsigma_a \omega_a s + \omega_a^2}$$

One can follow a similar procedure as in previous section and equate the coefficients of like powers of s of the characteristic equations $c(s)$ and the polynomial obtained from

$$C(s) = (s + A)(s^2 + 2\varsigma_c \omega_c s + \omega_c^2)(s^2 + 2\varsigma \omega s + \omega^2) \qquad \dots (4.70)$$

wherein we specify the dominant pole given by ς_c and ω_c and solve for the gains K_S and K_R. The algorithm is given in the box.

Fig. 4.10 gives the comparison of gains obtained

(1) using lateral acceleration equation (*i.e.*, incorporating z) and

(2) ignoring the lateral acceleration equation (*i.e.*, z = 0)

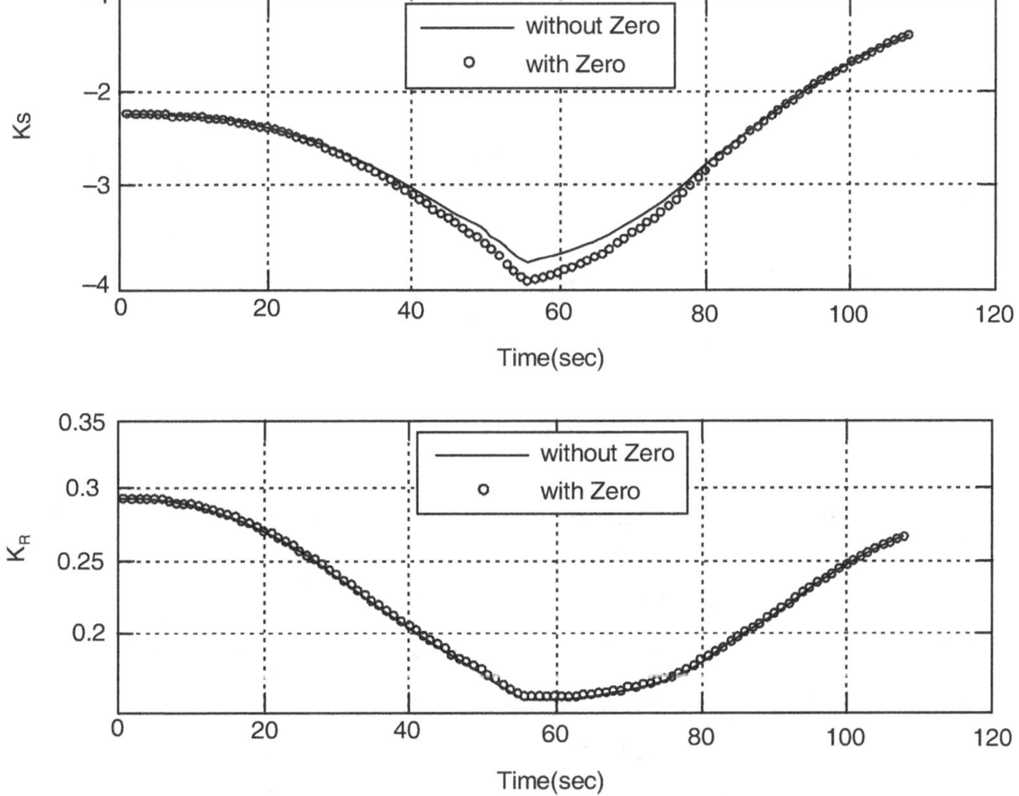

Fig. 4.10 Gain Schedules With and Without Zero

Fig 4.11 shows the comparison of stability margins when the vehicle dynamics for margin computation includes both the lateral acceleration equation as well as engine inertia effect. It is seen that the gains match well for low dynamic pressure region when the magnitudes of z and a_0 (or μ_α) are small but differ when magnitudes of z and a_0 are high. Also, the gains obtained by using fifth order polynomial (incorporating the effect of zero, z) give improved margins.

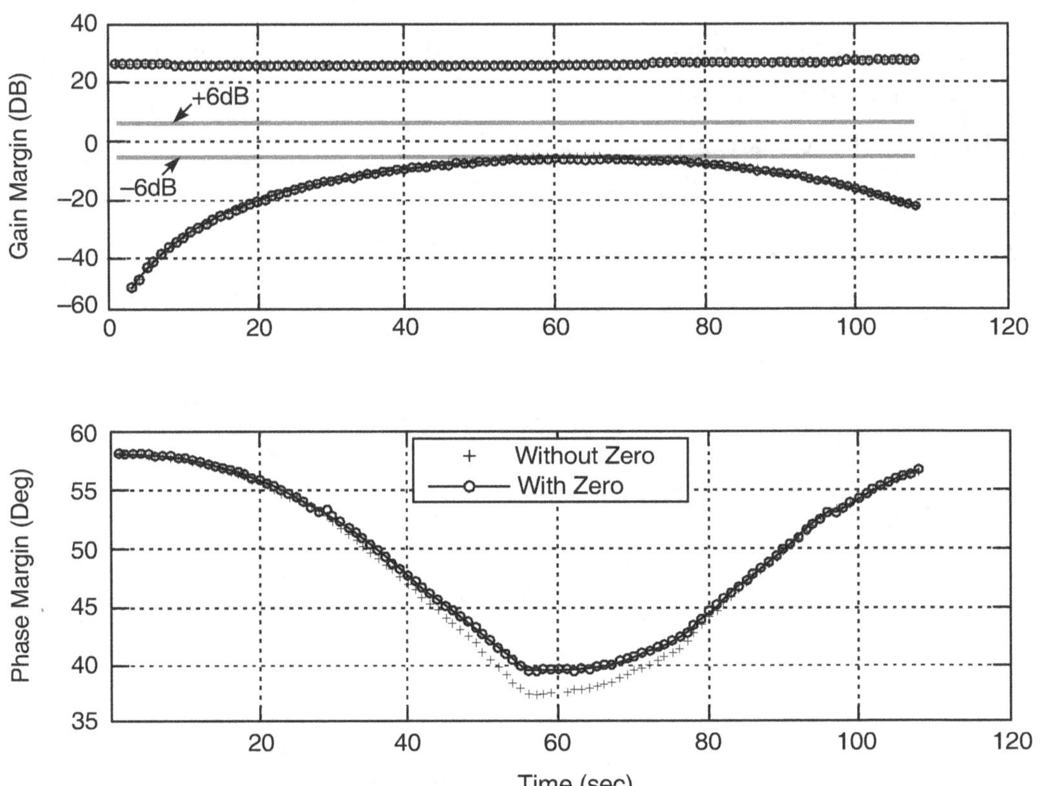

Fig. 4.11 Stablity Margins for two Cases of Gain Schedules

Lags Due to Digital Control System

The digital control system is associated with additional lag in the system due to

(1) computational delay T_c and

(2) the sample and hold feature due to sampling interval T_s

Fig. 4.12 shows a typical input signal and the dots show the sampling instants.

ALGORITHM FOR COMPUTING GAINS
(Lateral acceleration equation included)

$$a_1 = -(y_v + n_r)$$

$$a_0 = n_v u + y_v n_r - n_v y_r = -\mu_\alpha$$

$$z = -y_v + \frac{n_v y_\delta}{n_\delta}$$

$$\mu_c = n_\delta$$

Then

$$l_1 = 2\varsigma_a \omega_a - 2\varsigma_c \omega_c$$

$$l_2 = 2\varsigma_c \omega_c \cdot l_1$$

$$l_3 = \omega_a^2 - \omega_c^2 - l_2$$

$$K_1 = 2\varsigma_c \omega_c (a_0 + a_1 l_1 + l_3)$$

$$K_2 = \omega_c^2 (a_1 + l_1)$$

$$K_3 = K_1 + K_2 - a_1 \omega_a^2 - 2\varsigma_a \omega_a \cdot a_0$$

$$K_4 = \frac{K_3}{\omega_a^2}$$

$$K_5 = \frac{\omega_c^2}{\omega_a^2} \left[\frac{K_1}{2\varsigma_c \omega_c} \right] - a_0$$

$$K_s = \frac{1}{\mu_c} \left[\frac{K_5 - K_4}{1 - \dfrac{2\zeta_c}{\omega_c} \cdot z + \left(\dfrac{z}{\omega_c}\right)^2} \right]$$

$$K_R = \frac{z}{\omega_c^2} + \frac{K_4 \left[1 - \dfrac{2\zeta_c}{\omega_c} z + \left(\dfrac{z}{\omega_c}\right)^2 \right]}{(K_5 - K_4 z)}$$

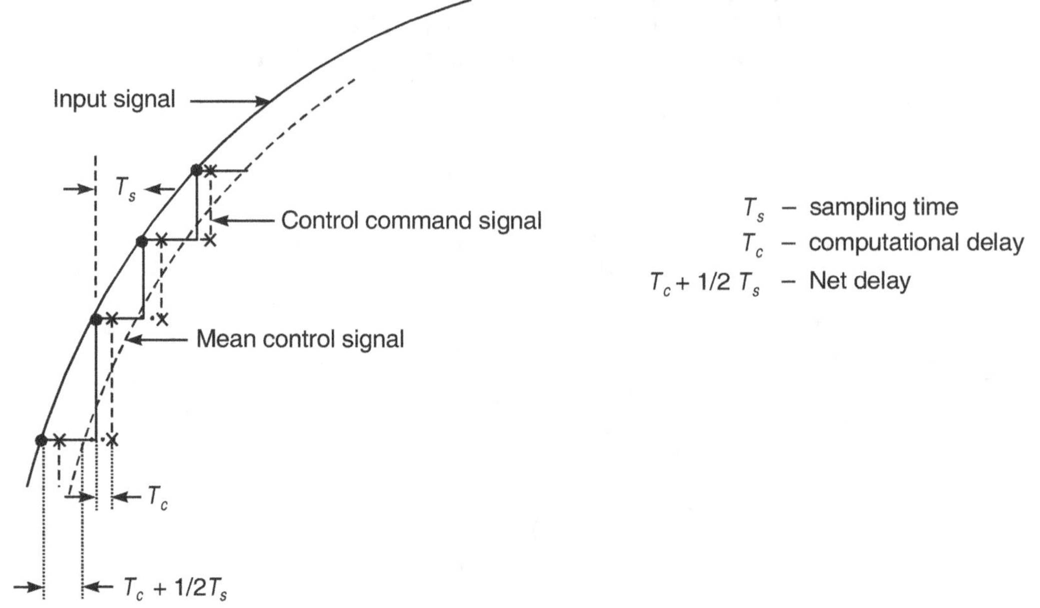

Fig. 4.12 Lag due to sample hold and computational delay

The digital control system takes the input and feedback at sampled times but takes finite time T_c to process it and compute the control signal. It is assumed here that the control signal is given to the actuators as soon as it is computed instead of waiting till the next sampling time. The crosses in the figure show the availability of computed control signal. This signal will continue till the next computed signal is available. If we draw a mean curve for the control signal, it will be seen that the control signal will have a net time shift of

$$D = T_c + \frac{1}{2} T_s \qquad \qquad \text{... (4. 71)}$$

from the ideal control, (*i.e.*, control which does not have sample and hold and computational delay.) This delay will result in degradation of stability margins. Hence, it will be instructive to compare the gains schedule obtained by incorporating the effect of this delay with that obtained by ignoring the same and study the improvement in stability margins if the gains are obtained by incorporating the delay.

For the present study, the effect of this delay is approximated by a first order lag.

Thus,
$$e^{-Ds} = \frac{1}{e^{Ds}} = \frac{1}{1+Ds} = \frac{d}{s+d} \qquad \qquad \text{... (4.72)}$$

where
$$d = \frac{1}{D} \qquad \qquad \text{... (4.73)}$$

The algorithm for computing gains K_s and K_R incorporating z and delay is given in the box.

ALGORITHM FOR COMPUTING GAINS
(INCORPORATING LATERAL ACCELRATION EQUATION
AND DELAY DUE TO DIGITAL CONTROL)

$$a_1 = -(y_v + n_r)$$

$$a_0 = n_v u + y_v n_r - n_v y_r = -\mu_\alpha$$

$$z = -y_v + \frac{n_v y_\delta}{n_\delta}$$

$$\mu_c = n_\delta$$

Then

$$l_2 = 2\varsigma_a \omega_a - 2\varsigma_c \omega_c$$

$$l_2 = 2\varsigma_c \omega_c \cdot l_1$$

$$l_3 = \omega_a^2 - \omega_c^2 - l_2$$

$$l_4 = a_0 + a_1 d + l_3 + (a_1 + d)l_1$$

$$c_1 = a_0 d\, \omega_a^2$$

$$c_2 = \omega_a^2(a_0 + a_1 d) + a_0 d \cdot 2\varsigma_a \omega_a$$

$$c_3 = 2\varsigma_a \omega_a(a_0 + a_1 d) + \omega_a^2(a_1 + d) + a_0 d$$

$$K_1 = c_3 - 2\varsigma_c \omega_c l_4$$

$$K_2 = c_2 - 2\varsigma_c \omega_c[K_1 - \omega_c^2(a_1 + d + l_1)] - \omega_c^2 l_4$$

$$K_3 = -c_1 + \omega_c^2[K_1 - \omega_c^2(a_1 + d + l_1)]$$

$$K_4 = \frac{K_2}{d\mu_c \omega_a^2}$$

$$K_5 = \frac{K_3}{d\mu_c \omega_a^2}$$

$$K_s = \frac{1}{\mu_c}\left[\frac{K_4 z - K_5}{1 - \dfrac{2\varsigma_c}{\omega_c}\cdot z + \left(\dfrac{z}{\omega_c}\right)^2}\right]$$

$$K_R = \frac{z}{\omega_c^2} + \frac{K_4\left[1 - \dfrac{2\varsigma_c}{\omega_c}z + \left(\dfrac{z}{\omega_c}\right)^2\right]}{(K_5 - K_4 z)}$$

Figure 4.13 shows the comparison of gains schedule. It is seen that both the gains K_s and K_R need significant adjustment throughout the operating region. It is also clear that K_R needs to be increased to compensate for the loss of phase margin due to delay.

Fig. 4.14 shows the comparison of stability margins for the two gains schedules.

It is observed that the negative gain margin and phase margin are barely acceptable at high dynamic pressure region but are quite good at lower dynamic pressure region. The positive gain margin is more than required. This indicates the possibility of increasing the forward gain to make the negative gain margin more comfortable and at the same time possibly select the optimum phase margin point.

The phase margin can also be enhanced by choosing somewhat higher value of K_R.

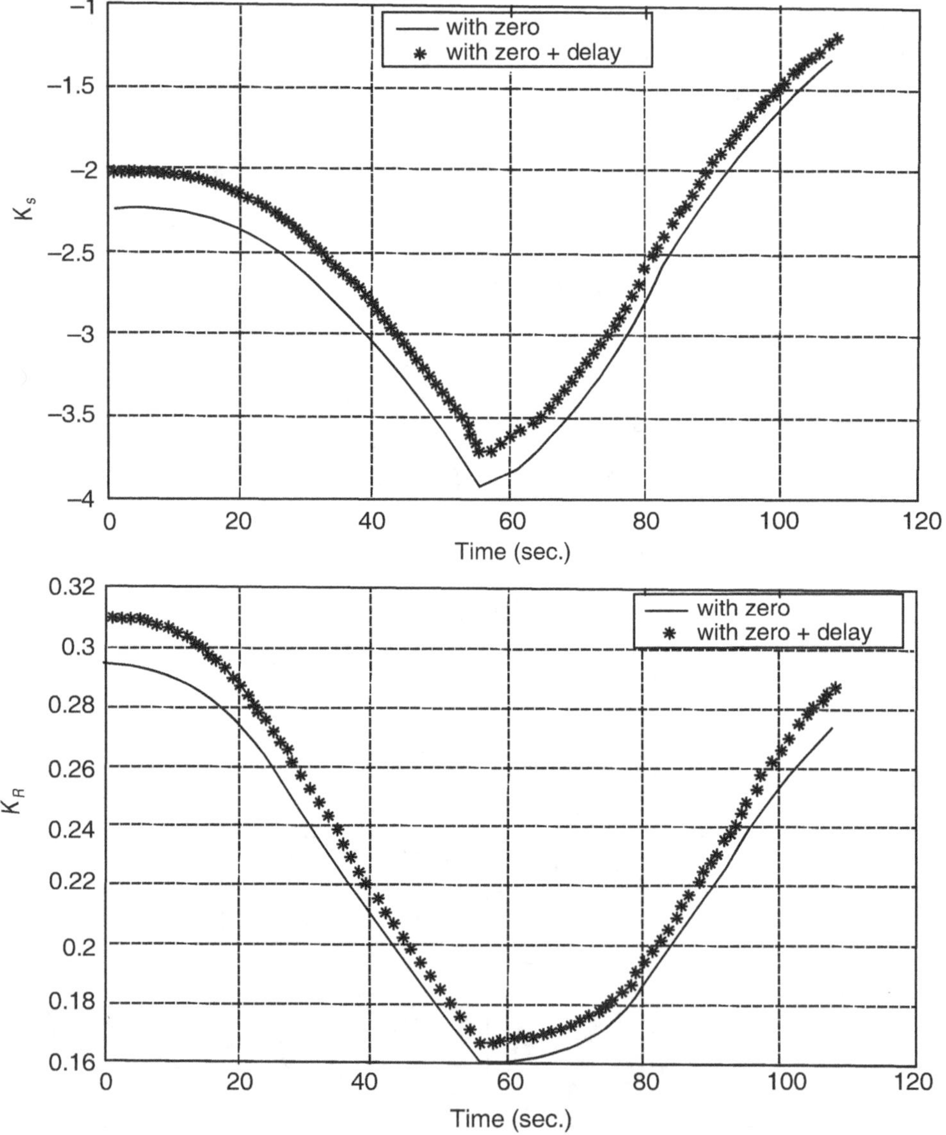

Fig. 4.13 Gain Schedules With and Without Delay

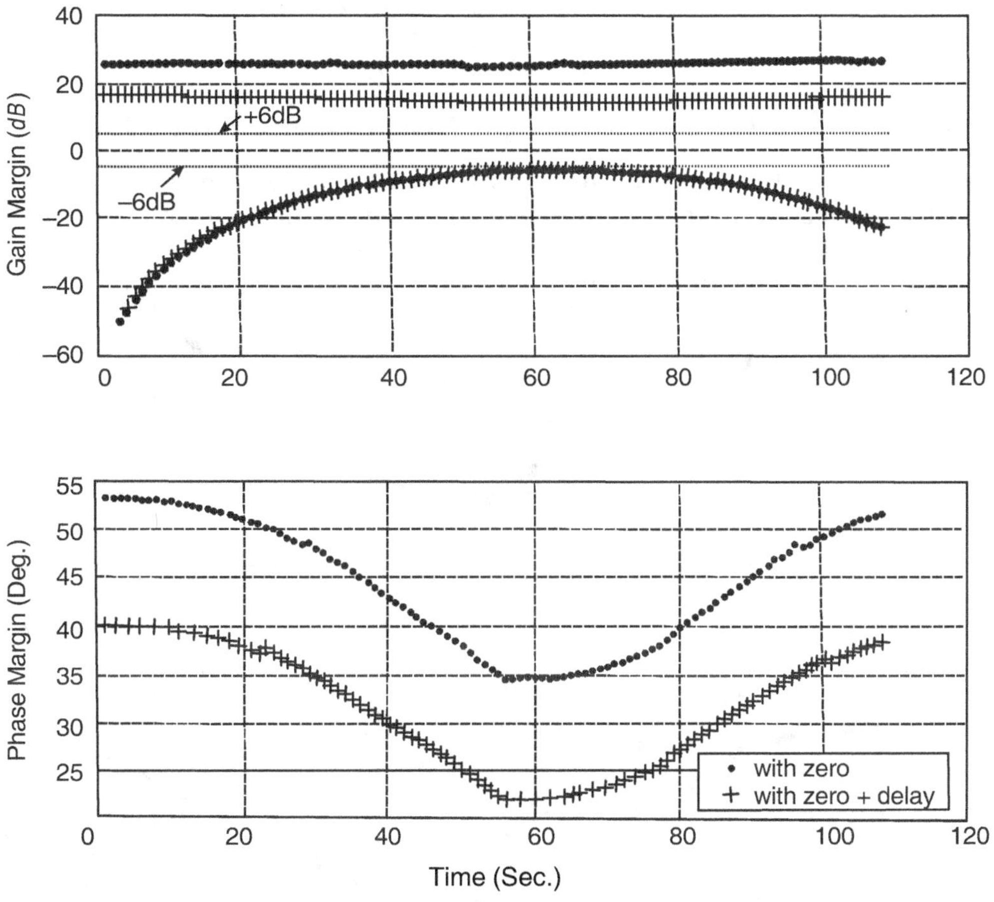

Fig. 4.14 Stability Margins for two Cases of Gain Schedules

The process of gain adjustment can be mechanized in different ways such as:

1. Alter the specification on desired pole (if other considerations do not put restrictions) such as

$$\omega_c = 6\,rad/s, \quad \zeta_c = 0.9$$

2. Make ω_c and ς_c function of dynamic pressure. Noting the fact that the maximum dynamic pressure is 30×10^3 N/m^2 at which we need $\omega_c = 6$ rad/sec and $\varsigma_c = 0.9$ one can write

$$\omega_c = 5 + 0.35 * 10^{-4}\,Q \qquad\qquad ...(4.74)$$

$$\varsigma_c = 0.7 + 0.7*10^{-5}\,Q \qquad\qquad ...(4.75)$$

3. Put a limit on lower value of K_R and ω_c as function of dynamic pressure. Then

$$\omega_c = 5 + 0.35 * 10^{-4} Q$$

and if $K_R < K_{Rlimit}$, use $K_R = K_{Rlimit}$ where K_{Rlimit} for the present example can be 0.2 to 0.25.

Fig. 4.15 shows the gains schedule and the corresponding stability margins for Rule No. 2. It is seen that the stability margins are satisfactory for complete time duration under study.

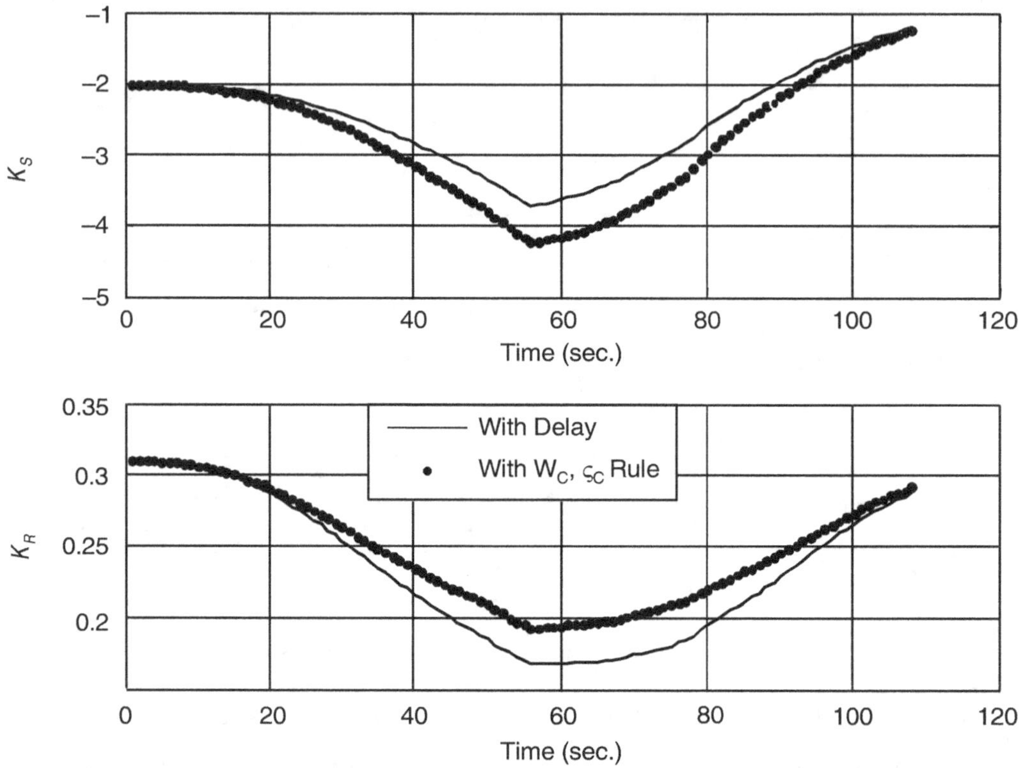

Fig. 4.15(a) Gains schedule

Fig 4.16 gives a typical step response at maximum dynamic pressure region.

One can observe the overshoot and a long tail in settling down characteristics. This is typical response for a highly unstable vehicle. When the step command is given and angle of attack is generated while changing the vehicle attitude, the aerodynamic disturbance torque also will be generated in the same direction as control torque and the response becomes fast. However, when the attitude becomes larger than the input command, control torque changes the direction but aerodynamic disturbance continues to push the attitude away from the zero position. Thus, only the residual torque (*i.e.*, control torque – aerodynamic torque) acts on the vehicle and as the control error becomes lower, control torque reduces causing a long tail in

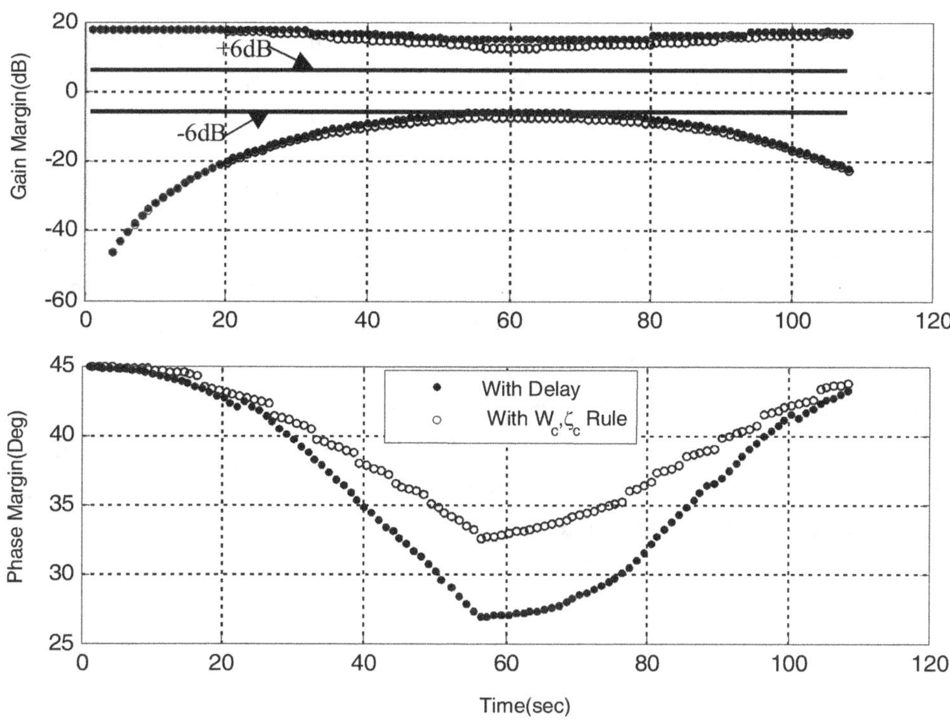

Fig 4.15(b) Stability Margins

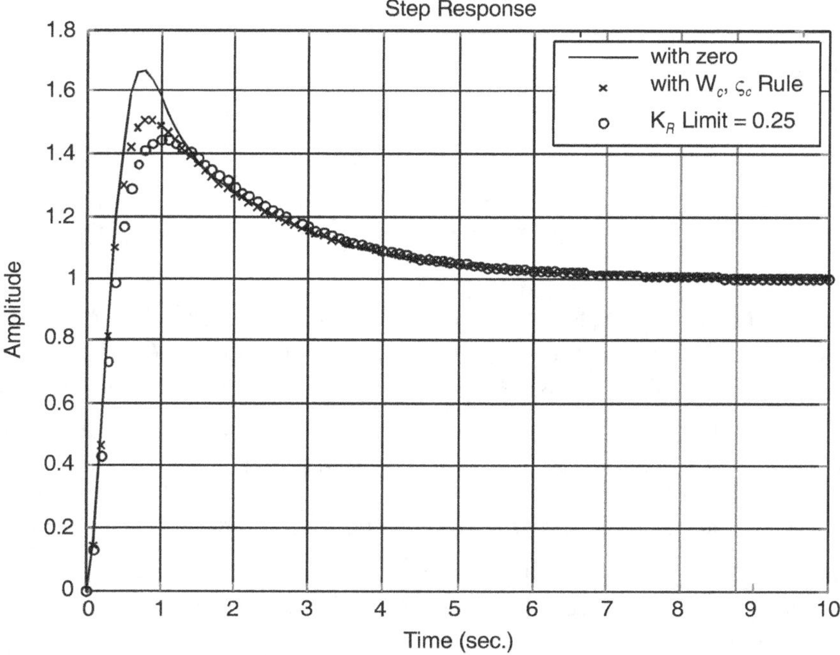

Fig 4.16 Typical Step Response

settling down response. Also the error will always settle down on a higher side of commanded attitude angle. The above tailing characteristics will be absent when the dynamic pressure is small or the vehicle is only marginally unstable (a_0 very small).

4.4.4 Secondary Injection Thrust Vector Control and Aerodynamic Control

In previous section, we have discussed the control provided by thrust vector control system either by gimballed nozzles or flexible nozzle where engine or nozzle inertia is significant. It has been observed that the engine inertia does not modify the gain schedule significantly since it is computed for fixing the low frequency dynamics. Hence, the parameter $K'_D = K_D / \mu_c$ in Eqns. 4.63 and 4.64 can be neglected. Further, the equations for $\mu_c K_s$ and K_R incorporating z and digital control lags remain unchanged. The gain schedule K_s, K_R for secondary injection thrust vector control (SITVC) and aerodynamic control can then be obtained by just computing appropriate value of μ_c and obtaining K_s from $\mu_c K_s$.

1. For SITVC

$$\mu_c = \frac{K_{DS} * l_c}{I_{yy}} \qquad \text{... (4.76)}$$

    where K_{DS} = side force per unit deflection of actuator

                         l_c = control moment arm

2. μ_c for aerodynamic control

   Assuming one pair of control surfaces give pitch control and another pair gives yaw control.

$$\mu_c = \frac{C_{N\delta} Q S l_c}{I_{yy}} \qquad \text{... (4.77)}$$

The side thrust for SITVC depends mainly on exhaust gas velocity in the nozzle and the flow rate of the injectant. As long as the stage motor is burning full, the K_{DS} parameter remains more or less constant. During tail off region of thrust, the value of K_{DS} becomes unreliable when the stage motor chamber pressure falls below a certain level. Therefore, it is desirable to switch off SITVC control when the chamber pressure falls below 15 to 20% of full level.

The parameter μ_c for aerodynamic control depends on the dynamic pressure Q and $C_{N\delta}$ which in turn depends on mach number. Since the dynamic pressure varies from zero at take off, reaches maximum value at certain flight time and altitude and again decreases at higher altitude when the air density becomes negligible. Therefore, this parameter for aerodynamic control varies over a very wide range. The forward gain K_s which varies inversely as μ_c therefore, also changes over a wide range. Thus, K_s becomes very large near take off region as well as at high altitude which only means that control surface deflection required per degree attitude error becomes very large. However, aerodynamic characteristics $(C_{N\delta})$ hold true only for control deflection of 20–25° and beyond which they become nonlinear and even decrease in value. Therefore, maximum allowed control surface deflections are around 25° and actuator strokes are designed to limit this maximum deflection value. Thus, high values of control gain K_s mean, the control surfaces will get saturated even for small attitude errors. Further, high gain also means large change in deflection for a small change in attitude error. Thus, the control surfaces will become highly oscillatory with large amplitude and angular rates due to sensor noise or digital control system quantization demanding high flow requirement of hydraulic oil in case of hydraulic actuator or large currents from electromechanical actuators. Hence, one must avoid large values of control gain K_S. This is normally done by following methods:

1. *Not to use aerodynamic control*

Aerodynamic control is not used near take off for pitch and yaw control for launch vehicles and long range missiles till the dynamic pressure builds upto a desirable level when the computed gains reach reasonable magnitudes. Till such time, another control system such as TVC (thrust vector control) or RCS or Jet vane control is used for controlling the vehicle attitude. The roll attitude errors being less critical in nature and not much disturbances are expected, roll control has been done using aerodynamic control in some launch vehicles and missiles and maximum value of forward gains is limited to avoid unnecessary oscillations.

Tactical missiles, are small missiles and do not afford the luxury of having two control systems. Such missiles also have high acceleration (25g to 30g) and hence quickly reach sufficient velocity and dynamic pressure so that the control becomes effective. Such missiles have inclined launch and use aerodynamic control even from start and quickly recover from the tip off errors when they leave the launcher.

2. *Reducing the desired control frequency (ω_c) and limiting the gain values*

This policy can be adopted for regions where aerodynamic control needs to be used because of other constraints. For example, many launch vehicles or missiles having multiple stages, have a coasting phase after the propulsion burn out to acquire reasonably low dynamic pressure for safe stage separation. For this phase, the thrust vector control is not possible. Hence, aerodynamic control needs to be used. In such case, the high gains can be avoided by reducing the control frequency (ω_c) by making it a function of dynamic pressure.

Referring to Eq. 4.74 where ω_c is made a function of dynamic pressure, our objective of making it a function of dynamic pressure was to increase the stiffness of the control system against the destabilization effect of a_0 or μ_α. The minimum frequency $\omega_c = 5r/s$ remains even for zero dynamic pressure since the control effectiveness of thrust vector control system is unaltered. The speed of control also depends on the available control torque. When control torque (which depends on aerodynamic pressure for aerodynamic control) itself is not available, one cannot think of higher control speed. Hence, the control frequency needs to be made directly dependant on dynamic pressure.

Noting that the angular acceleration θ due to control is proportional to μ_c and hence dynamic pressure Q

$$\theta = KQ \qquad \text{or} \qquad s^2\theta = KQ$$

This gives us a logical rule for ω_c dependence on Q and one can adopt a relation of type

$$\omega_c = K\sqrt{Q} \text{ where } K \text{ is a constant} \qquad \text{... (4.78)}$$

and use it for computing control gains so that the gains do not become unnecessarily high. A limit on higher value may also be imposed on the computed values of gains.

4.4.5   Design of Roll Control System

I. Gains Schedule

Referring to Eq. 4.14, the required transfer function for roll control system for computing gains schedule is given as follows:

$$G = \frac{\mu_{c_r}\left(1+\dfrac{K_{Dr}}{\mu_{cr}}s^2\right)}{s(s+a)} = \frac{\mu_{c_r}(1+K_D's^2)}{s(s+a)} \qquad \text{... (4.79)}$$

and
$$G_a = \frac{\omega_a^2}{s^2 + 2\varsigma_a\omega_a s + \omega_a^2}$$

For roll control system the parameter K_{Dr} is given by

$$K_{Dr} = \frac{2M_R L_R L_a}{I_{xx}} \qquad \text{... (4.79a)}$$

The inertial force $M_R L_R \delta_R$ and $-M_R L_R \delta_R$ due to two actuators located on opposite sides of central plane will form a couple with a total moment arm $= 2L_a$. The equal and opposite inertial torques $I_R \delta_R$ and $-I_R \delta_R$ due to two actuators are about axis parallel to yaw axis.

These will be borne by structural members and there will be no resultant roll torque on the vehicle due to I_R.

The gains schedule for roll control system can be obtained using the equations 4.63 to 4.65 by making $a = 0$, and $\mu_a = 0$. Similarly, the algorithms for incorporating the digital control system lags also can be used for roll control system. It is interesting to note that the control gains $\mu_{c_r} K_s$ and K_R can be solved only once since $\mu_\alpha = 0$ and all other parameters are constant (if ω_c is not made variable as is required for the case of aerodynamic control due to wide variation of dynamic pressure). The gain schedule K_s at any flight instant can then be obtained by computing μ_{c_r} as

$$\mu_{c_r} = \frac{2T_E l_a}{I_{xx}}, \qquad \qquad \dots (4.80)$$

and
$$K_s = \frac{(\mu_{c_r} K_s)}{\mu_{c_r}} = \left(\mu_{c_r} K_s \times \frac{I_{xx}}{2T_E l_a} \right) \qquad \dots (4.81)$$

If we assume that the engine thrust is constant and control moment arm also constant for roll control, the forward gain becomes only proportional to moment of inertia I_{xx}.

The example under consideration for this discussion is a case of vehicle with liquid propellant stage having two gimballed engines which are used for all the pitch, yaw and roll control. The liquid propellant inside the tanks is free to move and hence does not undergo the same roll angle motion as the missile. Hence, only small fraction of total propellant weight which is touching the tank walls really contributes to roll moment of inertia (due to viscous friction). Thus, the roll moment of inertia, in case of liquid propellant stages remains almost constant till the propellant is exhausted unless the propellant tanks are provided with cross baffles to partition the tanks to reduce slosh masses and increase the slosh frequency. For the case of partitioned tanks and cross baffles, the trapped volume of propellant contributes to roll moment of inertia.

For the case of solid propellant stages, the appropriate value of I_{xx} can be used for computing the roll control gains schedule.

It is also important to note that in case of roll control using gimballed engines, one must provide adequate attenuation for the hump in the open loop gain vs frequency plot for frequencies above TWD frequency to avoid TWD oscillations in the system.

Further, for the case of aerodynamic control, similar strategies as in case of pitch and yaw control for making ω_c a function of dynamic pressure and limiting the higher values of gains needs to be adopted.

II. Proportional Plus Integral Control and Integrator Saturation Limit

The proportional plus integral control is used to nullify the steady state errors due to constant disturbances acting on the system. For the example under consideration, the steady disturbances can be from following sources:

(*i*) plane of two engine not passing through the vehicle *CG*

(*ii*) unequal thrust of two engines in presence of equal engine deflection for yaw control

(*iii*) unequal engine deflection for same actuator commands

(*iv*) unequal roll moment arm for two engines

(*v*) actuator alignment errors.

Similarly, there could be rolling moment disturbance due to vehicle assymetries in external configuration which are represented by the aerodynamic coefficient C_L in which case the disturbance torque will depend on dynamic pressure as well as mach number.

Let the total disturbance moment from various sources be given by Td. The equivalent control deflection required to counter the disturbance moment is given by:

$$\delta_R = \frac{T_d}{2T_E l_a} \quad \text{for TVC} \qquad \qquad \ldots (4.82)$$

$$= \frac{T_d}{2C_{l\delta}QSd} \quad \text{For aerodynamic control (assuming four control} \qquad \ldots (4.83)$$

surfaces are used for roll control and $C_{l\delta}$ is rolling moment coefficient for one pair of surfaces)

The steady state error is given by

$$(\phi_c - \phi)_{ss} = \frac{\delta_R}{K_s} \qquad \qquad \ldots (4.84)$$

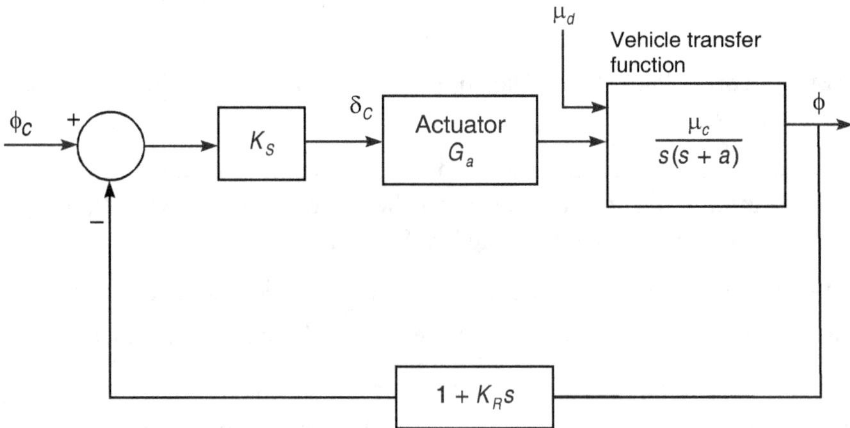

Fig. 4.17

For thrust vector control system, K_s is more or less constant and thus steady state error can be small. In case of aerodynamic control, the gain K_s becomes low at high dynamic pressure region and steady state error also becomes quite large. This needs to be compensated by providing integral control. The integrator output is limited to a value of maximum estimated steady state error in absence of integrator so that unnecessary delay in recovery from large integrator outputs in case of large transient errors are avoided. Consider the transfer function of proportional plus integral control:

$$G_I = 1 + \frac{K_I}{s} = \frac{K_1 + s}{s}$$

This has Gain $= \dfrac{\sqrt{K_1^2 + \omega^2}}{\omega}$ and Phase $= \left(\tan^{-1} \dfrac{\omega}{K_1} - 90° \right)$ ... (4.85)

Eq. 4.85 shows that larger the integrator gain, more will be the phase lag contributed by it and there will be corresponding degradation in phase margin of the control system. Larger gain of integrator also means faster correction of steady state error.

Since the integrator is provided for nullifying steady state errors due to steady disturbance, there need not be a high integrator gains and the gain can be mainly selected from the consideration of permissible degradation in phase margin of the system.

III. Sharing of Control Authority for Pitch, Yaw and Roll

For the example under consideration, the two engines are mounted in pitch plane. Hence pitch control is provided by deflecting engines in pitch plane. The yaw and roll control torque is provided by deflecting the engines unequally. Let δ_y and δ_R be the engine command required for yaw control and roll control respectively.

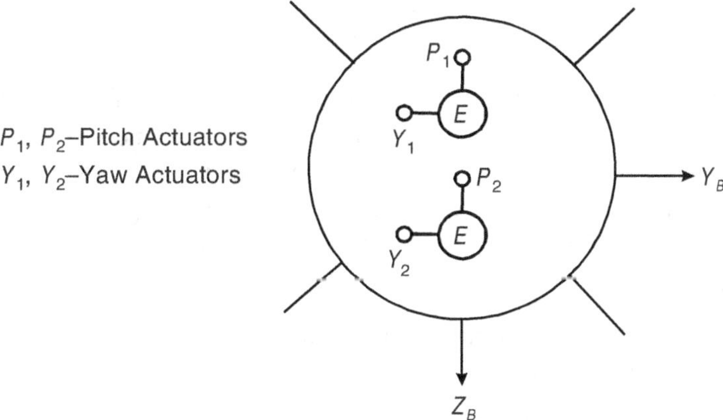

Fig. 4.18

Then the individual engines are commanded as

$$\delta_{y1} = \delta_y - \delta_R$$

$$\delta_{y2} = \delta_y + \delta_R \qquad \qquad ... (4.86)$$

Since the maximum engine deflection angle is limited, one must decide in advance, the maximum allowable deflection for roll and yaw and the commanded deflection accordingly limited before generating the actuator command as per Eq. 4.86.

Equation 4.82 gives the maximum deflection required to control the disturbance level. Allowing some additional deflection for transient control, the maximum level allowed for roll control can be fixed and remaining can be given for yaw control. The actual disturbance level can however be much less than the estimated worst case disturbance. Hence, it is possible to use a dynamic sharing of available control deflection by using a logic — 'if the actual deflection command is lower than the allowed limit, the difference can be allowed for yaw control as an extra available control deflection.' With the present day onboard computers, such logics and even more complex logics have become possible and have been successfully flight tested several times.

Similar sharing logic is also used for aerodynamic control where same four control surfaces are used for pitch, yaw and roll control systems.

4.4.6   Design of Lateral Acceleration Control System

Lateral acceleration control system is very commonly used for tactical missile in which case the required or demanded lateral acceleration is proportional to the line of sight turning rate λ. It has been established that the target moving with a constant velocity can be intercepted with a guidance law.

$$a_z = N'V_c\lambda_p \qquad \qquad ... (4.87)$$

where V_c is the resultant closing speed of the missile, λ_p is the angular rate of line of sight, N' is the navigation constant and a_z is the lateral acceleration demand in pitch plane. There are several variations to the above guidance law to ensure minimum miss distance for manoeuvring targets. The lateral acceleration control has also been frequently used in case of surface to surface missiles where guidance law enables the missile to follow a nominal trajectory which is designed to meet several constraints. The main considerations being

(*a*) The maximum angle of attack not to exceed a specified value.
    (This gives aerodynamic limit).

(*b*) The maximum control deflection not to exceed a specified limit
    (This gives control limit)

(c) The maximum load on the vehicle not to exceed a specified limit. (This gives structural limit).

The control system is configured in two different ways. We discuss first the configuration adopted by this author and used in case of a typical surface to surface missile.

Fig 4.19 gives the block diagram for latax control in yaw plane.

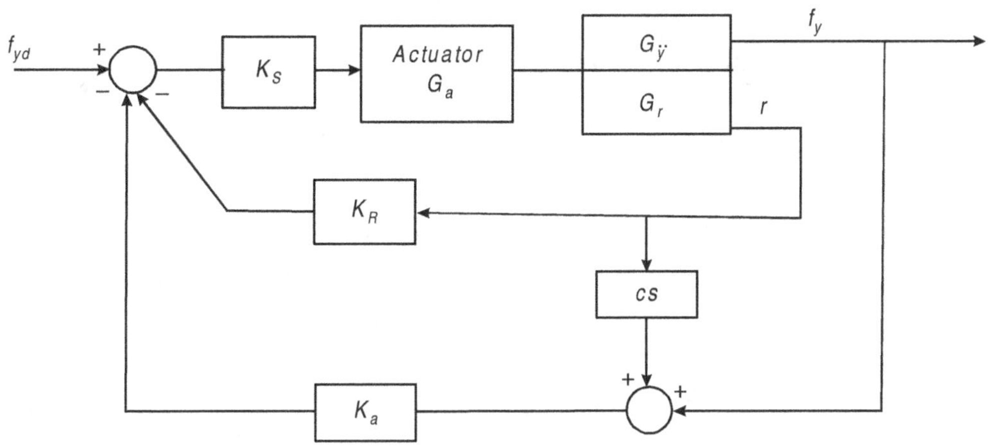

Fig. 4.19

Assuming aerodynamic control, the dynamic equations are (Ref. Eq. 4.7 and 4.8) given by:

$$v = y_v v + (y_r - u)r + y_\delta \delta \qquad \text{... (4.88)}$$

$$r = n_v v + n_r r + n_\delta \delta \qquad \text{... (4.89)}$$

Then
$$f_y = v + ur \qquad \text{... (4.90)}$$

and the control law is given by

$$\delta_c = K_s \left[f_{y_d} - K_a f_{ys} - K_R r \right] \qquad \text{... (4.91)}$$

(**Note:** For pitch control, the control law will be

$$\delta_c = K_s \left[f_{zd} - K_a f_{zs} + K_R q \right]) \qquad \text{... (4.92)}$$

where f_{ys} is the sensed acceleration and is given by

$$f_{ys} = f_y + cr \qquad \text{... (4.93)}$$

and c is the distance of sensor location from CG of the vehicle and is positive when measured towards nose side.

The closed loop transfer function (TF) of the system is given by

$$\frac{f_y}{f_{yd}} = \frac{K_s G_a G_y}{1 + (K_R + K_a cs) K_s G_a G_r + K_s K_a G_a G_y} \quad \ldots (4.94)$$

where

$$G_a = \text{Actuator } TF = \frac{\omega_a^2}{s^2 + 2\varsigma_a \omega_a s + \omega_a^2} \quad \ldots (4.95)$$

$$G_y = \frac{f_y(s)}{\delta(s)} = \frac{y_\delta s^2 - y_\delta n_r s - u(n_\delta y_v - n_v y_\delta)}{s^2 - (y_v + n_r)s + y_v n_r + u n_v - n_v y_r} \quad \ldots (4.96)$$

$$G_r = \frac{r(s)}{\delta(s)} = \frac{n_\delta s - y_v n_\delta + n_v y_\delta}{s^2 - (y_v + n_r)s + y_v n_r + u n_v - n_v y_r} \quad \ldots (4.97)$$

K_s, K_R and K_a are the control gains. Assuming $y_r = 0$, the characteristic equation is given by

$$s^4 + C_3 s^3 + C_2 s^2 + C_1 s + C_0 = 0 \quad \ldots (4.98)$$

where

$$C_3 = 2\varsigma_a \omega_a - (y_v + n_r) \quad \ldots (4.99)$$

$$C_2 = K_1 + K_2 K_a K_s \quad \ldots (4.100)$$

for

$$K_1 = \omega_a^2 - 2\varsigma_a \omega_a (y_v + n_r) + (u n_v + y_v n_r) \quad \ldots (4.101)$$

$$K_2 = (c n_\delta + y_\delta)\omega_a^2 \quad \ldots (4.\ 102)$$

$$C_1 = K_3 + K_4 K_R K_s + K_5 K_a K_s \quad \ldots (4.103)$$

for

$$K_3 = 2\varsigma_a \omega_a (u n_v + y_v n_r) - (y_v + n_r)\omega_a^2 \quad \ldots (4.104a)$$

$$K_4 = n_\delta \omega_a^2 \quad \ldots (4.104b)$$

$$K_5 = \omega_a^2 \left[c(n_v y_\varsigma - y_v n_\varsigma) - y_\varsigma n_r \right] \quad \ldots (4.105)$$

$$C_0 = K_6 + K_7 K_R K_s + K_8 K_a K_s \quad \ldots (4.106)$$

for

$$K_6 = (u n_v + y_v n_r)\omega_a^2 \quad \ldots (4.107)$$

$$K_7 = (n_v y_\varsigma - n_\varsigma y_v)\omega_a^2 \quad \ldots (4.108)$$

$$K_8 = K_7 u \quad \ldots (4.109)$$

This characteristic equation is represented by

$$(s^2 + 2\varsigma_c \omega_c s + \omega_c^2)(s^2 + 2\varsigma \omega s + \omega^2) = 0 \quad \ldots (4.110)$$

where ς_c and ω_c specify the desired poles locations.

Expanding and equating coefficients of like powers of s we get

$$C_3 = 2\varsigma\omega + 2\varsigma_c\omega_c \qquad\qquad \ldots(4.111)$$

$$C_2 = \omega^2 + \omega_c^2 + 2\varsigma\omega \cdot 2\varsigma_c\omega_c \qquad\qquad \ldots(4.112)$$

$$C_1 = 2\varsigma\omega\,\omega_c^2 + 2\varsigma_c\omega_c\omega^2 \qquad\qquad \ldots(4.113)$$

$$C_0 = \omega^2\omega_c^2 \qquad\qquad \ldots(4.114)$$

putting values of ω^2 from (4.114) and $2\varsigma\omega$ from 4.111 in eq. 4.112, one gets

$$C_2 = \frac{C_0}{\omega_c^2} + D_1 \qquad\qquad \ldots(4.115)$$

where $\qquad\qquad D_1 = \omega_c^2 + 2\varsigma_c\omega_c(C_3 - 2\varsigma_c\omega_c) \qquad\qquad \ldots(4.116)$

Using Eqs. 4.109 and 4.106 in (4.115), we get

$$K_R K_S K_7 + K_a K_s D_3 = D_2 \qquad\qquad \ldots(4.117)$$

where $\qquad\qquad D_2 = (K_1 - D_1)\omega_c^2 - K_6 \qquad\qquad \ldots(4.118)$

$$D_3 = (K_8 - K_2\omega_c^2) \qquad\qquad \ldots(4.119)$$

Using Eqs. 4.113, 4.111 and 4.114 in 4.107, one gets

$$K_R K_S D_6 + K_a K_s D_7 = D_5 \qquad\qquad \ldots(4.120)$$

where $\qquad\qquad D_4 = \omega_c^2(C_3 - 2\varsigma_c\omega_c) \qquad\qquad \ldots(4.121)$

$$D_5 = D_4\omega_c + 2\varsigma_c K_6 - K_3\omega_c \qquad\qquad \ldots(4.122)$$

$$D_6 = K_4\omega_c - 2\varsigma_c K_7 \qquad\qquad \ldots(4.123)$$

$$D_7 = K_5\omega_c - 2\varsigma_c K_8 \qquad\qquad \ldots(4.124)$$

Equations 4.117 and 4.120 give

$$K_a K_s = \frac{D_6 K_7 - D_2 D_6}{K_7 D_7 - D_3 D_6} \qquad\qquad \ldots(4.125)$$

$$K_R K_s = \frac{D_2 D_7 - D_3 D_5}{K_7 D_7 - D_3 D_6} \qquad\qquad \ldots(4.126)$$

K_s is then obtained by making the steady state value of closed loop gain unity.

Denoting steady state values of G_y and G_r by G_{yss} and $G_{r_{ss}}$ respectively and noting that $G_{a_{ss}} = 1.0,$ the equation 4.94 gives (for unity closed loop gain)

$$\left. \frac{f_y}{f_{yd}} \right|_{ss} = \frac{K_s G_{yss}}{1 + K_R K_s G_{rss} + K_a K_s G_{yss}} = 1.0$$

or

$$K_s = K_a K_S + K_R K_S \frac{G_{rss}}{G_{yss}} + \frac{1}{G_{yss}}$$

$$= K_a K_s + \frac{K_R K_S}{u} + \frac{1}{G_{yss}} \qquad\qquad ... (4.127)$$

Using Eqs. 4.127, 4.125 and 4.126, all the three gains can be computed.

The algorithm for computing the gains is given in the box. Fig. 4.20 shows a step response for the control gains obtained by using this algorithm for the following parameter values.

$$y_v = 0.080285 \qquad n_v = 0.003099 \qquad y_\delta = 1.991701 \qquad n_\delta = 1.045947$$

$$c = 3.118 \qquad K_s = -1.045947 \qquad K_R = 28.4149 \qquad K_a = 1.2130$$

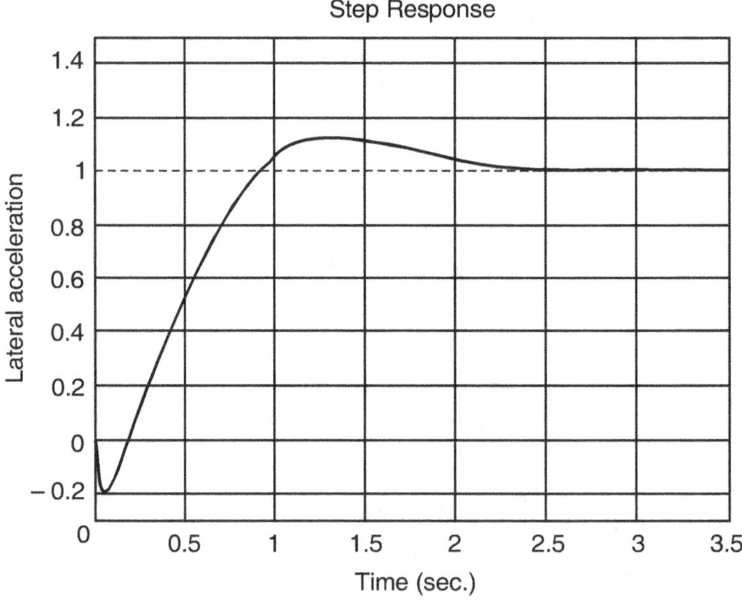

Fig. 4.20 Step Response of Latax Control system

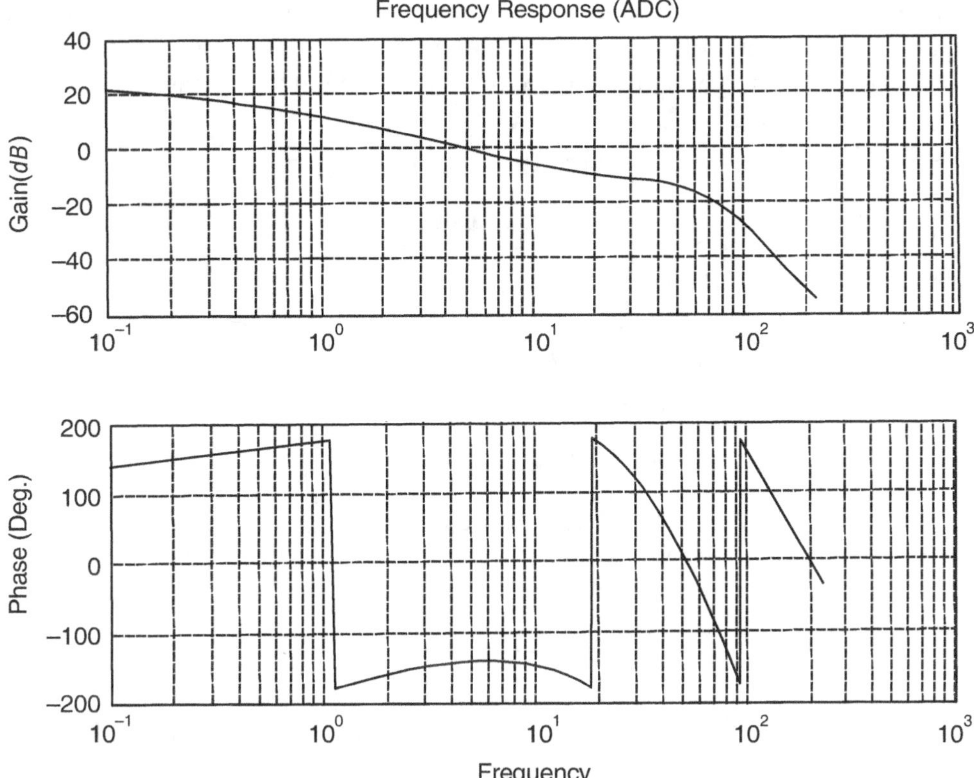

Fig. 4.21 Gain and Phase Vs Frequency

Fig. 4.21 shows the gain and phase plots against frequency which shows that the system is having stability margins as follows:

$$-\text{ve gain margin} = -11.6 \text{ dB}$$

$$\text{positive gain margin} = 10.15 \text{ dB}$$

$$\text{phase margin} = 37°$$

We have seen for attitude control system that the stability margins will degrade if we incorporate the effect of sample and hold and computational delays and the control gains will need some tuning to get better margins. Following a similar approach as for attitude control, we can incorporate the approximate effect of digital control lags by a first order transfer function $\dfrac{d}{s+d}$ where $\dfrac{1}{d} = D$ gives the approximate lag added by digital control system.

The characteristic equation will now become a fifth order equation and can be represented by

$$(s + A)\,(s^2 + 2\varsigma_c \omega_c s + \omega_c^2)\,(s^2 + 2\varsigma \omega s + \omega^2) = 0 \qquad \qquad \dots (4.128)$$

By equating coefficients of like powers of s, one can solve for the products $K_a K_s$ and $K_R K_S$ and again K_s can be obtained using Eq. 4.127 and K_R and K_a can then be obtained. The algorithm in the box also gives the steps required for computing the gains incorporating the digital control system lags.

ALGORITHM FOR COMPUTING GAINS

LATAX CONTROL

Assume

$$y_r = 0$$

$$a_0 = u\, n_v + y_v n_r$$

$$a_1 = -(y_v + n_r)$$

$$a_2 = c\, n_\delta + y_\delta$$

$$a_3 = n_v y_\delta - y_v n_\delta$$

$$K_1 = \omega_a^2 + 2\varsigma_a \omega_a a_1 + a_0$$

$$K_2 = (c\, n_\delta + y_\delta)\omega_a^2$$

$$K_3 = 2\varsigma_a \omega_a . a_0 + a_1 \omega_a^2$$

$$K_4 = n_\delta \omega_a^2$$

$$K_5 = \omega_a^2 [c\, a_3 - y_\delta n_r]$$

$$K_6 = a_0 \omega_a^2$$

$$K_7 = a_3 \omega_a^2$$

$$K_8 = K_7 u$$

$$d_0 = 2\varsigma_a \omega_a - 2\varsigma_c \omega_c + a_1$$

$$d_1 = \omega_c^2 + 2\varsigma_c \omega_c \cdot d_0$$

$$d_2 = (K_1 - d_1)\omega_c^2 - K_6$$

$$d_3 = K_8 - K_2 \omega_c^2$$

$$d_4 = \omega_c^2 d_0$$

$$d_5 = 2\varsigma_c K_6 - \omega_c^2 (K_3 - d_4)$$

$$d_6 = K_4 \omega_c - 2\varsigma_c K_7$$

$$d_7 = K_5 \omega_c - 2\varsigma_c K_8$$

GAINS WITHOUT DIGITAL DELAYS

$$den = K_7 d_7 - d_3 d_6$$

$$SKA = (d_5 d_7 - d_2 d_6)/den$$

$$SKR = (d_2 d_7 - d_3 d_5)/den$$

$$K_s = SKA + SKR/u + a_0/(u * a_3)$$

$$K_R = SKR/K_s$$

$$K_a = SKA/K_s$$

GAINS WITH EFFECT OF DELAYS

$$d_8 = -K_3 + d_4 + 2\varsigma_c \omega_c (K_1 - d_1)$$

$$d_9 = d_8 - [K_6 - \omega_c^2 (k_1 - d_1) + 2\varsigma_c \omega_c d_8]/d$$

$$d_{10} = -K_6 - \omega_c^2 (k_1 - d_1) - \omega_c^2 d_8/d$$

$$den = K_4(K_8 - K_2\omega_c^2) - K_7(K_5 - 2\varsigma_c \omega_c K_2)$$

$$SKA = (d_{10}K_4 - d_9 K_7)/den$$

$$SKR = [d_9(K_8 - dK_2) - d_{10}(K_5 - 2\varsigma_c \omega_c K_2)]/den$$

$$K_s = SKA + SKR/u + a_0/(ua_3)$$

$$K_R = SKR/K_s$$

$$K_a = SKA/K_s$$

4.4.7 An Alternate Configuration of Latax Control System

Some of the authors (Ref. 9,10,11) use a different configuration (which is sometimes called as 3 loop autopilot) for tactical missiles or homing missile. The configuration is shown in Fig. 4.22.

In this figure $\qquad\qquad \gamma_d = F_{zd}/u$ $\qquad\qquad\qquad\qquad$ ... (4.129)

And $\qquad\qquad\qquad \gamma = F_z/u$ $\qquad\qquad\qquad\qquad\qquad$ ... (4.130)

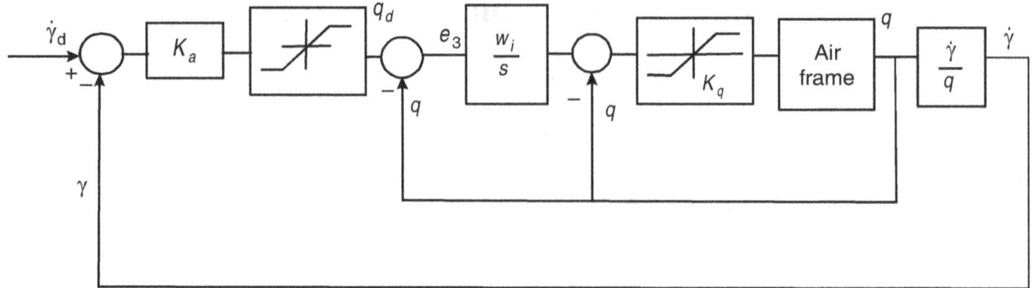

Fig. 4.22 3-loop latax autopilot

The inner two loops constitute a body rate demand loop. The authors claim that this configuration gives a direct control over the vehicle maximum body rate which can be controlled by adjusting the limiter after K_a block. It is also said that such a control over maximum body rate is necessary in case of homing missiles where the high body rates could lead to increased bore sight errors of the seeker due to coupling between the vehicle body rate and seeker stabilization loop and sometimes the seeker may lose the target if the bore sight error goes beyond the field of view.

Let the inner loops be redrawn as shown in Fig 4.23

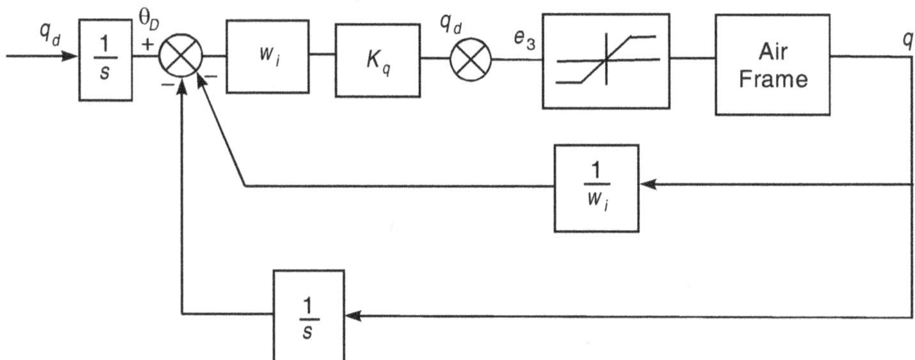

Fig. 4.23

It can thus be seen that the inner two loops can be conceptualized as an attitude control system which has been already discussed in detail. In this block diagram, we can identify control gains as

$$K_s = w_i K_q \qquad \qquad \text{... (4.131)}$$

$$K_R = \frac{1}{w_i} \qquad \qquad \text{... (4.132)}$$

A closer look at the two latax autopilot configurations (*i.e.*, Fig.4.19 and Fig. 4.22) leads to following observations:

1. Configuration of Fig. 4.19 (say, two-loop autopilot) requires sampling for feedback signals together and control calculations are also done at the same time. Hence, the specifications of ς_c and ω_c will automatically restrict the maximum body rate, sampling and computational delays will restrict increase in the autopilot bandwidth.(factors constraining the autopilot bandwidth are discussed in chapter 3).

2. Configuration of Fig. 4.22 (three Loop autopilot) has inner two loops independent of the outermost loop. For the same actuator and rate sensor transfer functions, it will be possible to design the rate demand loop with a higher closed loop bandwidth by using much higher sampling rate and lesser computational delays.

The higher closed loop bandwidth for the inner loops or rate demand loop will enable achieving higher bandwidth for the main latax loop or the outermost loop since it will be possible to allow higher rate build up using inner loops which is essential for achieving higher bandwidth.

It thus appears that the 3-Loop autopilot configuration will be more suitable for tactical missiles where higher speed of response is very much desirable.

4.4.8 Gain Schedule Adaptation

We have discussed the methodology for obtaining the desirable control gains at a given flight instant assuming that the vehicle parameters remain more or less constant over a short period of time. The study is then repeated at regular intervals of time to cover the entire flight trajectory and the gains schedule can thus be obtained. One would naturally feel that these gains can be stored as a function of time in tabular form. At any flight instant the control gains can then be interpolated and used for control calculations. This gives the gains as function of time.

I. Gains as a function of time

This scheme can be used in case of launch vehicles where the vehicle trajectory is very well defined. The gains schedule can be designed for nominal performance of the vehicle propulsion.

In actual practice, the vehicle state at any flight instant can be different from the nominal state due to following factors:

1. Thrust time characteristics can have a dispersion within the specification band.

2. Vehicle weight can be different from the nominal value within the specified band.

3. And vehicle drag can be different within the specified band.

The control system stability margins, therefore, will be different from the nominal stability margins. It will be desirable that the designer verifies the margins for worst case combinations such as:

1. High thrust, low drag and low weight and

2. Low thrust, high drag and high weight.

Further, one may study the performance with static margin dispersions within the specified band for each of the above cases.

The stability margins with these extreme cases could be lower than the desirable specifications and one may accept degradation upto 4db gain margin and 25° phase margin and also check the vehicle performance in 6 degree of freedom trajectory simulation.

II. Gains schedule as function of velocity or dynamic pressure

Some designers have preferred to schedule the gains as a function of vehicle relative velocity or dynamic pressure.

Since aerodynamic force is a function of dynamic pressure, scheduling gains as a function of dynamic pressure would give desirable features in case of aerodynamic control. The gain schedule as a function of velocity can suffer at higher altitudes where velocity can be high but dynamic pressure can be quite low due to low air density at higher altitudes.

Both the parameters, *i.e.*, velocity and dynamic pressure are really not adequate to make the gains fully adaptive. For example, during initial phase of flight, the vehicle velocity and dynamic pressures are low but the inertial properties of the vehicle such as mass, moment of inertia and distance of centre of gravity from control force location will be quite different than in case of coast phase after propellant burn-out even though dynamic pressure may be same at two instants. Thus, the gains will be multi-valued function of dynamic pressure and is not a fully satisfactory solution.

III. Generalised adaptation of the gains schedule (Ref. 12)

A look at the gains computation scheme makes it clear, that the vehicle transfer function depends on the following parameters:

(*i*) Inertial properties of the vehicle such as mass, moment of inertia and centre of gravity of the vehicle.

   These can be fairly represented as a function of time assuming a nominal propellant burn rate.

(*ii*) Air density of the atmosphere (ρ) and velocity of sound

   These are functions of vehicle altitude from sea surface and can be obtained from standard atmosphere models.

(*iii*) Aerodynamic coefficients for the vehicle such as $C_{N\alpha}, C_p, C_{N\delta}, C_{l\delta}$ etc.

   These are mainly the functions of mach number. The mach number can be obtained by knowing the relative velocity of the vehicle and velocity of sound (v_s) which is again function of altitude and can be obtained from the standard atmosphere model.

The inertial navigation system (INS) available onboard the vehicle gives the data regarding t (time), h (altitude) and v (relative velocity) of the vehicle. Then following steps can be adopted for computing the gains on board the vehicle at a required time interval:

1. Get t, h, v from *INS*
2. Find mass, I_y, *CG* and I_{xx} either from a table as a function of time or as pre-fitted polynomial of time (t)
3. Find ρ and v_s

   This can be obtained as a function of altitude from standard atmosphere and can be stored as a table or pre-fitted polynomials of altitude (h).

   Compute mach number and aerodynamic coefficients.
4. Compute y_v, y_δ and n_v, n_δ. Assume $n_r \approx 0$
5. Compute control gains K_s, K_R and Ka using algorithm given in previous sections.
6. Verify stability margins for the entire flight profile
7. Design a suitable filter/compensator to get the required augmentation to margins and attenuation at higher frequencies to cater for structural flexibility, propellant sloshing, engine inertia etc.

The present day onboard computers have made the above scheme quite feasible and has been successfully implemented in case of a surface to surface missile. It may be noted that the trajectory of surface to surface missile is not known in advance and will be decided based on the actual ground range of the target from the missile launch point. It will be very difficult to adapt the gain schedule in any other way for a satisfactory solution. The above scheme has been found to be giving completely satisfactory performance for any given range trajectory selected within the capability of the missile and can thus be said to be a generalized adaptation scheme for control gains schedule.

The inertial properties of the vehicle depend on the propellant burn rate and hence the actual values will be different from the values obtained from table or polynomial in 't'. The control system stability margins are expected to absorb this dispersion in data. Further, the actual burn out of the vehicle needs to be sensed and parameters can be accordingly chosen by providing the required logical statements in computer program.

The vehicles having liquid propulsion will have a thrust cut off based on the required velocity for a given range trajectory. In such cases, there will be left over propellant inside the tanks. Moreover, the left over propellants would shift towards the head end of the propellant tanks due to drag deceleration. This will introduce jumps in vehicle centre of gravity and moments of inertia. This needs to be appropriately modelled based on the time of thrust cut off and quantity of left over propellant.

The models for ρ and v_s which can be used in the design of control system are given below in the box. The inertial parameters and aerodynamic coefficients are vehicle dependant and hence not given here.

Models for Air Density (ρ) and Sound Velocity (V_S)

Altitude Range	Parameter
$h < 11$ Km	$\rho = 1.151943 - 0.1014426\,h + 0.0027712\,h^2$
$11 < h \leq 25$ Km	$\rho = 1.007382 - 0.0732885\,h + 0.0013867\,h^2$
$25 < h \leq 32$ Km	$\rho = 0.3934705 - 0.0221768\,h + 0.000322\,h^2$
$32 < h \leq 43$ Km	$\rho = 0.1578482 - 0.0072362\,h + 8.47962 * 10^{-5}\,h^2$
$43 < h \leq 60$ Km	$\rho = 2.8624*10^{-2} - 9.5336*10^{-4}\,h + 8.0464 * 10^{-6}\,h^2$
$60 < h \leq 80$ Km	$\rho = 4.6289*10^{-3} - 1.1638*10^{-4}\,h + 7.3447 * 10^{-7}\,h^2$
$h \leq 16$ Km	$Vs = 350.0907 - 4.12537\,h$
$16 < h \leq 49$ Km	$Vs = 258.8023 + 1.53088\,h$
$49 < h \leq 75$ Km	$Vs = 429.972 - 1.956\,h$
$75 < h \leq 80$ Km	$Vs = 282.9$

IV. A further refinement to generalized adaptation of gains schedule

As stated in above paragraphs, the Generalized adaptation scheme of gains schedule is found to be giving satisfactory performance for the entire operational zone of the missile. However, it is to be noted that the gains are worked out onboard based on the estimated parameters of the vehicle both inertial (such as mass, moment of inertia and distance of CG from nose tip of the vehicle) and aerodynamic (such as $C_{N\alpha}$, $C_{N\delta}$ and distance of Centre of Pressure CP from nose tip). If during flight the actual parameters match with the estimated parameters, the control system stability margins would be as per prediction. But if the parameters differ from the estimated parameters due to either the propulsion variation during the flight or aerodynamic angle of attack profile, the stability margins during flight can defer or degrade from the predicted values. This deviation of stability margins can now a days (thanks to the very high computational power available on board) be minimized by devising more sophisticated schemes for estimating the vehicle parameters in case the degradation in margins is beyond tolerable limits for the expected worst case of dispersions in vehicle parameters.

(a) Inertial Parameters

The instantaneous values of inertial parameters (mass, CG and moment of Inertias) depend on the burnt out propellant mass during stage burning. One may pre compute these parameters and tabulate at every 10% of propellant burning. The actual inertial parameters at any instant can then be found out by interpolation after estimating the consumed propellant till that instant. The consumed propellant is estimated as follows.

Repeat the following steps at regular interval Δt.

(*i*) Read forward acceleration from INS and pass it through a low pass filter to get a smoothened value.

(*ii*) Find r, V_s from atmosphere model for the height obtained from INS and estimate Mach no., drag coefficient and drag force,

$$D = \frac{1}{2}\rho V^2 C_D QS$$

(*iii*) Thrust $T = D + Ma$; M– Vehicle instantaneous mass

$\qquad\qquad\qquad\qquad\qquad a$ – Acceleration sensed by INS (without gravity compensation)

$$m = \frac{T}{I_{S_P}};\qquad\qquad I_{S_P} - \text{Specific Impulse (assumed constant)}$$

(*iv*) Consumed propellant mass

$$M_{PC} = M_{PC} + m\Delta t$$

If computed M_{PC} is greater than the stored propellant mass, Limit M_{PC} to stored value.

(*v*) Vehicle Mass = $M_o - M_{PC}$

(*vi*) Interpolate from table the vehicle parameters cg, Iy, Ix for this mass.

In the above process, I_{S_P} is assumed constant and propulsion thrust variation is assumed to be due to environmental temperature, propellant composition etc. The uncertainty of drag value due to C_D estimation error is ignored as it is a small fraction of the actual thrust.

(b) Aerodynamic Parameters

The generalized adaptation scheme of gains schedule described earlier assumes aerodynamic parameters (C_{N_α}, C_{N_δ}, C_P etc.) to be only a function of Mach number. Further refinement can be done to it by estimating these parameters as a function of Mach number and angle of attack. The angle of attack can be obtained from vehicle altitude and flight path angle using INS information about direction cosine matrix and velocity. The angle of attack so obtained should, however, be passed through an appropriate low pass filter before aerodynamic coefficients are interpolated from two dimensional tables for each parameter (e.g. rows for mach number and columns for angle of attack).

4.4.9 Effect of Sample and Hold and Computational Delay

4.4.9.1 Z-domain Transfer Function

In section 4.1.1, the system equations have been put in a standard form. Writing the time varying nature of the states more explicitly and adding output equation, the systems equations can be written as:

$$\dot{X}(t) = FX(t) + GU(t) \qquad\qquad \text{... (4.133)}$$

$$Y(t) = HX(t)$$

The control variable U is a function of input commands and measured states of the system. In digital control systems, the measurements are done at sampling instants and no information is available about states between the two sampling instants. The controller processes these measurements and works out the commands to be issued to the actuators. The computer takes finite time to process this information before the command is issued to actuator. This is the computational delay. The actuator command continues till the controller updates the commands based on information available at the next sampling time. We need a tool to find stability margins of the system incorporating the effect of sample and hold and computational delay. The following development is based on Ref. 13.

The dynamic equations (4.133) are modified to account for computational delay as

$$\dot{X}(T) = FX(T) + GU(T - \lambda)$$

$$Y(T) = HX(T) \qquad\qquad \text{... (4.134)}$$

Solving this equation over one sampling interval gives

$$X(KT + T) = e^{FT} X(KT) + \int_{KT}^{KT+T} e^{F(KT+T-\tau)} GU(\tau - \lambda)d\tau \qquad \text{... (4.135)}$$

Substituting $\qquad \eta = KT + T - \tau \qquad\qquad \text{... (4.136)}$

$$d\eta = -d\tau$$

$$X(KT + T) = e^{FT} X(KT) + \int_{0}^{T} e^{F\eta} GU(KT + T - \lambda - \eta)d\eta \qquad \text{... (4.137)}$$

Let $\qquad\qquad \lambda = T - m \qquad\qquad \text{... (4.138)}$

The variations of η from 0 to T corresponds to variation of time from $(KT + T)$ to KT. For the duration $0 < \eta < m$, i.e., for $(KT + T) > t > (KT + T - m)$ or $(KT + T) > t > (KT + \lambda)$, the updated control based on the input command and missile state sampled at $t = KT$ is available. This control is designated as $U(KT)$.

For the duration $m < \eta < T$ *i.e.*, for $(KT + \lambda) > t > KT$, the updated control is not available due to computational delay and old value of control variable is used. This control is designated as $U(KT - T)$.

Now, splitting the integral in Eq. (4.137)

$$X(KT + T) = e^{FT} X(KT) + \int_0^m e^{F\eta} Gd\eta\, U(KT) + \int_m^T e^{F\eta} Gd\eta\, U(KT - T)$$

OR $\qquad X(KT + T) = \Phi X(KT) + \Gamma_1 U(KT - T) + \Gamma_2 U(KT) \hfill \ldots (4.139)$

where $\qquad\qquad \Phi = e^{FT} \hfill \ldots (4.140)$

$$\Gamma_1 = \int_m^T e^{F\eta} Gd\eta \hfill \ldots (4.141)$$

$$\Gamma_2 = \int_0^m e^{F\eta} Gd\eta \hfill \ldots (4.142)$$

To eliminate $U(KT - T)$ from the right hand side of Eq. (4.139), define a new state

$$X_{n+1}(KT) = U(KT - T)$$

where n is the dimension of the original system. Then, the new system equations can be written as

$$\begin{bmatrix} X(KT + T) \\ X_{n+1}(KT + T) \end{bmatrix} = \begin{bmatrix} \Phi & \Gamma_1 \\ 0 & 0 \end{bmatrix} \begin{bmatrix} X(KT) \\ X_{n+1}(KT) \end{bmatrix} + \begin{bmatrix} \Gamma_2 \\ 1 \end{bmatrix} U(KT)$$

and $\qquad\qquad Y(KT) = [H \quad 0] \begin{bmatrix} X(KT) \\ X_{n+1}(KT) \end{bmatrix} \hfill \ldots (4.143)$

This gives the standard form of system equations in discrete domain *i.e.*,

$$X(K + 1) = \Phi X(K) + \Gamma U(K) \hfill \ldots (4.144)$$

$$Y(K) = HX(K)$$

Taking the Z transform, the Z-domain transfer function is given by

$$\frac{y(z)}{u(z)} = H[zI - \Phi]^{-1}\Gamma \hfill \ldots (4.145)$$

For computation of stability margins, one needs open loop transfer function *GH*. Fig. 4.24 shows a typical block diagram of a feedback control system.

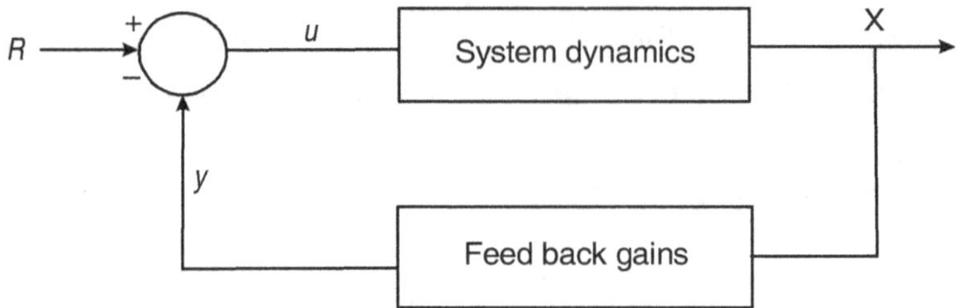

Fig. 4.24

If we define the control u and output y signal as shown in the figure, by using appropriate matrix H, equation 4.145 gives the required open loop gain for computing the stability margin.

For example, for an attitude control system with state vector $\left[\theta, \dot\theta, \delta, \dot\delta\right]^T$ and control law

$\delta = K_s(\theta_c - \theta - K_R\dot\theta)$, we can define output matrix as

$$y = \theta + K_R\dot\theta = \begin{bmatrix} 1 & K_R & 0 & 0 \end{bmatrix}\begin{bmatrix} \theta \\ \dot\theta \\ \delta \\ \dot\delta \end{bmatrix} \qquad \text{... (4.146)}$$

The forward gain K_s can be used as a multiplying factor to the gain plot.

4.4.9.2 Computation of Φ, Γ_1, Γ_2 etc.

We have following equations for computing Φ, Γ_1, Γ_2 :

$$\Phi = e^{FT}$$

$$\Gamma_1 = \int_m^T e^{F\eta}G d\eta$$

$$\Gamma_2 = \int_m^T e^{F\eta}G d\eta$$

Substituting $\sigma = \eta - m$, eqn. for Γ_1 can be written as

$$\Gamma_1 = e^{Fm}\int_0^{T-m} e^{F\sigma}d\sigma\, G \qquad \text{... (4.147)}$$

Defining for any positive scalar a,

$$\Phi(a) = e^{Fa} \qquad \text{... (4.148)}$$

$$\Psi(a) = \frac{1}{a}\int_0^a e^{F\sigma}d\sigma \qquad \text{... (4.149)}$$

One can write

$$\Gamma_1 = \Phi(m)\,.\,\Psi(T-m)\cdot(T-m)\cdot G \qquad \ldots(4.150)$$

$$\Gamma_2 = \Psi(m)\,.\,m\,.\,G \qquad \ldots(4.151)$$

We can write $\Phi(a)$ as

$$\Phi(a) = I + \sum_{k=1}^{\infty} \frac{F^k a^k}{k!} \qquad \ldots(4.152)$$

and writing the integrand of Eq. 4.149 in series form and integrating term by term, one gets

$$\Psi(a) = \sum_{k=0}^{\infty} \frac{F^k a^k}{(k+1)!} \qquad \ldots(4.153)$$

Rewriting $\Phi(a)$ as

$$\Phi(a) = I + \sum_{k=1}^{\infty} \frac{F^k a^k}{k!}$$

Substituting $k = j + 1$, we can write

$$\Phi(a) = I + \sum_{j=0}^{\infty} \frac{F^{j+1} a^{j+1}}{(j+1)!}$$

$$= I + \sum_{j=0}^{\infty} \frac{F^j a^j}{(j+1)!} Fa$$

$$= 1 + \Psi(a)\cdot a \cdot F \qquad \ldots(4.154)$$

Thus, $\Phi(a), \Gamma_1, \Gamma_2$ can be easily obtained once $\Psi(a)$ is known.

Ref. 13 has suggested that following form for $\Psi(a)$ gives good numerical properties.

$$\Psi(a) = I + \frac{Fa}{2}\left(I + \frac{Fa}{3}\left(I + \frac{Fa}{4}\left(I + \frac{Fa}{5}\left(\ldots + \frac{Fa}{N-1}\left(1 + \frac{Fa}{N}\right)\right)\right)\right)\right) \qquad \ldots(4.155)$$

The detailed procedure for incorporating the effect of sample and hold and computational lag can now be summarized as follows:

1. Take A, B, C, D matrices (or F and G matrices if D matrix is absent).

2. Compute F and G matrices.

3. Take appropriate H matrix for loop gain calculation.

4. Compute Φ, Γ_1 *and* Γ_2 if computational delay is present. (Φ and Γ if delay is absent).

5. Compute z transform of loop gain *i.e.*,

$$\frac{y(z)}{u(z)} = H[zI - \Phi]^{-1}\Gamma$$

6. Transform to w domain using

$$z = \frac{1+w}{1-w}$$

7. If compensator is present, multiply the loop gain transfer function by the compensator transfer function in w - plane.

8. Compute the frequency response *i.e.*, gain and phase versus frequency or use other continuous linear control system analysis techniques.

4.4.9.3 Transformation to W-domain

In order to use familiar techniques in s-domain, the transfer function in z-domain (eq. 4.145) is transformed to w-domain using the transformation.

$$z = \frac{1+w}{1-w} \qquad \qquad ...\,(4.156)$$

To appreciate the correlation between w-domain frequency and s-domain frequency, we can write

$$z = \frac{1+w}{1-w}$$

or

$$\frac{z+1}{z-1} = \frac{1}{w}$$

or

$$w = \frac{z-1}{z+1} = \frac{e^{Ts}-1}{e^{Ts}+1}$$

$$= \frac{e^{\frac{Ts}{2}} - e^{-\frac{Ts}{2}}}{2j} \times \frac{2j}{e^{\frac{Ts}{2}} + e^{-\frac{Ts}{2}}}$$

Writing $s = j\omega$ and noting that

$$\sin\frac{T\omega}{2} = \frac{e^{\frac{jT\omega}{2}} - e^{-\frac{jT\omega}{2}}}{2j}$$

and

$$\cos\frac{T\omega}{2} = \frac{e^{\frac{jT\omega}{2}} + e^{-\frac{jT\omega}{2}}}{2}$$

we get

$$w = j\tan\frac{\omega T}{2} \qquad\qquad \text{... (4.157)}$$

It is known that for small angles $\tan\theta \approx \theta$. Therefore, for small frequency, the w-domain frequency is simply the scaled s-domain frequency (with a scale factor $=\frac{T}{2}$). However, at higher frequencies, the w-domain frequency is a tangent function of $\frac{\omega T}{2}$ and approaches infinity as $\frac{\omega T}{2}$ approaches 1.57 radians or 90°.

This condition occurs at half the sampling frequency.

Since

$$f_s = \frac{1}{T}$$

At half sampling frequency

$$\omega = 2\pi\left(\frac{f_s}{2}\right) = \pi f_s$$

Then

$$\frac{\omega T}{2} = \pi f_s \frac{T}{2} = \frac{\pi}{2} \qquad\qquad \text{... (4.158)}$$

This fact needs to be kept in mind since w-domain frequencies will be very large as we approach half the sampling frequency and the numerical computations will be unreliable.

4.4.9.4 *Frequency Aliasing*

It will be interesting to know what happens if the s-domain frequency is higher than half the sampling frequency or $\frac{\omega T}{2} > \frac{\pi}{2}$.

Eq. 4.157 gives *w*-domain frequency negative which is rather confusing. But, if we have a system with a low damped dynamics and if we vary the frequency starting from lower than half the sampling frequency and going up beyond half the sampling frequency, we will encounter the effect of frequency aliasing. This is a real scenario in case of flight control systems where we have a number of low damped structural mode frequencies which vary during flight due to burning of the propellant and designer has to cater for it. Several text books have explained the effect of aliasing due to sampling. In bode plot, we will observe the aliasing effect as follows:

For illustration, let the sampling frequency be 50Hz and *f* be the frequency of low damped dynamics. Then, the peak in gain versus frequency plot in *w*-domain will be observed as follows:

As the frequency approaches sampling frequency, we will get peak in Bode plot at low frequency (*i.e.*, difference of sampling frequency and actual frequency). This will pose very difficult problem for attenuating the mode and get adequate stability margins. The aliasing is the effect of sampling. Hence, if there are dynamic modes with frequencies nearer the sampling frequencies, one must attenuate them adequately using analog filters and then only the sensor signal should be sampled for control calculations. This will avoid significant peaks occurring in lower frequency range which otherwise would deteriorate the stability margins.

TABLE 4.4

Hz	*S-domain frequency (rad)*	*W-domain frequency*
10	$2\pi \times 10$	$\tan 2\pi \times 10 \cdot \dfrac{T}{2} = \tan 10\pi T$
15	$2\pi \times 15$	$\tan 15\pi T$
20	$2\pi \times 20$	$\tan 20\pi T$
25	$2\pi \times 25$ (half the sampling freq.)	∞ since $25\pi T = \dfrac{\pi}{2}$
30	$2\pi \times 30$	$\tan(fs - 30)\pi T = \tan 20\pi T$
40	$2\pi \times 40$	$\tan(fs - 40)\pi \cdot T = \tan 10\pi T$
50	$2\pi \times 50$ (sampling freq.)	0
60	$2\pi \times 60$	$\tan(60 - fs)\pi T = \tan 10\pi T$
75	$2\pi \times 75$ (1½ times sampling freq.)	$\infty \quad 25\pi T = \dfrac{\pi}{2}$
80	$2\pi \times 80$	$\tan(2fs - 80) = \tan 20\pi T$
	and so on	

4.4.9.5 Updatement of Compensator Z-transfer Function due to Change in Sampling Interval

Present day flight control systems are all digital autopilots and digital computer is used for all data processing needs. Due to this, the designers are able to use very complex algorithms for computing the autopilot gains so that the autopilot gives good performance in the entire zone of missile operation. The control law uses outputs from various sensors sampled at a given time instant. We have seen earlier that the sampling interval (or the fixed interval between two successive samples) causes a phase lag and degrades the stability margins. Also, the computer takes a finite time to process all the data and finally compute the actuator command. This computation delay further degrades the stability margins. Hence, the designer would like to minimize the length of sampling interval. The minimum length however will be decided by the time required for the computer to sample all the sensors, complete the data processing and send the actuator commands.

Every one is familiar with the great advancement in digital computer technology. Computation speed which was unthinkable 15-20 years before has now been possible. The onboard computers have since been being updated and shorter sampling intervals are being adopted.

This necessitates appropriate changes in the autopilot compensators and/or filters whose z-transfer function is based on the sampling interval. As the compensator/filters are designed in w-domain by first obtaining the frequency response of the system in w-domain, we intend to change the w-domain filter by keeping s-domain filter same.

Let the filter be indicated in w-domain for a sampling interval T and v-domain for a sampling interval of T'. Then from Eq. 4.157, one can write

$$w = j \, \tan \frac{\omega T}{2}$$

And
$$v = j \, \tan \frac{\omega T'}{2} \quad \text{where s} = j\,\omega$$

Or assuming small angle approximations (the stability margins occur at relatively low frequencies):

$$w = \frac{T}{T'} v$$

Therefore to get a new filter for sampling interval T' from existing filter for sampling interval T :

1. Write the w-domain transfer function of compensator/filter

2. Replace w by $\dfrac{T}{T'} v$

3. Convert v-domain transfer function to z-domain transfer function using

$$v = \frac{z-1}{z+1}$$

4. Check the stability margins in v-domain.

Table gives the conversion formulae from w domain to z domain and vice versa as a ready reference for first order and second order compensator segments which are normally used as a part of total compensator.

Fig 4.25 shows the w-domain and v-domain plots for a typical unsymmetrical (or modified) notch filter when sampling time is changed from 10 msecs to 5 msecs. Note that the notch frequency has shifted by a factor of 2.

Caution: Since the process of conversion from W to Z etc. requires normalization of terms like $\frac{z-1}{z+1}$, one should not attempt to find formulae for first order by putting zero value to coefficients a_2, b_2 of 2nd order terms.

W to Z Transformation

compensator order	w-domain transfer function	z-domain transfer function
Ist	$\dfrac{a_0 + a_1 w}{b_0 + b_1 w}$	$\dfrac{(a_0 + a_1) + (a_0 - a_1)z^{-1}}{(b_0 + b_1) + (b_0 - b_1)z^{-1}}$
IInd	$\dfrac{a_0 + a_1 w + a_2 w^2}{b_0 + b_1 w + b_2 w^2}$	$\dfrac{(a_0 + a_1 + a_2) + 2(a_0 - a_2)z^{-1} + (a_0 - a_1 + a_2)z^{-2}}{(b_0 + b_1 + b_2) + 2(b_0 - b_1)z^{-1} + (b_0 - b_1 + b_2)z^{-2}}$

Z to W Transformation

compensator order	z-domain transfer function	w-domain transfer function
Ist	$\dfrac{c_0 + c_1 z^{-1}}{d_0 + d_1 z^{-1}}$	$\dfrac{(c_0 + c_1) + (c_0 - c_1)w}{(d_0 + d_1) + (d_0 - d_1)w}$
IInd	$\dfrac{c_0 + c_1 z^{-1} + c_2 z^{-2}}{d_0 + d_1 z^{-1} + d_2 z^{-2}}$	$\dfrac{(c_0 + c_1 + c_2) + 2(c_0 - c_2)w + (c_0 - c_1 + c_2)w^2}{(d_0 + d_1 + d_2) + 2(d_0 - d_2)w + (d_0 - d_1 + d_2)w^2}$

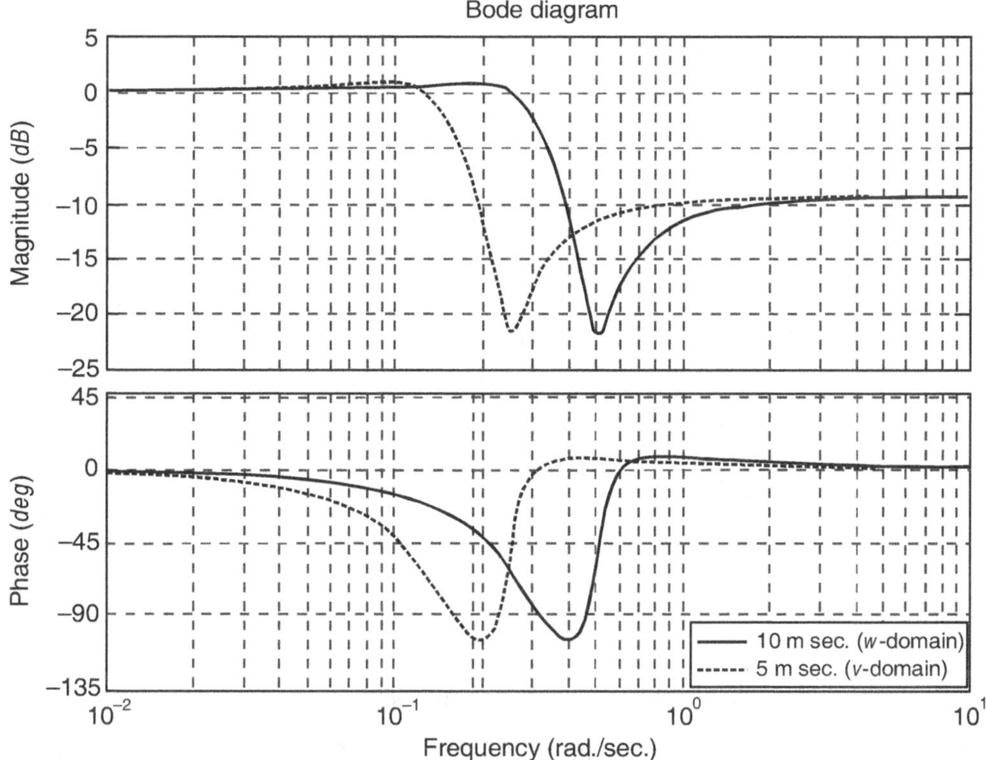

Fig. 4.25 Filter Frequency Response in *W*-domain and *V*-domain

4.5 PROPELLANT SLOSHING

Any one having an experience of carrying partially filled liquid containers would agree with the difficulty of carrying a partially filled container in comparison with a filled container of similar weight. This is because the liquid in a partially filled container gets into an oscillatory motion and hence causes an oscillatory force on the containers. It is said that there have been accidents of tankers partially filled with liquids due to heavy sloshing of the liquid resulting in overturning of the tankers.

Many launch vehicles use liquid propellant stages due to their high specific impulse. As the propellant gets depleted, there would be free volume available and with the disturbances or control forces acting on the vehicle, the propellant gets into an oscillatory motion called propellant sloshing. The frequency of the oscillatory motion and the sloshing mass depends on the acceleration, tank geometry and the depth of the liquid column. The frequencies, most of the time, are very near the control frequency and hence pose problems for stabilization.

The shapes of propellant tanks in case of small vehicles are circular cylinders with domes on either ends. In case of large vehicles such as Saturn, various configurations for tanks are used. Figure 4.26 shows some typical configurations such as clustered tanks or partitioned tanks.

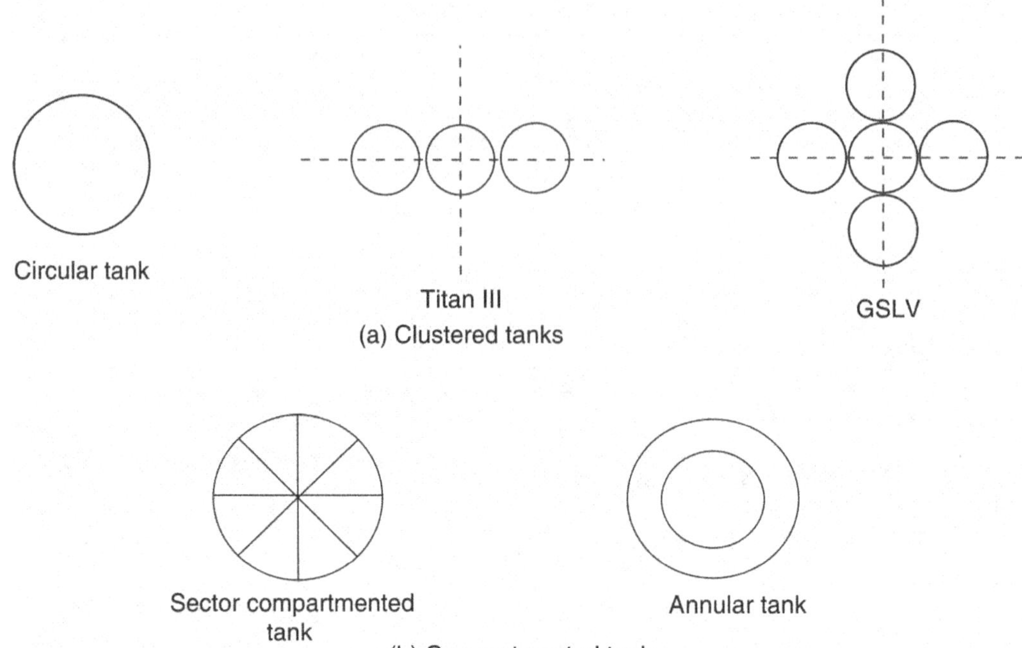

Fig. 4.26 Some typical tank configurations

The objective of clustering or partitioning of the tanks is to increase the frequency of oscillation and reduce the sloshing mass. This, however, results in many more slosh modes each one with a smaller slosh mass and higher frequency. In addition, the position of the slosh mass also changes resulting into altered disturbance force and moment acting on the vehicle.

The hydrodynamic motion of the liquid under acceleration field has been analysed in References 1, 14, 15 and equivalent pendulum models and spring mass models have been given. The generalized model given in Chapter two uses a pendulum model (which can also be easily modified to use spring mass model). This model is quite general in nature and can incorporate slosh pendulums in tanks clustered around main central tanks or excitation of the slosh motion in peripheral tanks due to roll motion etc. in addition to pitch plane or yaw plane motion.

The force and torque acting on the vehicle due to each slosh mass can be incorporated in the equations of motion by appropriately indicating the position of slosh mass by (y_i and z_i) for the ith tank in Equations 2.45 and 2.46 with 2.47 of chapter-2.

Following observations can be made for cylindrical tanks with circular cross section from the results given in Ref. 15.

1. The ratio of slosh mass to total fluid mass (M_s/M_p) increases as the ratio of fluid depth to tank radius (h/R) decreases indicating that the sloshing becomes more pronounced as the liquid depth decreases.

2. Slosh mass for the second mode is very small compared to first mode indicating that one slosh mode per propellant tank may be good enough for stability studies.

The natural damping to slosh motion due to viscous friction and tank wall friction is very low and hence various types of baffles are provided inside the tank to enhance the damping of the slosh modes as an effort to improve the slosh mode stability. Studies (Ref. 15) indicate that the liquid upto a depth of $1/4^{th}$ the tank diameter from the surface of undisturbed liquid surface only participates in the slosh motion. However, since the liquid column height keeps continuously changing due to depletion of propellant, the baffles are normally provided in the entire tank length at a uniform spacing distance. The main considerations in the baffle design are:

Minimum weight and volume penalty and high damping effect. Fig. 4.27 shows a typical ring baffle configuration.

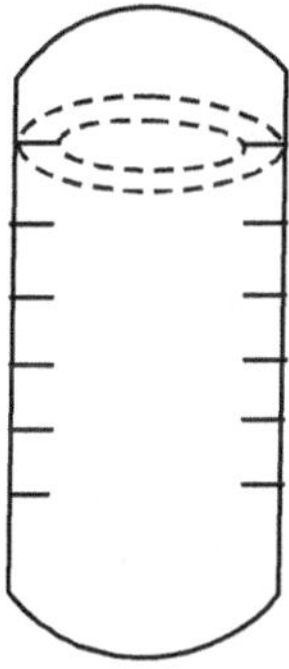

Fig. 4.27 Schematic ring baffle configuration

The slosh amplitude even when the mode becomes unstable in case of a flight vehicle grows very slowly with time. When the free surface comes near the next baffle zone the damping increases and the rising trend will get altered. Hence, one must see the effect of slosh mode instability on the vehicle attitude and angular rate and tolerate the instability if its effect on the vehicle state is insignificant. When the free surface reaches near the bottom of the tank, the mode itself may be found unstable. However, the sloshing mass will not have enough energy so as to affect the vehicle attitude or rate and hence may not be looked at seriously from vehicle control consideration. However, heavy sloshing during this region has a possibility of uncovering the tank pipe outlet and pressurized air getting into propellant feed line and creating propulsion disturbances.

4.6 STRUCTURAL FLEXIBILITY

The tactical missiles in general are designed for certain manoeuvrability and are stiffer vehicles compared to launch vehicles. The launch vehicles are highly flexible and their modal frequencies are quite low. For example, the first mode frequencies for launch vehicles are of the order of 1 to 2 Hz and even lower for very large vehicles whereas the first mode frequencies for some of the missiles are 15 to 20 Hz and even 50–60 Hz for smaller missiles.

The structural modes can get excited due to external forces such as thrust, control force, aerodynamic force, wind gusts etc. When the vehicle flexes, the local angle of attack along the vehicle length changes which alters the aerodynamic loading on the vehicle in such a way that the bending moments are further increased, thus, introducing structure-aerodynamic coupling. Control sensors are mounted on the vehicle. The sensor senses the vehicle state at sensor location which consists of the rigid body vehicle state plus the effect of vehicle flexing. The actuator input will thus be based on not only the rigid vehicle state but also the contribution due to vehicle flexibility. Thus, the control-structure coupling gets introduced into the control loop through sensors and actuators. The control system design needs to ensure that this aero-structure-control interaction is stable and does not lead to diverging oscillation. There have been instances of unstable structure control interaction giving rise to diverging oscillations and ending up with the structural failure of the vehicle. Hence, the stability analysis of flexible modes is an extremely important part of the flight control systems design and is discussed here in more details.

4.6.1 Models for Stability Analysis

In the present discussion, the vehicle is assumed symmetric in pitch and yaw planes and hence stability analysis is discussed only for pitch plane and the results are assumed to be applicable for yaw plane also. In case, the vehicle is not symmetric and the control design is also different for the two planes, the analysis will have to be carried out for yaw plane also separately.

Referring to Equations 2.120 to 2.126, one can see that one needs to evaluate several integrals to account for the distributed aerodynamic load along the vehicle length to make the model general to incorporate n-bending modes. These integrals are given in Table 4.5.

The details of matrix model to incorporate m-slosh modes and n-bending modes are given in Appendix 1.

Defining the vehicle state variable as

$$[\alpha\ \theta\ \dot{\theta}\ \delta\ \dot{\delta}\ \lambda_1\ \dot{\lambda}_1 \,..\lambda_m\ \dot{\lambda}_m\ q_1\ \dot{q}_1....q_n\ \dot{q}_n] \qquad ...(4.159)$$

one can write the state equation as

$$\dot{\bar{X}} = A\bar{X} + D\dot{\bar{X}} + BU$$

or

$$\dot{\bar{X}} = (I - D)^{-1}A\bar{X} + (I - D)^{-1}BU$$
$$= F\bar{X} + GU \qquad ...(4.160)$$

where U is a control variable which in the present case is the actuator command δ_c and is obtained as in following sections for different control laws:

TABLE 4.5: Integrals required for Linear Model

$I(1)$	$\displaystyle\int_0^L \frac{\partial C_{N\alpha}(l)}{\partial l}\, dl$
$I(2)$	$\displaystyle\int_0^L \frac{\partial C_{N\alpha}(l)}{\partial l}(l_{cg}-l)\, dl$
$I(3)$	$\displaystyle\int_0^L \frac{\partial C_{N\alpha}(l)}{\partial l}(l_{cg}-l)^2\, dl$
$I(3+i)$ $i=1$ to n	$\displaystyle\int_0^L \frac{\partial C_{N\alpha}(l)}{\partial l}\cdot\phi_i(l)\, dl$
$I(3+n+i)$ $i=1$ to n	$\displaystyle\int_0^L \frac{\partial C_{N\alpha}(l)}{\partial l}\cdot(l_{cg}-l)\phi_i(l)\, dl$
$I(3+2n+i)$ $i=1$ to n	$\displaystyle\int_0^L \frac{\partial C_{N\alpha}(l)}{\partial l}\frac{\partial\phi_i(l)}{\partial l}\, dl$
$I(3+3n+i)$ $i=1$ to n	$\displaystyle\int_0^L \frac{\partial C_{N\alpha}(l)}{\partial l}(l_{cg}-l)\frac{\partial\phi_i(l)}{\partial l}\, dl$
$I(3+4n+(i-1)n+j)$ $j=1$ to $n,\ i=1$ to n	$\displaystyle\int_0^L \frac{\partial C_{N\alpha}(l)}{\partial l}\phi_i(l)\phi_j(l)\, dl$
$I(3+4n+n^2+(i-1)n+j)$ $j=1$ to $n,\ i=1$ to n	$\displaystyle\int_0^L \frac{\partial C_{N\alpha}(l)}{\partial l}\phi_i(l)\frac{\partial\phi_j(l)}{\partial l}\, dl$

4.6.1.1 *Attitude Control Law*

The δ_c for attitude control law is given by

$$\delta_c = K_s(\theta_c - \theta_s - K_R\theta_s) \qquad\qquad \text{... (4.161)}$$

In smaller missiles, the angle and angular rate sensors are located at the same place or many times one signal is derived from another signal using mathematical processing. However, in large missiles and launch vehicles, more than one sensors are possible. Hence, as a general case,

$$\delta_c = K_s(\theta_c - \theta_s(PG) - K_R\theta_s(RG))$$

Where *PG* indicates the location of angle sensor and *RG* indicates the location of angular rate sensor.

Then δ_c can be written as

$$\delta_c = -H\overline{X} + K_s\theta_c \qquad \text{... (4.162)}$$

where *H* matrix is given in Appendix 4.1 (Also see the following sub-section)

Equation 4.151 can then be re-written as

$$\overline{X} = (F - GH)\overline{X} + GK_s\theta_C \qquad \text{... (4.163)}$$

The eigen values of (*F-GH*) then give the stability characteristics of the attitude control system. A root locus plot also can be obtained for a given K_R and various values of K_s by finding the roots of the characteristic equation of (*F–GH*).

To study the gain and phase margins, one can find the loop gain by defining the output signal as

$$y = \theta_s + K_R\theta_s = H\overline{X} \qquad \text{... (4.164)}$$

Then the loop gain is given by

$$GH = H(sI - F)^{-1} GK_s \qquad \text{... (4.165)}$$

The bode plot can now be plotted by varying the frequency $s = j\omega$. *GH* in equation 4.165 can be updated to account for any filter or compensator in the loop.

4.6.1.2 *Latax Control Law*

The latax control law in pitch plane (Ref. Fig. 4.19 and eq. 4.92) is given by

$$\delta_c = K_s[f_{zd} - K_a f_{zs} + K_R q_s] \qquad \text{... (4.166)}$$

where f_{zs} is the sensed acceleration at accelerometer location and q_s is the sensed angular rate at rate gyro location. These signals are obtained as follows for stability analysis.

$$f_{zs} = f_{zcg} - c\theta + \sum_{i=1}^{n} \phi_i(l_{acc})q_i \qquad \text{... (4.167)}$$

and
$$q_s = q + \sum_{i=1}^{n} \frac{\partial\phi_i(RG)}{\partial l} q_i \qquad \text{... (4.168)}$$

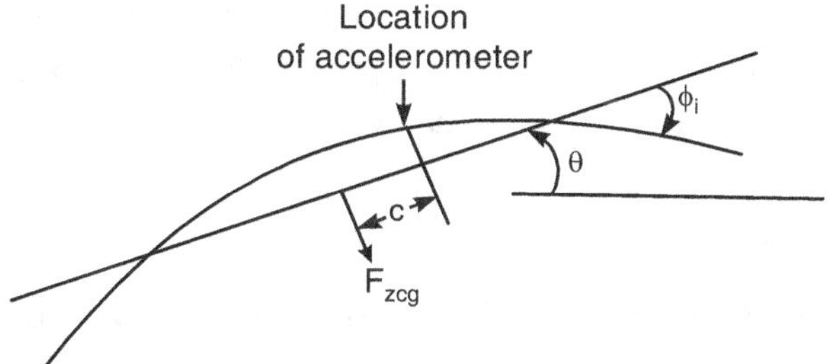

Fig. 4.28

Now,

(*i*) $\quad f_{zcg} = w - u\theta = u\alpha + u\alpha - u\theta$

$$= u \sum_{i=1}^{5+2m+2n} a_{1,\,i}\, x_i + u\alpha - u\theta \qquad \qquad \text{... (4.169)}$$

(*ii*) $\quad \theta = \sum_{i=1}^{5+2m+2n} a_{3,i}\, x_i \qquad \qquad \text{... (4.170)}$

(*iii*) $\quad q_i = \sum_{i=1}^{5+2m+2n} a_{5+2m+2n,\,i}\, x_i \qquad \qquad \text{... (4.171)}$

Define output matrix H (for computing open loop gain GH) as

$$H\overline{X} = -K_R\,\theta_s + K_a f_{zs}$$

$$= -K_R\!\left[\theta_r + \sum_{i=1}^{n} \frac{\partial \phi_i(RG)}{\partial l}\, q_i\right]$$

$$+ K_a\left\{ u\alpha - u\theta + u \sum_{i=1}^{5+2m+2n} a_{1,i}\, x_i - \left[c \sum_{i=1}^{5+2m+2n} a_{3,i}\, x_i \right] \right.$$

$$\left. + \sum_{i=1}^{n} \phi_i\,(l_{acc}) \left(\sum_{j=1}^{5+2m+2n} a_{5+2m+2i,\,j}\, x_j \right) \right\} \qquad \text{... (4.172)}$$

H matrix computation is given in Appendix 4.1

The open loop gain GH is then given by eqs. 4.165, *i.e.*,

$$GH = H(sI - F)^{-1}\, GK_s$$

This open loop gain can be further updated to include any compensator or filter inside the loop. The open loop gain can now be used for getting bode plot by varying the frequency $s = j\omega$ and the stability margins can be obtained.

4.6.2 Sensor Location

The attitude control law uses sensed angle and angular rates as feedback signal.

The latax control law uses the sensed acceleration and angular rates as feedback signals. The inertial sensors sense the vehicle state at the location of the sensor. When the vehicle is flexible the sensor senses not only the rigid body state but also contribution due to vehicle flexibility at the location of the sensor. The control law processes the sensor signals and computes the actuator commands. The net phase lag contributed by the sensor dynamics, computational delay, sample and hold and the actuator dynamics should not be such as to make some of the flexible modes unstable. The sensed signals by various sensors are given as follows:

1. Angle sensor : Location l_{PG} from nose tip

$$\theta_s = \theta_r + \sum_{i=1}^{n} \frac{\partial \phi_i(l_{pg})}{\partial l} q_i \qquad \text{... (4.173)}$$

2. Angular rate sensor : Location l_{RG} from nose tip

$$\theta_s = \theta_r + \sum_{i=1}^{n} \frac{\partial \phi_i(l_{RG})}{\partial l} q_i \qquad \text{... (4.174)}$$

3. Acceleration sensor : Location l_{acc} from nose tip

$$f_s = f_{cg} - c\theta + \sum_{i=1}^{n} \phi_i(l_{acc}) q_i \qquad \text{... (4.175)}$$

where
$$c = l_{cg} - l_{acc} \qquad \text{... (4.176)}$$

If there were only one mode of vibration and it were invariant with time, one would like to position angle sensor and angular rate sensor at a place where the slope of the mode $\dfrac{\partial \varphi_i}{\partial l}$ is zero (or anti node) and accelerometer at a place where ϕ_i is zero (*i.e.*, the node point). In such a case, the sensor will not sense any effect of flexibility and any vibration which gets excited due to external forces will gradually damp out due to inherent structural damping though the decay rate will be quite slow. In actual practice, the structural flexibility consists of large number of modes and the nodes and anti nodes for each mode will be at different place. Hence, it is practically impossible to locate sensors so as to exclude the effect of flexibility.

Experience shows that for most of the vehicles, inclusion of two to three flexible modes in the stability analysis would be adequate. In addition, the first mode frequency is the closest frequency to the control frequency and it would present maximum difficulty in attenuating the same through filters. Hence, one must attempt to minimize the first mode signal in the control law by positioning the sensor suitably.

The vehicle attitude is mostly sensed using inertial measurement unit which is used for navigation purpose. This is a sophisticated unit and mostly a single unit is used for the entire mission and hence needs to be necessarily located in the uppermost stage and not much freedom is available to locate it to exclude the effects of vehicle flexibility. The angular rates and accelerations are sensed by rate gyros and accelerometers and more than one packages can be used if required for stabilizing the modes.

Fig. 4.29 shows a root locus study for a four stage satellite launch vehicle in which the actuator is assumed to be a first order, angle sensor is placed at a fixed location and four possible rate gyro locations are considered *viz.*,

(1) same location as IMU, interstage between 3rd and 4th stages

(2) interstage between 2nd and 3rd stages

(3) interstage between 1st and 2nd stages

(4) base shroud region of first stage.

It is seen that the closed loop pole corresponding to first mode moves towards the imaginary axis for first two location. For the 3rd rate gyro location, the closed loop pole for the designed gain is farther from the imaginary axis than the open loop pole and it is still farther from imaginary axis in case of location no. 4. This indicates that the closed loop damping for the first mode in case of locations 3 and 4 is much better than the pure structural damping in open loop mode. This is a desirable feature and hence the locations are preferred choices. However, the severe noise environment due to engine at lift off and the acoustic environment may drive the sensors to saturation and hence NASA launch vehicle design criteria (Ref. 3) recommends not to use this place for locating the sensors.

Fig. 4.29 shows that this location gives slight reduction in the closed loop damping for the second and third modes. However, there is quite large gain margin before the mode becomes unstable. Further, it is possible to design suitable filters to provide attenuation for higher modes. The scenario that a particular location is most desirable for one mode but not desirable for another mode will always occur in practice and hence the designer has to decide the location based on the most critical mode which otherwise is presenting maximum difficulty in stabilizing it by compensation or filtering.

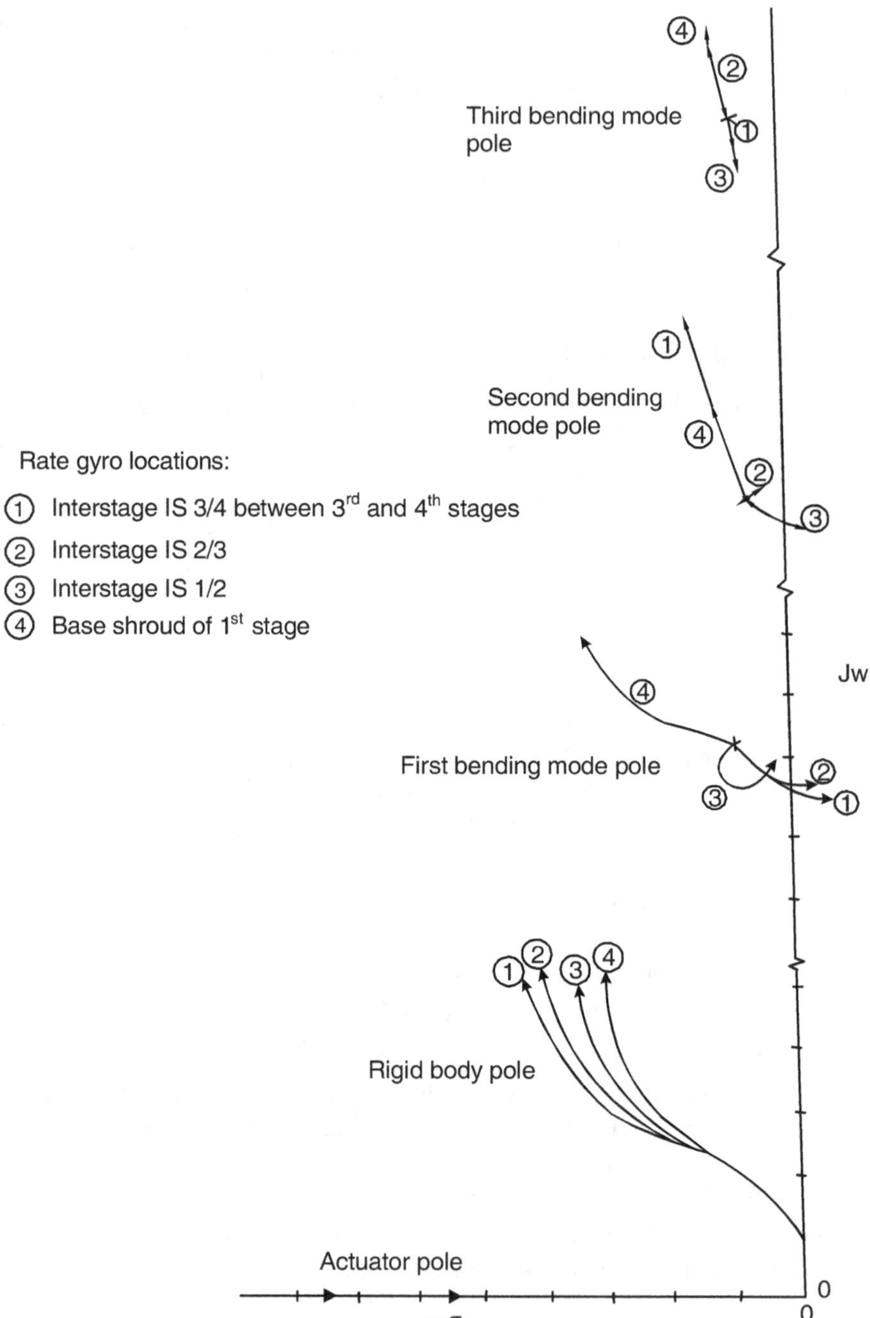

Fig. 4.29 Root loci for different rate gyro locations

4.6.3   Methods of Stabilization of Structural Modes

The loop gain *GH* when plotted against frequency $j\omega$ will show a peak at each low damped mode dynamics. The slosh modes being very close to control frequency, the peaks will occur within the crossover frequency or near the crossover frequency and it will be very difficult to provide any compensation for them. In case, they are found to be unstable, one can resort to partitioning of the tanks to increase the frequency and add baffles or other damping devices inside the propellant tanks to increase the damping and reduce the incremental *GH* swing in the GH plot (*i.e.* $(GH_{max} - GH_{min})$ at slosh mode frequency) and thus make it stable.

Fig. 4.30 shows gain and phase plots for a typical vehicle where one slosh mode for oxidizer tank and one mode for fuel tank is considered. The plots show that the modes are stable.

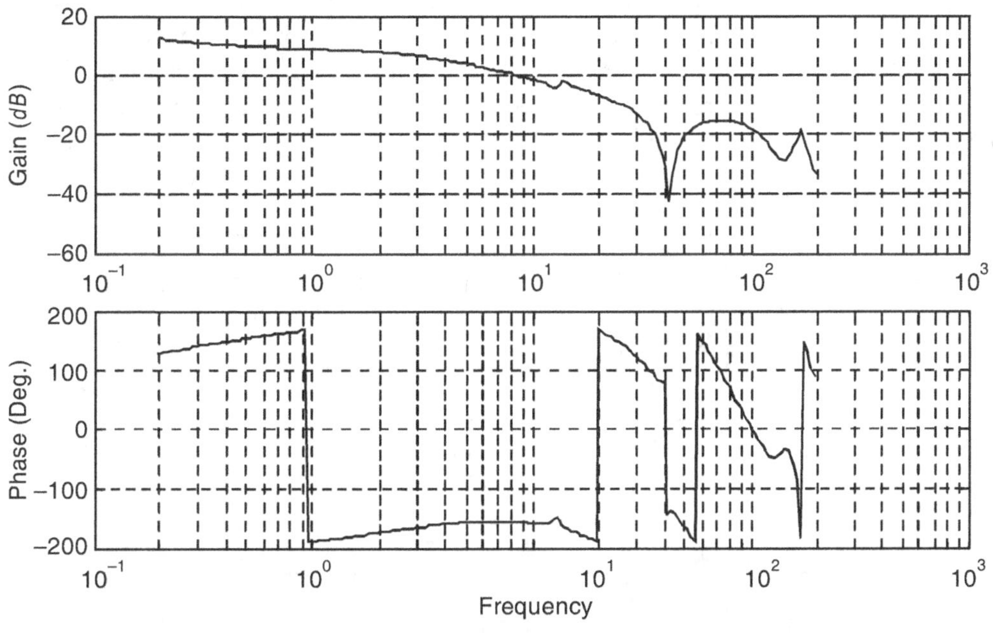

Fig. 4.30

The structural damping being very low, there will be very sharp peaks occurring in *GH* plot at these frequencies and the amplitudes of the peaks may be well above the zero *db* line. Hence, it is essential to study their stability in detail and ensure adequate stability margins. Following are some of the methods used for stabilizing modes.

1. Sensor Location. This has been already discussed in detail in the previous section.

2. Compensation and Filtering.

Gain Stabilization

Whenever possible, the designer should provide adequate attenuation for the flexible modes. It is essential that the inertia effects of the control effectors (Tail-wags-dog effect) is included in the model since it gives pronounced hump in the *GH* plot beyond TWD frequency (See Fig. 4.9) and frequencies which are well attenuated in absence of inertia effect may in fact become unstable when this effect is included. It is comparatively easier to attenuate frequencies which are well separated from the control frequency meaning the frequencies higher than 5 times the control frequency. The method of stabilizing the mode by providing adequate attenuation is called "Gain stabilization" of the modes. It may be noted that the mode peak depends upon the structural damping ratio assumed for the modes. One must consider the lowest possible structural damping ratio while gain stabilizing the modes. Further, the prediction uncertainty on modal frequency also increases for higher modes. Hence, adequate caution needs to be used whenever notch filter is used for attenuation of a particular mode. There have been number of cases wherein the actual modal frequency during flight is found to be outside the notch even though the notch frequency was set after conducting experiments on the structure. It will be desirable to provide an unsymmetrical notch by which lowest possible frequency is provided with a required attenuation and there is a

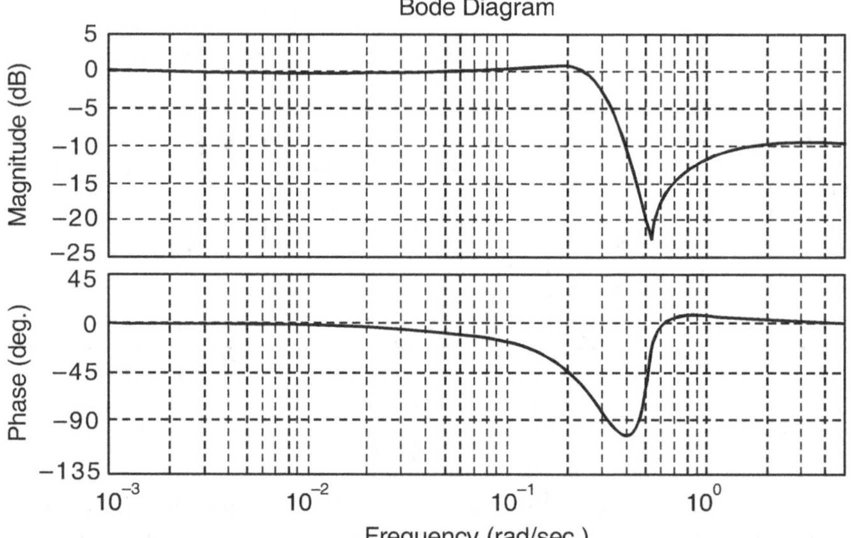

Fig. 4.31 Asymmetric Notch Characteristics

definite attenuation above the notch frequency. Fig. 4.31 shows such a notch and Eqn. 4.177 gives the required transfer function of the notch.

$$F_{notch} = \left(\frac{s^2 + 2\varsigma_1\omega_1 s + \omega_1^2}{s^2 + 2\varsigma_2\omega_2 s + \omega_2^2} \right) \frac{\omega_2^2}{\omega_1^2} \qquad \text{... (4.177)}$$

Phase Stabilization

First one or two modes in case of launch vehicles are quite close to control frequency. Any attempt in attenuation of these modes is associated with a large phase lag at control frequency, thus, presenting difficulties in providing adequate gain and phase margins at the rigid body frequency (or control frequency). Hence, such modes are usually phase stabilized which means an adequate phase margin is provided at each crossover. Each mode which has a peak more than zero db, will have two gain crossovers and thus two phase margin points will occur. If possible, it is preferable to position the peak towards the right half of (*RHP*) *GH* plane. This will ensure higher closed loop damping of the mode than the pure structural damping. (See. Fig. 4.29 for root locus plots). Fig. 4.32 shows a mode peak placed in the *RHP* of *GH* plane and the two phase margin points ϕ_1 (lag phase margin) and ϕ_2 (lead phase margin).

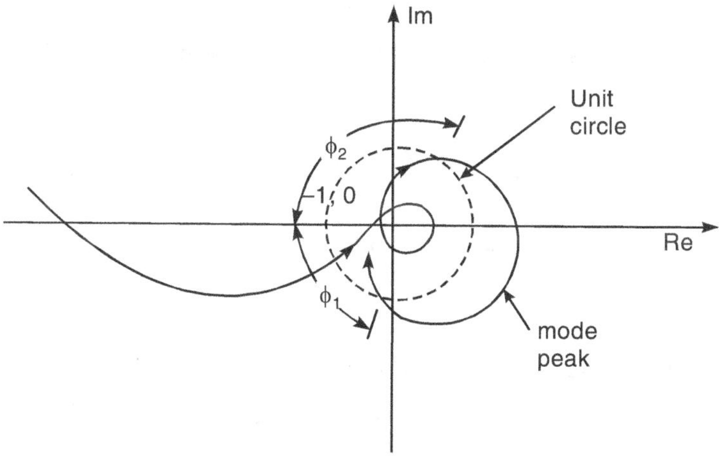

Fig. 4.32

The phase margins to be provided for the phase stabilized modes must be much higher than the 30° phase margin suggested in Section 4.3.1 and following factors need to be considered for deciding on the required phase margin:

1. Modal frequency changes with burning of the propellant.

2. Uncertainty band on the prediction of frequencies and mode shape data.

3. Variation of mode shape and slope data with time at the sensor location and uncertainty of prediction on the same.

4. Any phase uncertainty for actuator or sensor at the required modal frequency.

The phase margin of 45° to 50° may be initially considered but its adequacy needs to be verified after checking for the above uncertainties.

3. Blending of the Sensor Signals

The first mode signal in the control loop can be minimized by locating the rate gyros at anti nodes provided such a location is feasible. If such a location is not available, one would place it at a minimum possible slope or at a place of desirable sign of the slope. If one is finding difficulty in phase stabilizing or gain stabilizing such a mode, additional sensors may be used in order to minimize the signal due to the mode.

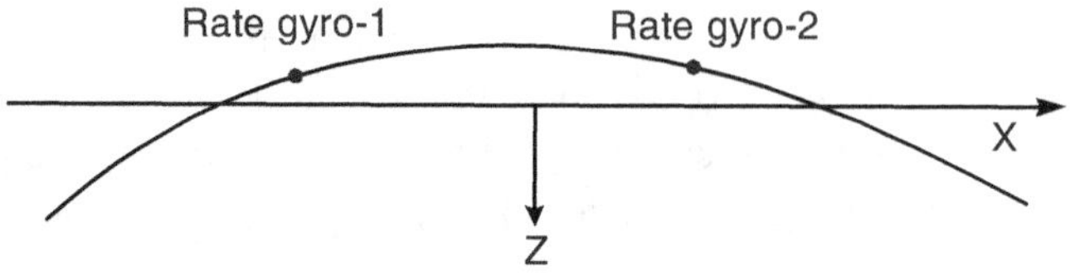

Fig. 4.33

Fig. 4.33 indicates two sensors located at places where the mode slope is given by σ_1 and σ_2. The sensed rates are given by:

$$\theta_{s1} = \theta_r + \frac{\partial \phi_1(1)}{\partial l} q_1 = \theta_r + \sigma_1 q_1 \qquad \ldots (4.178)$$

$$\theta_{s2} = \theta_r + \frac{\partial \phi_1(2)}{\partial l} q_1 = \theta_r + \sigma_2 q_1 \qquad \ldots (4.179)$$

Let these signals be mixed as

$$K\theta_{s1} + (1-K)\theta_{s2} = K[\theta_r + \sigma_1 q_1] + (1-K)(\theta_r + \sigma_2 q_1)$$

$$= \theta_r + (K\sigma_1 + \sigma_2 - K\sigma_2) q_1$$

The modal signal can be minimized by selecting K such that

$$K\sigma_1 + \sigma_2 - K\sigma_2 = 0$$

or
$$K = \frac{\sigma_2}{\sigma_2 - \sigma_1} \qquad \ldots (4.180)$$

The locations for two rate gyros now can be chosen based on the availability and where the slopes are well defined (*i.e.*, the slopes are not nearly zero, which can cause large errors). Further, the sign of effective mode slope can also be chosen as per the requirement for stabilizing the mode by choosing K such as to make the effective slope

$$\sigma' = K(\sigma_1 - \sigma_2) + \sigma_2 \qquad \ldots (4.181)$$

positive or negative thus enabling one to phase stabilize the mode in addition to providing some attenuation to the mode peak by selecting the magnitude of σ'. The factor K can also be varied as a function of time to account for mode shape variation in flight as a function of left over propellant.

4. Tracking Notch Filters

Present day onboard processors provide adequate computational power and very complex algorithms can now be easily implemented in the design. Tracking notch filters is one such feature to avoid higher frequency flexible modes entering into the control loop.

Transfer function of a notch filter is given as

$$G = \frac{s^2 + 2\xi_1\omega_n s + \omega_n^2}{s^2 + 2\xi_2\omega_n s + \omega_n^2} = \frac{1 + 2\xi_1 S + S^2}{1 + 2(k\xi_1)S + S^2}$$

where $\quad S = \dfrac{s}{\omega_n}$ and $\xi_2 = k\xi_1$

The gain and phase plots of this notch are given in Figure 4.34 for various values of ξ_1 and

$k = 10$. At the notch frequency the value of transfer function is given by $G = \dfrac{\xi_1}{\xi_2}$ or gain

$$= 20 \log\left(\frac{\xi_1}{\xi_2}\right) db.$$

Thus a ratio of 0.1 gives 20 *db* notch depth.

The variation in structural frequency of the missile is due to following factors:

1. Change in vehicle mass due to burning of the propellant and
2. Prediction uncertainty due to dispersion in the stiffness distribution and joint rotation constants used for predicting the frequencies and mode shapes.

Prediction uncertainty can be minimized by carrying out the Ground Resonance Test (GRT) on the vehicle with propellant filled condition and the empty condition and tuning the joint rotation constants to match the prediction frequency with the GRT frequency. Once the model is tuned, the structural frequency and mode shapes can be determined for different amount of propellant remaining in the missile. The actual frequency at any flight instant can then be obtained by linear interpolation after computing the instantaneous mass as described in section 4.4.8 on further refinement of Generalized adaptation scheme for Gains Schedule. The same frequency can be used for tracking notch filter. The width of the notch filter for a given amount of attenuation can be fixed depending on the prediction uncertainty of the mode shape frequency. The width can be adjusted either by adjusting the damping ratios ξ_1, ξ_2 or by using additional notch filter with slightly different notch frequency provided the degradation in phase margin is tolerable.

The notch filtering scheme does not give any attenuation for frequencies outside the notch zone. Hence it is desirable to attempt a low pass filter to get the desirable attenuation as a first choice since it provides attenuation for all higher frequencies which one might have

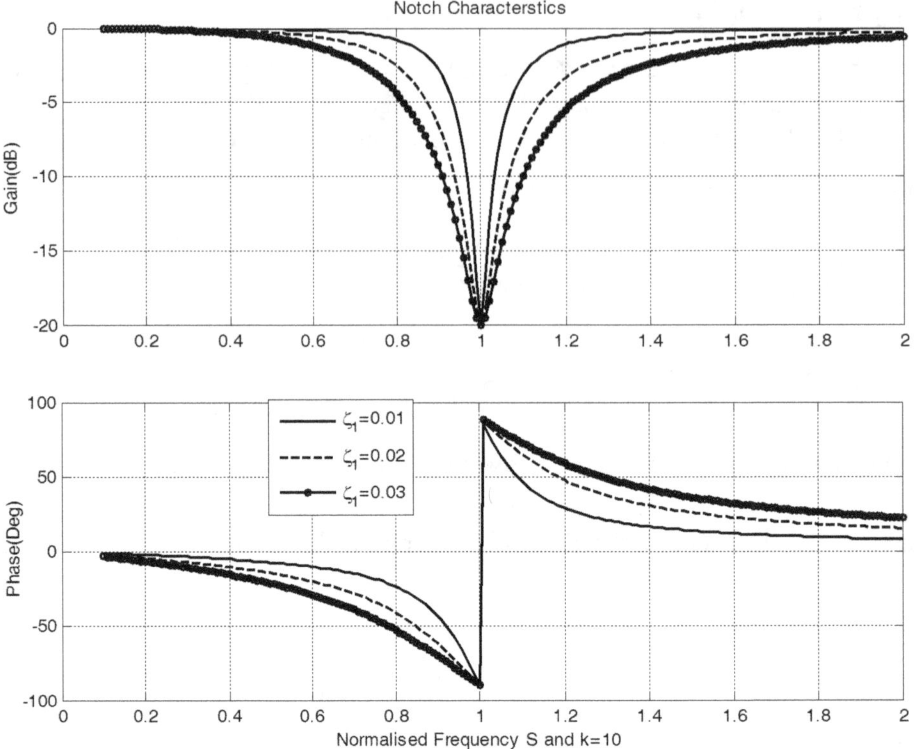

Fig. 4.34 Notch Characteristics

ignored in the stability analysis.

4.6.4 Compensator Design Using Mathematical Programming

4.6.4.1 Defining Objective Function and Constraints

Design of compensator for the system when number of flexible modes are present is a somewhat complicated task. The system must have good stability margins in the region of rigid body frequency and lower flexible modes and adequate attenuation at higher mode frequencies. When one is adjusting compensator coefficients by trial and error method, some constraints will improve but some other may get degraded. Hence, it is desirable to have a systematic algorithm by which the compensator coefficients are changed in such a way that all the constraints get improved.

A flight control system block diagram can be considered to be consisting of number of loop gain channels *e.g.* attitude loop, angular rate loop, lateral acceleration loop etc. Each loop can be considered to be having a separate compensator. We consider here compensator to be present only in the forward path and common to all the loop gain channels. We also consider here that the compensators are first order or second order so that their stability is guaranteed if the coefficients are constrained to be positive. The compensated frequency response of the system can then be written as

$$GH = (c + jd)\prod_{i=1}^{k} G_{ci}(s) = e + jf \qquad \qquad \text{... (4.182)}$$

where k gives the number of compensators in the loop and $(c + jd)$ gives the uncompensated GH as a function of frequency.

The objective functions are variously defined by different authors as follows:

1. Coffey[16] defined objective function as

$$J = \left| (y* - \hat{y}*)^T ww^T (y - \hat{y}) \right| \qquad \qquad \text{... (4.183)}$$

where $\hat{y}$ is a vector of desired frequency response points, y is a vector of frequency response points as a function of compensator coefficients, w is the diagonal weighting matrix and (*) denotes the conjugate.

Gradient search algorithm is used to improve the compensator coefficients to minimize the objective function. This method puts unnecessarily high requirements on compensation since the desired frequency response many times need not be strictly as specified. Further, the selection of weighting matrix also needs to be done before a satisfactory compensator is arrived at.

2. Stear and Page[17] have defined the objective function as

$$J = \sum_{i=1}^{N} K_i (1 - s_i^a / s_i^d)^2 \qquad \qquad \text{... (4.184)}$$

where N is the number of specifications considered, s_i^d is the desired value of specification and s_i^a is the actual value of specification as a function of compensator coefficients and K_i is the weighting constant. The K_i is selected as positive value when the $s_i^a < s_i^d$ (*i.e.*, the specification is not satisfied) and zero when $s_i^a \geq s_i^d$ (*i.e.*, specification is satisfied).

Then, search techniques are used to select the compensator coefficients so as to drive the objective function to zero.

3. Jerrel Mitchell and McDaniel[19,20] defined the problem as a constraints satisfaction problem instead of maximization or minimization of an objective function. For this, the concept of generalized stability margin as defined in Section 4.3.1 has been used where the stability margin is given by the minimum radial distance of the GH plot from the critical point $(-1, j0)$. In general, the various constraints can be defined as the radial distance from any point of interest $(-k, j0)$. Thus stability margin is defined by distance from $(-1, j0)$ and attenuation margin is defined by distance from $(0, j0)$.

Then the constraints are given by

Stability margins

$$d_i(\omega) \geq D_i \quad \text{for } i = 1, \ \ldots\ldots. n \qquad \ldots (4.185)$$

Attenuation margin

$$d_j(\omega) \leq D_j \quad \text{for } j = 1, \ldots\ldots. m \qquad \ldots (4.186)$$

Some of the constraints may be already satisfied by the initial compensator and one should ensure that they do no get violated when the compensator coefficients are being changed. The constraints which are not satisfied are called 'active constraints' and one needs to change the compensator coefficients in such a way that there is an improvement in the actual constraint value. The process is continued till all the constraints are satisfied or no further improvement is possible. The later case happens if adequate number of compensator coefficients are not available for perturbation. In such case, one has to add additional compensator transfer functions making more number of coefficients available for perturbations.

Ref. 20 has shown that in order to move n frequency points of GH plot in n optimal directions by realizable compensator it is sufficient to have $2n$ independent compensator coefficients and in order to move n points in directions within $\pm\dfrac{\pi}{2}$ of the optimal directions, it is sufficient to have n independent compensator coefficients.

4.6.4.2 Compensator Improvement Algorithm

Let $\overline{X}^T$ denote a vector of all coefficients of a compensator which gives a stable control system to start with and let all constraints be given by

$$g_i(\overline{X}^T) \geq b_i \quad \text{where } i = 1, \ \ldots\ldots. m \qquad \ldots (4.187)$$

Let $K_1, K_2, \ldots\ldots, K_n$ be the indices of active constraints. Then a direction vector D is given by

$$D = a_1 \nabla g_{k1} + a_2 \nabla g_{k2} + \ldots\ldots. a_n \nabla g_{kn} \qquad \ldots (4.188)$$

where $\nabla g_{k1}, \nabla g_{k2}$ etc. are the gradients of the constraints with respect to compensator coefficients. If we select coefficients $a_i's$ such that

$$D \cdot \nabla g_{ki} > 0 \quad \text{for } i = 1, \ \ldots\ldots. n \qquad \ldots (4.189)$$

we will ensure improvement in each constraint value by traveling in the direction of vector D.

Let
$$D \cdot \nabla g_{ki} = c_i \quad i = 1, \ldots\ldots.n \qquad \text{... (4.190)}$$

where $c_i's$ are positive coefficient, then one gets,

$$(\nabla g^T \cdot \nabla g)\,\bar{a} = \bar{c} \qquad \text{... (4.191)}$$

where ∇g is a matrix whose columns are gradients of active constraints. If all the gradients are linearly independent, one can get

$$\bar{a} = [\nabla g^T \cdot \nabla g]^{-1}\,\bar{c} \qquad \text{... (4.192)}$$

$a_i's$ give the desired dot product of direction vector D and gradient of ith active constraint.

The compensator coefficients are then updated as follows:

$$\bar{X}_{k+1}^T = \bar{X}_k + h\nabla g\,\bar{a} \qquad \text{... (4.193)}$$

where ∇g is the matrix with gradients as columns and are computed with compensator coefficients at k^{th} iteration. 'h' is a small positive step size and is selected as in usual optimization methods using gradients such that the $(k+1)$th iteration gives a better value of constraints than the kth value.

The iterations are continued till all the constraints are inactive which means that the required compensator has been found. The process can also end up either in some of the gradients of the active constraints becoming zero or linearly dependent on other gradients. In such a case, one may have to accept the solution or restart the process with a new starting point.

4.6.4.3 Computation of Gradient

We give here a procedure assuming that the compensator is common for all channels. The more general treatment for different compensators in different channels is given in Ref. 20.

Let
$$GH = e + jf = (a + jb)\prod_{i=1}^{k} G_{ci}$$

$$= (c + jd)G_{cq} \qquad \text{... (4.194)}$$

where $a + jb$ = uncompensated GH as a function of frequency, s

$$(c + jd = (a + jb)\prod_{\substack{i=1 \\ i \neq q}}^{k} G_{ci} \qquad \text{... (4.195)}$$

Let the constraint be defined by a distance $d(\omega)$ from a general point p given by $(-A + jB)$ (*e.g.*, the point p for stability margin is given by $(-1, j\,0)$). Then

$$d(\omega) = \left| A + jB + (c + jd)G_{cq} \right| \qquad \text{... (4.196)}$$

The general transfer function for G_{cq} is given by

$$G_{cq} = \frac{x_0 + x_1 s + x_2 s^2}{y_0 + y_1 s + y_2 s^2} \qquad \text{... (4.197)}$$

Ref. 20 assumes a more general case of compensator with numerator and denominator orders n and m respectively. However, it is suggested here to restrict the compensator segments to 1st or 2nd order due to ease of ensuring stable poles. The square of the distance *i.e.*, $d^2(\omega)$ is given by:

$$D(\omega) = d^2(\omega) = \frac{\left[A(y_0 - y_2\omega^2) - By_1\omega + c(x_0 - x_2\omega^2) - dx_1\omega \right]^2 + \left[B(y_0 - y_2\omega^2) + Ay_1\omega + d(x_0 - x_2\omega^2) + cx_1\omega \right]^2}{(y_0 - y_2\omega^2)^2 + (y_1\omega)^2} \qquad \text{... (4.198)}$$

The gradient of $d(\omega)$ with compensator coefficients can now be easily obtained by differentiating with each coefficient of the compensator. The general form of the expression for nth order numerator and mth order denominator compensator is given by[19]

$$D(\omega) = \frac{\left(\sum_{i=0}^{n} C_i x_i + A\sum_{j=0}^{k} E_{2j}Y_{2j} - B\sum_{j=0}^{p} E_{2j+1}Y_{2j+1}\right)^2 + \left(\sum_{i=0}^{n} D_i x_i + A\sum_{j=0}^{p} E_{2j+1}Y_{2j+1} + B\sum_{j=0}^{k} E_{2j}Y_{2j}\right)^2}{\left(\sum_{j=0}^{k} E_{2j}Y_{2j}\right)^2 + \left(\sum_{j=0}^{p} E_{2j+1}Y_{2j+1}\right)^2} \qquad \text{... (4.199)}$$

where $\qquad k = \dfrac{m}{2}$ and $p = \dfrac{m}{2} - 1$ if m is even

and $\qquad k = \dfrac{m-1}{2}$ and $p = \dfrac{m-1}{2}$ if m is odd

$$[C_0, C_1, C_2, C_3, C_4 ....] = [c, -d\omega, -c\omega^2, d\omega^3, c\omega^4, -d\omega^5 ......] \qquad \text{... (4.200a)}$$

$$[D_0, D_1, D_2, D_3, D_4 ....] = [d, c\omega, -d\omega^2, -c\omega^3, d\omega^4 ......] \qquad \text{... (4.200b)}$$

$$[E_0, E_1, E_2, E_3, E_4 ....] = [1, \omega, -\omega^2, -\omega^3, \omega^4, \omega^5 ......] \qquad \text{... (4.200c)}$$

General partial differential coefficient with compensator coefficient is then given by

  (*i*) For *i* even,

$$\frac{\partial D(\omega)}{\partial y_i} = \frac{2[FD(A \cdot FN_1 + B \cdot FN_2) - FN \cdot FD_1]E_i}{(FD)^2} \qquad \text{... (4.201)}$$

(*ii*) For *i* odd,

$$\frac{\partial D(\omega)}{\partial y_i} = \frac{2[FD(-B \cdot FN_1 + A \cdot FN_2) - FN \cdot FD_1]E_i}{(FD)^2} \qquad \text{... (4.202)}$$

(*iii*) and for *i* even or odd

$$\frac{\partial D(\omega)}{\partial x_i} = \frac{2[FN_1 \cdot C_i + FN_2 \cdot D_i)]}{FD} \qquad \text{... (4.203)}$$

Where

$$FN_1 = \left[\sum_{i=0}^{n} C_i x_i + A \sum_{j=0}^{k} E_{2j} Y_{2j} - B \sum_{j=0}^{p} E_{2j+1} Y_{2j+1} \right] \qquad \text{... (4.204)}$$

$$FN_2 = \left[\sum_{i=0}^{n} D_i x_i + A \sum_{j=0}^{k} E_{2j+1} Y_{2j+1} + B \sum_{j=0}^{p} E_{2j} Y_{2j} \right] \qquad \text{... (4.205)}$$

$$FD_1 = \sum_{j=0}^{k} E_{2j} Y_{2j} \qquad \text{... (4.206)}$$

$$FD_2 = \sum_{j=0}^{p} E_{2j+1} Y_{2j+1} \qquad \text{... (4.207)}$$

$$FN = (FN_1)^2 + (FN_2)^2 \qquad \text{... (4.208)}$$

$$FD = (FD_1)^2 + (FD_2)^2 \qquad \text{... (4.209)}$$

4.6.4.4 General Observations

The use of mathematical programming for improving the compensator so as to meet the specifications on stability margins and attenuation margins for various frequency zones appears quite systematic and Ref. 20 has illustrated the use of compensator improvement program (CIP) by arriving at a compensator after several iterations which gives significantly improved constraints. However, there are several factors which discourage straightforward use of this program.

For launch vehicles and missiles, the open loop transfer function keeps changing with time due to:

1. Burning of propellant (mass, cg and Iyy)

2. Vehicle state *i.e.*, velocity, mach no. and altitude (aerodynamic characteristics)

3. Structural modes (frequency, mode shape and generalized mass)

4. Slosh mode parameters.

The compensator arrived at for one flight instant after large number of iterations need to be checked for different flight instants and it is not clear how well it would perform. Further, it is desirable to check the stability margins for various perturbations cases of vehicle parameters *e.g.* nominal, upper bound and lower bound data regarding inertial parameters, aerodynamic characteristics and structural mode shape and frequency data as a robustness check for the design.

It appears quite reasonable to expect that the compensator which is found optimum at one flight instant may not be satisfactory at other flight instant. Hence, a different compensator needs to be obtained for different flight instants. A natural question arises as to whether the coefficients will vary in an orderly fashion with time or with some other variable so that these can be programmed and used after interpolating at various flight instants. Most of the mathematical programming problems give local optimum unless the objective function is convex or concave. The objective functions for compensator design are neither convex nor concave and hence the compensators obtained using mathematical programming are likely to be the local optimum.

Therefore, one is not confident enough that the compensator improvement program when used at different flight instants will give an orderly variation of the compensator coefficients.

However, in complex situations when it is difficult to fix parameters of a large order compensator, this program will serve as a valuable tool to get an initial feel about the compensator and the designer can then judiciously set the coefficients or rules for varying the coefficients so that the compensator gives satisfactory stability margins for the entire zone of operations.

4.7 CONCLUDING REMARKS

In this chapter, we have mainly addressed the autopilot loop design which includes gain schedule and compensator design. It starts with simplification of vehicle dynamics which is required for designing the gains schedule which mainly controls the overall behaviour of vehicle. It discusses various steps involved in the design and main objectives of autopilot design.

The autopilot design is then discussed in detail first assuming rigid body dynamics. It discusses how the pole location influences the step response for a 3rd order system and then discusses the effects of engine inertia (tail-wag-dog effect), transfer function zero and digital control system lags due to sample and hold and computational delay. It covers attitude control loop for pitch/yaw and roll control systems and also latax control system and gives various schemes of gain adaptation including generalized adaptation scheme. It then discusses method of incorporating sample and hold and computational delay and problem of frequency aliasing arising due to sampling.

The chapter then discusses higher order dynamics namely the propellant sloshing and structural flexibility which includes problem of propellant sloshing and methods of managing them. It also discusses system model for incorporating structural flexibility for both attitude control and latax control system and methods of stabilizing the structural modes such as sensor location, compensation or filtering, blending of sensor signals etc.

We have also discussed the use of mathematical programming for arriving at an improved compensator and limitations of the methodology.

The chapter gives only the classical methods of control system design. There have been several attempts for autopilot design by various modern techniques such as fuzzy logic and neural networks. Some of them have also given reasonably good simulated performance for a sample case. However, these techniques appear heuristic in nature and without a theoretical background about the robustness of performance and the only confidence one can gain is with repeated and large number of simulation runs. On the other hand, the classical design approach has a sound theoretical basis and one can predict the system performance under various cases of disturbance or parameter variation. The flight trial being a very costly affair, many times involving life of human beings, autopilot designers of flight control systems have been using mainly classical design methods.

REFERENCES

1. **A.L. Greensite:** Control Theory: Vol. II—Analysis and Design of Space Vehicle Flight Control Systems. Spartan Books. 1970 (Chapter IV).

2. **NV Kadam:** Revised models for aerodynamic-structure-control-slosh interaction studies. No. RCI-3300.1019.513 Jan. 1991.

3. **Anonymous:** Effects of structural flexibility on launch vehicle control systems. NASA Space Vehicle Design Criteria (Guidance and Control), NASA SP - 8036.

4. **R.F. Ringland:** Dynamic stability of space vehicles, Vol. X—Exit stability.

5. **B.C.Kuo:** Feedback control systems.

6. **Truxal:** Feedback control systems.

7. **G.A. Jones:** On the step response of a class of Third Order Linear Systems. IEEE transactions on Automatic Control (Corresp) Vol. AC-12, Jun. 1967, P. 341.

8. **Dirk V. Duytschaever:** Comment "On the step response of a class of third order linear systems". IEEE Trans on Auto Cont. (Corresp) Feb. 1968, pp. 134-135.

9. **Nesline FW, Nabbefield N.C.:** Design of digital autopilots for homing missiles. AGARD CP-270.

10. **R.N. Bhattacharjee, T.K. Goshal, N. Sarkar:** Three loop autopilot, JPTR/One/T-11, 1986.

11. **R.N. Bhattacharjee, Arun Saha:** Design relations, Methodology and related subsystem specifications for common missile Autopilots, Vol. I and Vol. II.

12. **N.V. Kadam:** Practical design of flight control systems—Some problems and their solutions. Defence Science Journal—Special Issue on Guidance & Control of Missiles, Vol. 55 No. 3. Jul. 2005, pp. 211-221.

13. **J. F. Franklin, J.D. Powell:** Digital control of dynamic systems. Addition - Wesley Publishing Company, 1980.

14. **H.N. Abramson (Editor):** The dynamic behaviour of liquids in moving containers. NASA SP 106, 1966.

15. **J.R. Roberts, E.R. Basuro, Pei-Ying Chen:** Slosh Design Handbook I. NASA CR 406, May 1966.

16. **T. C. Coffey:** Automated frequency-domain synthesis of multi loop control systems. AIAA Journal, Vol. 8, No. 10, 1970.

17. **Edwin B. Stear, John A. Page:** Automated design of multivariable control systems. Proc. of 1971 IEEE Decision and Control Conference, Univ. of Florida, Gainsville, pp. 192-196.

18. **Edwin B. Stear, Charles P. Lefkowitz:** Automated design of space booster control systems International Journal and Control,1972, Vol. 16 No. 5 pp. 849-868.

19. **Jerrel R. Mitchell, Willie L. McDaniel Jr.:** A computerized compensator Design algorithm with launch vehicle Applications. IEEE Transactions on automatic control June 1976, pp. 366-371.

20. **J.R. Mitchell, Willie Mc Daniel Jr.:** An innovative approach to compensator design. NASA CR-2248, May 1973.

APPENDIX 4.1	COMPUTATION OF SYSTEM MATRICES A, B, D AND H

1. INTRODUCTION

The system matrix elements are given below for the model given by equations 2.120 to 2.127 of chapter 2 and the H matrix is given for equations 4.162 and 4.172 of chapter 4. The model assumes total vehicle parameters and mode shapes are also assumed to be obtained for total vehicle. Computation of elements uses integrals for aerodynamic load distribution as given in Table 4.4.

The state vector is defined as

$$\bar{X}^T = [\alpha \; \theta \; \theta \; \delta \; \delta \; \lambda_1 \; \lambda_1 \ldots \lambda_m \; \lambda_m \; q_1 \; q_1 \ldots q_n \; q_n]$$

A-MATRIX:

$$A(1,1) = -\frac{u}{u} - \frac{QS}{Mu} I(1)$$

$$A(1,2) = -\frac{g \sin \gamma_0}{u}$$

$$A(1,3) = 1 + \frac{QS}{Mu^2} I(2)$$

$$A(1,4) = -\frac{2T_E}{Mu}$$

$$A(1,5) = 0$$

$$A(1, 4+2k) = 0$$

$$A(1, 5+2k) = 0 \qquad\qquad k = 1, \ldots\ldots m,$$

$$A(1, 4+2m+2i) = -\frac{2T_E \dfrac{\partial \phi_{zi}(G)}{\partial l}}{Mu} - \frac{GS}{Mu} I(3+2n+i)$$

$$i = 1, \ldots\ldots, n$$

$$A(1, 5+2m+2i) = -\frac{QS}{Mu^2} I(3+i)$$

$$A(2, i) = 0 \qquad\qquad i = 1, \ldots\ldots, 5+2m+2n$$

$$A(2, 3) = 1.0$$

$$A(3,1) = \frac{QS}{Iyy}I(2)$$

$$A(3,2) = 0.0$$

$$A(3,3) = -\frac{QS}{I_{yy}u}I(3)$$

$$A(3,4) = -\frac{2T_E L_C + 2M_R L_R(u + g\sin\gamma_0)}{I_{yy}}$$

$$A(3,4+2j) = -\frac{M_{sj}L_{sj}(u + g\sin\gamma_0)}{I_{yy}}$$

$$A(3,5+2j) = 0.0 \qquad\qquad j = 1, \ldots, m$$

$$A(3,4+2m+2i) = \frac{2T_E\left[\phi_{zi}(G) - L_C\dfrac{\partial\phi_{zi}(G)}{\partial l}\right]}{I_{yy}} + \frac{QS}{I_{yy}}I(3+3n+i)$$

$$i = 1, \ldots, n$$

$$A(3,5+2m+2i) = \frac{QS}{I_{yy}u}I(3+n+i)$$

$$A(4,i) = 0 \qquad\qquad i = 1, \ldots, 5+2m+2n$$

$$A(4,5) = 1.0$$

$$A(5,1) = \frac{-M_R L_R u}{I_R}$$

$$A(5,2) = \frac{-M_R L_R\, g\sin\gamma_0}{I_R}$$

$$A(5,3) = \frac{M_R L_R u}{I_R}$$

$$A(5,4) = -\omega_n^2 - \frac{M_R L_R(u + g\sin\gamma_0)}{I_R}$$

$$A(5,5) = -2\varsigma_n\omega_n$$

$$A(5,4+2k) = 0$$

$$A(5,5+2k) = 0 \qquad\qquad k = 1, \ldots, m$$

$$A(5,4+2m+2i) = -\frac{M_R L_R(u + \sin\gamma_0)}{I_R}\frac{\partial\phi_{zi}(G)}{\partial l} \qquad\qquad i = 1, \ldots, n$$

$$A(5,5+2m+2i) = 0$$

$A(4+2k, i) = 0$ $A(4+2k, 5+2k) = 1.0$	$i = 1, \ldots\ldots, n$ $k = 1, \ldots\ldots, m$
$A(5+2j,1) = -\dfrac{u}{L_{sj}}$ $A(5+2j,2) = \dfrac{-g\sin\gamma_0}{L_{sj}}$ $A(5+2j,3) = \dfrac{u}{L_{sj}}$ $A(5+2j,4) = 0$ $A(5+2j,5) = 0$ $A(5+2j,4+2j) = -\omega_{pj}^2$ $A(5+2j,5+2j) = -2\varsigma_{pj}\omega_{pj}$ $A(5+2j,4+2m+2i) = 0$ $A(5+2j,5+2m+2i) = 0$	$j = 1, \ldots., m$ $i = 1, \ldots., n$
$A(4+2m+2i, j) = 0, \quad j = 1, \ldots\ldots, (5+2m+2n)$ $A(4+2m+2i, 5+2m+2i) = 1.0$	$i = 1, \ldots., n$
$A(5+2m+2i,1) = \dfrac{-QS}{m_i}I(3+i)$ $A(5+2m+2i,2) = 0$ $A(5+2m+2i,3) = \dfrac{QS}{m_i u}I(3+n+i)$ $A(5+2m+2i,4) = -\dfrac{2T_E\phi_{zi}(G)}{m_l}$ $A(5+2m+2i,5) = 0$ $A(5+2m+2i,4+2k) = 0$ $A(5+2m+2i,5+2k) = 0$	 $i = 1, \ldots\ldots, n$ $k = 1, \ldots\ldots, m$

$$A(5+2m+2i, 4+2m+2j) = -\omega_i^2 \delta_{ij} - \frac{2T_E \phi_{zi}(G)\dfrac{\partial \phi_{zi}(G)}{\partial l}}{m_i}$$

$$-\frac{QS}{m_i} I(3 + 2n + n^2 + (i-1)n + j)$$

$$A(5+2m+2i, 5+2m+2j) = -2\varsigma_i \omega_i \delta_{ij} - \frac{QS}{m_i u} I(3 + 4n + (i-1)n + j)$$

$$\delta_{ij} = 0 \qquad\qquad \text{for } i \neq j$$

$$= 1 \qquad\qquad \text{for } i = j$$

$$j = 1, \ldots\ldots, n$$

B-MATRIX:

$$B(i) = 0, \text{ for } i = 1, \ldots\ldots, (5+2m+2n)$$

$$B(5) = \omega_n^2$$

D-MATRIX:

$$\text{Initialise } D(i,j) = 0 \qquad\qquad \text{for } i = 1, \ldots\ldots, (5+2m+2n)$$
$$j = 1, \ldots\ldots, (5+2m+2n)$$

Only non-zero elements are listed below.

$$D(1,5) = -\frac{2M_R L_R}{Mu}$$

$$D(1,5+2j) = -\frac{M_{sj} L_{sj}}{Mu} \qquad\qquad \text{for } j = 1, \ldots\ldots, m$$

$$D(3,5) = -\frac{2(I_R + M_R L_R L_C)}{I_{yy}}$$

$$D(3,5+2j) = \frac{M_{sj} X_{sj} L_{sj}}{I_{yy}} \qquad\qquad \text{for } j = 1, \ldots\ldots, m$$

$$D(5,1) = -\frac{M_R L_R u}{I_R}$$

$$D(5,3) = -1 - \frac{M_R L_R L_C}{I_R}$$

$$D(5, 5+2m+2i) = -\frac{\partial \phi_{zi}(G)}{\partial l} - \frac{M_R L_R \phi_{zi}(G)}{I_R}, \qquad \text{for } i = 1, \ldots\ldots, n$$

$$D(5+2j,1) = -\frac{u}{L_{sj}}$$

$$D(5+2j,3) = \frac{X_{sj}}{L_{sj}}$$

$$D(5+2j,\, 5+2m+2i) = -\frac{\phi_i(sj)}{L_{sj}}, \ \text{for} \ i = 1, \ldots\ldots, n$$

$$j = 1, \ldots\ldots, m$$

$$D(5+2m+2i,5) = -\frac{2M_R L_R \left(\phi_{zt}(G) + L_R \dfrac{\partial \phi_{zi}(G)}{\partial l} \right)}{m_i}$$

$$D(5+2m+2i,5+2j) = -\frac{M_{sj} L_{sj} \phi_{zi}(sj)}{m_i}$$

$$\text{for } j = 1, \ldots\ldots, m$$
$$\text{for } i = 1, \ldots\ldots, n$$

H-MATRIX

Initialise $H(i) = 0$ for $i = 1, \ldots\ldots, (5+2m+2n)$

Only Non-zero elements are given below.

CASE I: ATTITUDE CONTROL LAW

Ref. eqns. 4.161 and 4.162

$$H(2) = 1.0$$

$$H(3) = K_R$$

$$H(4+2m+2i) = \frac{\partial \phi_{zi}(PG)}{\partial l}$$

$$H(5+2m+2i) = K_R \frac{\partial \phi_{zi}(RG)}{\partial l}$$

$$i = 1, \ldots,$$

n **CASE II** : LATAX CONTROL LAW

Ref. eqns. 4.166 to 4.172

1. Compute

$$H(k) = K_a \left\{ uA(1,k) - c\,A(3,k) + \sum_{i=1}^{n} \phi_{zi}(l_{acc})[A(5+2m+2i,k)] \right\}$$

$$k = 1, \ldots, (5+2m+2n)$$

2. Update $H(k)$ Computed in step1 as follows:

$$H(1) = H(1) + u\,K_a$$

$$H(3) = H(3) - K_R - u\,K_a$$

$$H(5+2m+2i) = H(5+2m+2i) - K_R \frac{\partial \phi_{zi}(RG)}{\partial l}, \qquad \text{for } i = 1, \ldots, n$$

Control Systems Design–3: Analysis and Design of ON-OFF Reaction Control Systems

5.1 INTRODUCTION

For launch vehicle and missile trajectories out of atmosphere, the ON-OFF reaction control system (RCS) is one of the candidates during powered phase and the only available candidate for the coast phase. Launch vehicles have a long duration of coast phases to gain altitude. For long range and medium range missiles also there will be a significantly long duration of free flight and on-off reaction control is required to keep the desired orientation for the re-entry phase of flight. In some of the vehicles such as SLV-3. ASLV, the reaction control system is used even during powered phases.

There is only limited literature available on the analysis and design of on-off control systems and hence, this topic is discussed in this chapter in some detail. Different types of reaction control systems and their general layout is described briefly in Chapter I Section 1.3.4 and hence, not repeated here.

The on-off reaction control systems have a limited maximum force level. Hence, it is necessary that the disturbance force is properly estimated. This feature of the on-off reaction control system in addition to its pulsating nature makes it unsuitable for using in high aerodynamic pressure region. The pulsating force will be having higher harmonic frequencies and is likely to have a tendency to excite the vehicle vibration modes. The limited maximum force feature can lead to a situation that the transient atmospheric disturbances can lead to a situation that the disturbances exceed the control force level which is catastrophic for unstable vehicles. However, it is possible to use the RCS when the dynamic pressure is very low. For example; at the take off-phase of a missile, till the aerodynamic pressure builds for aerodynamic control to become effective, or at a higher altitude when the small dynamic pressure is present but will become negligible in a short duration of flight.

The digital simulation is one of the popular tools for the analysis and design of RCS system. However, the simulation does not give an insight into the behaviour of the system even though one will be able to fix the design parameters of the control system. Hence, the subject is dealt with in detail in this chapter. The analysis mainly concentrates on out-of-atmosphere phase where aerodynamic torque is totally neglected. Experience shows that the same analysis can be effectively used to study the system behaviour when the aerodynamic torque is very small compared to the control torque and the limit cycle characteristics obtained using the analytical method match quite well with the characteristics obtained using simulation. However, Section 5.10 gives summary discussion when aerodynamic moment is also considered. Thus, we discuss RCS for three conditions:

  (*i*) A constant disturbance torque

 (*ii*) Zero disturbance torque and

(*iii*) Variable disturbance torque (torque proportionate to aerodynamic angle of attack).

5.2 CONTROL SYSTEM CHARACTERISTICS

Fig. 5.1 gives the block diagram of a typical RCS control system.

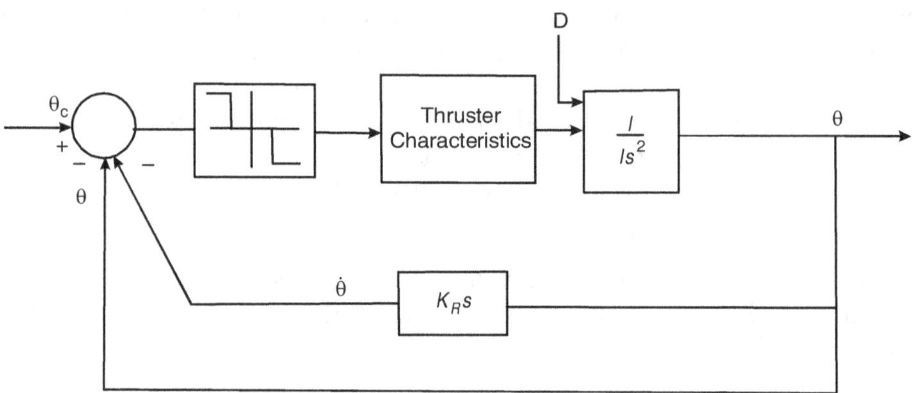

Fig. 5.1

The control law is given by

$$e' = \theta_c - \theta - K_R \theta$$

or
$$e = -\theta_c + \theta + K_R \theta \qquad\qquad \dots (5.1)$$

Then the control torque is given by

$$u = U \quad \text{if } e > dz$$
$$= -U \quad \text{if } e < -dz$$
$$= 0 \quad \text{if } |e| \le d_z \qquad\qquad \dots (5.2)$$

D is the constant disturbance moment acting on the system and U is the maximum control torque level.

The representative thruster characteristics is given in Fig. 5.2

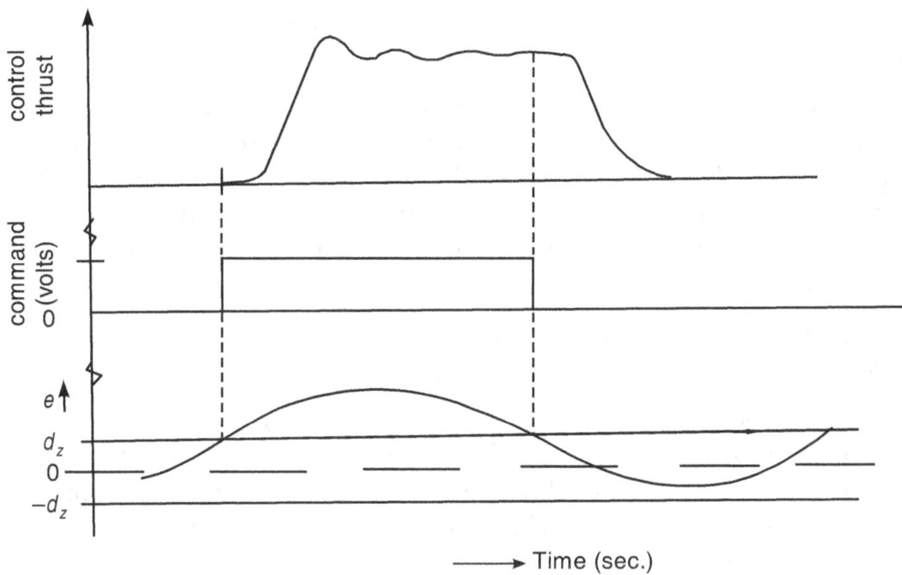

Fig. 5.2

For the simplicity of analysis, the thrust characteristics are modelled in different ways by different authors and are briefly described below:

1. Ref. 1 has modelled the characteristics as a second order transfer function.

Thus
$$\frac{F}{m} = \frac{F_c\omega_n^2}{s^2 + 2\varsigma_n\omega_n s + \omega_n^2} \qquad \ldots (5.3)$$

Where m is the output of the nonlinearity block.

He uses describing function approach to study the limit cycle characteristics. The on-delay and cut-off delay can be separately considered. However, the same characteristics is used for rise time and fall time of thrust. The damping ratio of, say, 0.6 gives a overshoot while rising which can be considered as realistic but it also gives same undershoot during thrust fall, which is never the case in actual practice. Moreover, in practical systems, a considerable difference is observed in the rise time and fall time of the thrust which cannot be represented in this model. The describing function approach becomes further complicated if one intends to use different second order model for rise time and fall time. Further, the approach gives a limit cycle frequency which is independent of disturbance torque.

The authors have observed that there is a good agreement between the results of describing function analysis and analog simulation study as far as duty cycle, steady state error and fuel consumption is concerned for varying thrust misalignment but the agreement is not good for the limit cycle frequency because the analytical results predict that the frequency of limit cycle does not change with thrust misalignment while the test data show a definite increase in frequency with increasing misalignment.

2. Ref. 2, 3 and 4 consider only the on-delay and cut-off delay in the model and ignore rise time and fall time. Ref. 2 considers a constant disturbance case (due to thrust misalignment) and out of atmosphere trajectory. Ref. 3 considers the atmosphere phase for both unstable and stable vehicle characteristics but ignores the constant disturbance torque due to thrust misalignments. Ref. 4 considers the case where the disturbance torque is absent.

3. Ref. 5 (the present author) models the thrust characteristics as valve delay (t_{d1}), cut-off delay (t_{d2}), a linear rise time (t_r) and linear fall time (t_f).

It also discusses special cases wherein the control command pulse is too short for the thruster to rise to full thrust level or off command is for a too short duration for the thruster to fall to zero thrust level. These cases occur when the control torque is too high for a given disturbance torque and the control torque is not adequately higher than the disturbance torque level.

Ref. 2 as well as 5 indicate that the limit cycle frequency is a function of the ratio of control torque to disturbance torque and thus, better represents the limit cycle characteristics than the describing function approach. However, the results of this analysis cannot be used if the disturbance torque is zero.

We first give a complete analysis of the limit cycle oscillations, then give the conditions to ensure stable limit cycle and finally give a step by step procedure for designing a control system.

5.3   ASSUMPTIONS

The following analysis assumes that the vehicle (launch vehicle or missile) is out of sensible atmosphere. Hence, the aerodynamic torque acting on the vehicle is neglected. The vehicle is assumed to be having a disturbance moment due to thrust misalignment whose magnitude may be varying slowly but is unidirectional. The time slice approach discussed in Chapter 4 is used and the vehicle parameters are assumed to remain constant over a short period of time. The analysis can then be repeated at regular intervals of time and the characteristics can be obtained for the entire powered phase.

The control system is assumed to have a constant pitch angle command θ_c and the system is expected to hold the vehicle attitude within the reasonable accuracy against a constant disturbance torque D. The case of varying θ_c at a constant pitch rate $\dot{\theta}_c$ is subsequently briefly discussed.

Consider the following process. Initially, vehicle is having an attitude of θ_c and the control thruster is off. The vehicle attitude and rate will build up due to the disturbance torque D. When the error function exceeds a preset value of dead zone d_z, the control motor is fired.

The control torque reverses the rate $\dot{\theta}$ and reduces the angle and the error function reduces to less than the dead zone. The control motor is commanded to shut off. However, the angle and rates will be further reduced due to the cut off delay and fall characteristics of thruster. Following two possibilities will occur:

1. The dead zone value is small enough, so that the error function crosses the other boundary of dead zone *i.e.*, $-d_z$. The opposite control motor is fired to bring back the vehicle state.

2. The dead zone is large enough so that the error function does not cross the opposite boundary of dead zone $(-d_z)$ before the thrust is reduced to zero. Once the control thrust is reduced, the disturbance force will automatically bring back the vehicle state and again exceed the dead zone boundary d_z.

The process repeats thereafter for the entire powered phase.

It is only logical to think that, the designer should select the dead zone, in such a way, that the error function oscillates about only one boundary of the dead zone and torque is activated only in a direction to oppose the disturbance torque, since

(*i*) disturbance torque is anyway going to bring back the vehicle state and there is no need to waste control fuel to do the same thing.

(*ii*) if the opposing control torque is also activated, the net torque in the direction of disturbance torque will be much higher and will give much higher angular rates and will require further more fuel to oppose the disturbance torque.

Since the error function is assumed to oscillate about one boundary of dead zone, there will be an average error over a cycle of oscillation and this average error will be larger if deadzone value is made larger. Therefore, an optimum value of deadzone would be that value for which the error function just touches the opposite boundary. In this case, the average error will be the smallest without spending any extra control fuel.

5.4 EXPRESSIONS FOR VEHICLE STATE AT DIFFERENT SALIENT POINTS DURING ONE OSCILLATION

Let I be the moment of inertia of the vehicle and

$$\alpha_c = \frac{U}{I} = \text{Angular acceleration due to control thruster} \qquad \dots (5.4)$$

$$\alpha_d = \frac{D}{I} = \text{Angular acceleration due to disturbance torque} \qquad \text{... (5.5)}$$

Consider one full cycle of oscillation as shown in Fig. 5.3.

Then $\qquad\qquad \theta = \alpha'$ $\qquad\qquad\qquad\qquad\qquad\qquad\qquad\qquad$ ... (5.6)

where α' is the resultant angular acceleration at any instant.

Integrating the Eq. 5.7 successively, one gets

$$\theta = \theta_0 + \int_0^t \alpha' dt \qquad \text{... (5.7)}$$

$$\theta = \theta_0 + \int_0^t \left[\theta_0 + \int_0^\eta \alpha' d\eta \right] dt$$

$$= \theta_0 + \theta_0 t + \int_0^t \int_0^\eta \alpha' d\eta \, dt \qquad \text{... (5.8)}$$

α' will have following expressions for various cases.

Case I: Control thrust is zero

$$\alpha' = \alpha_d \qquad \text{... (5.9)}$$

Case II: Control thrust is rising $\qquad\qquad\qquad\qquad\qquad\qquad$ ... (5.10)

$$\alpha' = \alpha_d - n_1 t, \ \ 0 \le t \le t_r, n_1 \text{ is the slope of rise of control torque}$$

Case III: Control thrust is maximum $\qquad\qquad\qquad\qquad\qquad$ ... (5.11)

$$\alpha' = \alpha_d - \alpha_c$$

Case IV: Control thrust is falling $\qquad\qquad\qquad\qquad\qquad\qquad$ ... (5.12)

$$\alpha' = \alpha_d - \alpha_c + n_2 t, \ \ 0 \le t \le t_f, n_2 \text{ is the slope of falling control torque}$$

Depending on the case under consideration, one gets following expressions for θ and θ.

Case I : $\qquad\qquad t_0 \le t \le t_2 \qquad\qquad\qquad\qquad t_6 \le t \le t_7$

$$\theta = \theta_0 + \alpha_d t \qquad\qquad\qquad \theta = \theta_6 + \alpha_d t \qquad \text{... (5.13)}$$

$$\theta = \theta_0 + \theta_0 t + \frac{1}{2}\alpha_d t^2 \qquad\qquad \theta = \theta_6 + \theta_6 t + \frac{1}{2}\alpha_d t^2 \qquad \text{... (5.14)}$$

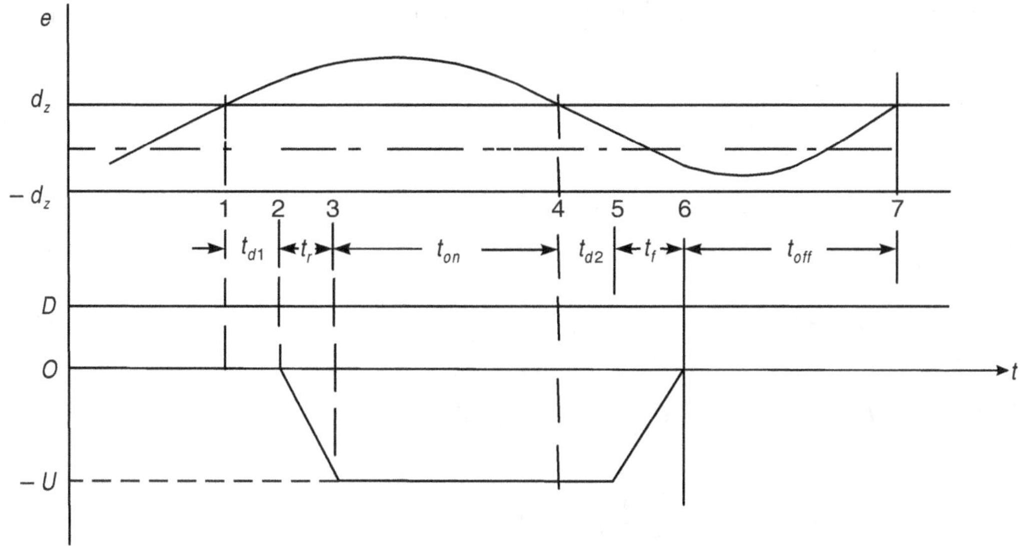

Fig. 5.3 Error function and Control Torque Vs. Time

Case II: $t_2 \leq t \leq t_3$

$$\theta = \theta_2 + \alpha_d t - \frac{1}{2} n_1 t^2 \tag{5.15}$$

$$\theta = \theta_2 + \theta_2 t + \frac{1}{2}\alpha_d t^2 - \frac{1}{6} n_1 t^3, \text{ for } 0 \leq t \leq t_r \tag{... (5.16)}$$

Case III : $t_3 \leq t \leq t_5$

$$\theta = \theta_3 + (\alpha_d - \alpha_c)t \tag{... (5.17)}$$

$$\theta = \theta_3 + \theta_3 t + \frac{1}{2}(\alpha_d - \alpha_c)t^2 \tag{... (5.18)}$$

Case IV : $t_5 \leq t \leq t_6$

$$\theta = \theta_5 + (\alpha_d - \alpha_c)t + \frac{1}{2} n_2 t^2 \tag{... (5.19)}$$

$$\theta = \theta_5 + \theta_5 t + \frac{1}{2}(\alpha_d - \alpha_c)t^2 + \frac{1}{6} n_2 t^3 \text{ for } 0 \leq t \leq t_f \tag{... (5.20)}$$

Using appropriate equations and initial conditions and using

$$n_1 = \frac{\alpha_c}{t_r} \text{ and } n_2 = \frac{\alpha_c}{t_f}, \tag{... (5.21)}$$

one can write down the expressions for the vehicle state at various salient points in Fig. 5.3 during one oscillation as follows:

$$\theta_2 = \theta_1 + \dot\theta_1 t_{d1} + \frac{1}{2}\alpha_d t_{d1}^2 \qquad \ldots (5.22)$$

$$\dot\theta_2 = \dot\theta_1 + \alpha_d t_{d1} \qquad \ldots (5.23)$$

$$\theta_3 = \theta_1 + \dot\theta_1(t_{d1} + t_r) + \frac{1}{2}\alpha_d(t_{d1} + t_r)^2 - \frac{1}{6}\alpha_e t_r^2 \qquad \ldots (5.24)$$

$$\dot\theta_3 = \dot\theta_1 + \alpha_d(t_{d1} + t_r) - \frac{1}{2}\alpha_c t^2 r \qquad \ldots (5.25)$$

$$\theta_4 = \theta_1 + \dot\theta_1(t_{d1} + t_r + t_{on}) + \frac{1}{2}\alpha_d(t_{d1} + t_r + t_{on})^2 - \frac{1}{2}\alpha_c\left[\left(\frac{1}{2}t_r + t_{on}\right)^2 + \frac{1}{12}t_r^2\right]$$
$$\ldots (5.26)$$

$$\dot\theta_4 = \dot\theta_1 + \alpha_d(t_{d1} + t_r + t_{on}) - \alpha_c\left(\frac{1}{2}t_r + t_{on}\right) \qquad \ldots (5.27)$$

$$\theta_5 = \theta_1 + \dot\theta_1\,(t_{d1} + tr + t_{on} + t_{d_2})^2 + \frac{1}{2}\alpha_d(t_{d1} + t_r + t_{on} + t_{d_2})^2$$
$$-\frac{1}{2}\alpha_c\left[\left(\frac{1}{2}t_r + t_{on} + t_{d2}\right)^2 + \frac{1}{12}t_r^2\right] \qquad \ldots (5.28)$$

$$\dot\theta_5 = \dot\theta_1 + \alpha_d(t_{d1} + tr + t_{on} + t_{d_2}) - \alpha_c\left(\frac{1}{2}t_r + t_{on} + t_{d2}\right) \qquad \ldots (5.29)$$

$$\theta_6 = \theta_1 + \dot\theta_1(t_{d1} + t_r + t_{on} + t_{d2} + t_f) + \frac{1}{2}\alpha_d(t_{d1} + t_r + t_{on} + t_{d2} + t_f)^2$$
$$-\frac{1}{2}\alpha_c\left[\left(\frac{1}{2}t_r + t_{on} + t_{d2} + \frac{1}{2}t_f\right)^2 + \left(\frac{1}{2}t_r + t_{on} + t_{d2} + \frac{1}{2}t_f\right)t_f + \frac{1}{12}\left(t_r^2 - t_f^2\right)\right]$$
$$\ldots (5.30)$$

$$\dot\theta_6 = \dot\theta_1 + \alpha_d(t_{d1} + t_r + t_{on} + t_{d2} + t_f + t_{off}) - \alpha_c\left(\frac{1}{2}t_r + t_{on} + t_{d2} + \frac{1}{2}t_f\right) \qquad \ldots (5.31)$$

$$\theta_7 = \theta_1 + \dot\theta_1(t_{d1} + t_r + t_{on} + t_{d2} + t_f + t_{off}) + \frac{1}{2}\alpha_d(t_{d1} + t_r + t_{on} + t_{d2} + t_f + t_{off})^2$$
$$-\frac{1}{2}\alpha_c\left[\left(\frac{1}{2}t_r + t_{on} + t_{d2} + \frac{1}{2}t_f\right)^2 + \left(\frac{1}{2}t_r + t_{on} + t_{d2} + \frac{1}{2}t_f\right) + \frac{1}{12}\left(t_r^2 - t_f^2\right)\right]$$
$$-\alpha_c\left[\frac{1}{2}t_r + t_{on} + t_{d2} + \frac{1}{2}t_f\right]t_{off} \qquad \ldots (5.32)$$

$$\dot\theta_7 = \dot\theta_1 + \alpha_d\,(t_{d1} + t_r + t_{on} + t_{d2} + t_f + t_{off}) - \alpha_c\left[\frac{1}{2}t_r + t_{on} + t_{d2} + \frac{1}{2}t_f\right] \qquad \ldots (5.33)$$

5.5 EXPRESSION FOR CONTROL IMPULSE

For steady state oscillation (with θ_c = constant) the state of the vehicle at points (1) and (7) must be same. Hence

$$\theta_1 = \theta_7 \qquad\qquad \text{... (5.34)}$$

and

$$\theta_1 = \theta_7 \qquad\qquad \text{... (5.35)}$$

Using Equation 5.35 in 5.33, one gets

$$\alpha_d(t_{d1} + t_r + t_{on} + t_{d2} + t_f + t_{off}) = \alpha_c\left[\frac{1}{2}t_r + t_{on} + t_{d2} + \frac{1}{2}t_f\right] \qquad \text{... (5.36)}$$

Let $\quad T_p = t_{d1} + t_r + t_{on} + t_{d2} + t_f + t_{off} =$ Period of limit cycle $\qquad$ ... (5.37)

And $\quad T_{on} = \dfrac{1}{2}tr + t_{on} + t_{d2} + \dfrac{1}{2}t_f =$ Effective on time of control motor $\qquad$ ... (5.38)

If we multiply both sides of Eq. 5.36 by Moment of Inertia I, one gets

$$DT_p = UT_{on} \qquad\qquad \text{... (5.39)}$$

If we consider entire powered phase to consist of 'n' such oscillations and multiply both sides of Eq. 5.39 by 'n', LHS of Eq. 5.39 gives area under the disturbance torque vs. time curve and right hand side as control torque impulse.

Thus, we get a very important expression for control torque impulse which simply states that

Control torque-impulse = Disturbance torque-impulse $\qquad\qquad$ ... (5.40)

Control force-impulse can then be simply calculated by dividing the control torque impulse by the control moment arm. Since the control moment arm will be continuously changing due to variation of vehicle *CG* due to propellant burning, one should calculate instantaneous equivalent control force by dividing the disturbance torque by the control moment arm and then find area under the equivalent control force vs. time curve.

The duty cycle for RCS during the oscillation is defined as effective on time divided by the period of oscillation. Therefore, using Eq. 5.39

$$\text{Duty cycle} = \frac{T_{on}}{T_p} = \frac{D}{U} = \frac{\textit{Disturbance torque}}{\textit{Control torque}} \qquad\qquad \text{... (5.41)}$$

It must be noted that these expressions for control impulse and duty cycle are correct only when the error function oscillates about one boundary of the dead zone and the control motor on the other side (in aiding direction to disturbance torque) does not get activated. The control impulse will increase rapidly if the other direction motor gets activated.

5.6   LIMIT CYCLE CHARACTERISTICS

5.6.1   Expressions for θ_1 and $\dot{\theta}_1$ and $\theta_{max}, \dot{\theta}_{max}$, etc.

Using Eq. 5.32, 5.33, 5.37 and 5.38, one gets

$$\theta_1 T_p + \frac{1}{2}\alpha_d T_p^2 - \frac{1}{2}\alpha_c\left[T_{on}^2 + t_{on}t_f + \frac{1}{12}(t_r^2 - t_f^2)\right] - \alpha_c T_{on}t_{off} = 0$$

Using the value of α_c Ton from Eq. 5.36 and simplifying one gets

$$\theta_1 = \frac{1}{2}\alpha_d\left[T_{off} + \frac{1}{2}t_f - \left(t_{d1} + \frac{1}{2}t_r\right)\right] + \frac{1}{24}\alpha_c\frac{\left(t_r^2 - t_f^2\right)}{T_p} \qquad \text{... (5.42)}$$

θ_{max} will occur during rise portion when

$$\theta = \alpha_d - n_1 t = 0 \qquad \text{or} \qquad t = \frac{\alpha_d}{n_1} \qquad \text{... (5.43)}$$

$$\therefore \qquad \theta_{max} = \theta_2 + \frac{1}{2}\frac{\alpha_d^2}{n_1} \qquad \text{... (5.44)}$$

Similarly, negative maximum of angular rate (or θ_{min}) will be occurring during control thrust fall duration and is given by

$$\theta_{min} = \theta_5 - \frac{1}{2}\frac{(\alpha_c - \alpha_d)^2}{n_2} \qquad \text{... (5.45)}$$

θ_2 and θ_5 can be computed once θ_1 and various time segments are known. θ_1 is then obtained from

$$\theta_1 = \theta_c + d_z - K_R\theta_1 \qquad \text{... (5.46)}$$

θ_{max} will be reached when $\theta = 0$, which can occur either between points 2 and 3 or between 3 and 4 depending on whether θ_3 is negative or positive.

After knowing θ_1 and all time segments, compute the vehicle states (*i.e.*, θ_5 and $\dot{\theta}_5$) at all the six points. If θ_3 is positive, θ_{max} will reach between 3 and 4 and if θ_3 is negative, θ_{max} will reach between 2 and 3. The $t_{\theta max}$ can be obtained using appropriate equation for θ from eqn. 5.15 or 5.17.

Thus, if θ_3 is negative, one gets

$$t_{\theta\max} = \frac{\alpha_d + \sqrt{\alpha_d^2 + 2n_1\theta_2}}{n_1} \qquad \ldots (5.47)$$

$$\theta_{\max} = \theta_2 + \theta_2 t_{\theta\max} + \frac{1}{2}\alpha_d t_{\theta\max}^2 - \frac{1}{6}n_1 t_{\theta\max}^3 \qquad \ldots (5.48)$$

If θ_3 is positive, one gets

$$t_{\theta\max} = \frac{\theta_3}{(\alpha_c - \alpha_d)} \qquad \ldots (5.49)$$

$$\theta_{\max} = \theta_3 + \frac{1}{2}\frac{\theta_3^2}{(\alpha_c - \alpha_d)} \qquad \ldots (5.50)$$

where θ_2, θ_3 can be computed once the value of d_z is known.

Similarly, $\theta_{\min}$ is obtained as follows:

(1) If $\theta_6 > 0$

$$t_{\theta\min} = \frac{(\alpha_c - \alpha_d) + \sqrt{(\alpha_c - \alpha_d)^2 - 2n_2\theta_5}}{n_2} \qquad \ldots (5.51)$$

$$\theta_{\min} = \theta_5 + \theta_5 t_{\theta\min} + \frac{1}{2}(\alpha_c - \alpha_d)t_{\theta\min}^2 + \frac{1}{6}n_2 t_{\theta\min}^3 \qquad \ldots (5.52)$$

(2) If $\theta_6 < 0$

$$t_{\theta\min} = -\frac{\theta_6}{\alpha_d} \qquad \ldots (5.53)$$

$$\theta_{\min} = \theta_6 - \frac{1}{2}\frac{\theta_6^2}{\alpha_d} \qquad \ldots (5.54)$$

5.6.2 Expressions for t_{off}, t_{on} and T_p

Use the vehicle state at Point 4 to get expressions for time segments. This point satisfies, the deadzone boundary as

$$-\theta_c + \theta_4 + K_R\theta_4 = d_z, \qquad \ldots (5.55)$$

Using , $-\theta_c + \theta_1 + K_R\theta_1 = d_z$, expression for θ_1 from Eq. 5.42 and t_{on} in terms of t_{off} using Eq. 5.36 and after simplification one gets

$$(t_{off} + \frac{1}{2}t_f)^2 + B(t_{off} + \frac{1}{2}t_f) + C = 0 \qquad \ldots (5.56)$$

$$\text{where } B = (t_{d1} + \frac{1}{2}t_r + t_{d2} + \frac{1}{2}t_f) - \frac{K_t(t_{d2} + \frac{1}{2}t_f)\left[K_R - \frac{1}{2}(t_{d2} + \frac{1}{2}t_f)\right] + \dfrac{t_r^2 - t_f^2 - k_t t_r^2}{24}}{K_R - \frac{1}{2}(t_{d1} + \frac{1}{2}t_r + t_{d2} + \frac{1}{2}t_f)}$$

$$\dots (5.57)$$

$$C = (t_{d1} + \frac{1}{2}t_r)(t_{d2} + \frac{1}{2}t_f)\left[1 - \frac{K_t\left[K_R - \frac{1}{2}(t_{d2} + \frac{1}{2}t_f)\right]}{K_R - \frac{1}{2}(t_{d1} + \frac{1}{2}t_r + t_{d2} + \frac{1}{2}t_f)}\right]$$

$$+ \frac{1}{24} \frac{K_t(t_{d1} + \frac{1}{2}t_r)t_f^2 + K_t(t_{d2} + \frac{1}{2}t_f)(t_r^2 - t_f^2) - (t_{d2} + \frac{1}{2}t_f)(t_r^2 - t_f^2)}{K_R - \frac{1}{2}(t_{d1} + \frac{1}{2}t_r + t_{d2} + \frac{1}{2}t_f)} \quad \dots (5.58)$$

where K_t = (Control torque / Disturbance torque) $= \dfrac{U}{T}$ $\qquad \dots (5.59)$

The solution of this equation is given by

$$t_{off} + \frac{1}{2}t_f = -\frac{B}{2} \pm \frac{1}{2}\sqrt{B^2 - 4C} \qquad \dots (5.60)$$

If we ignore $\dfrac{t_r^2}{24}$ and $\dfrac{t_f^2}{24}$ which are quite small, $(B^2 - 4C)$ becomes a perfect square and we get two values as

$$\left(t_{off} + \frac{1}{2}t_f\right) = -\left(t_{d1} + \frac{1}{2}t_r\right)$$

$$\text{OR}$$

$$= \frac{K_t(td_2 + \frac{1}{2}t_f)\left[K_R - \frac{1}{2}(td_2 + \frac{1}{2}t_f)\right]}{K_R - \frac{1}{2}(td_1 + \frac{1}{2}t_r + td_2 + \frac{1}{2}t_f)} - (td_2 + \frac{1}{2}tf)$$

The first value being negative, is not admissible. Then, the value of $(t_{off} + \frac{1}{2}t_f)$, $(t_{on} + \frac{1}{2}t_r)$ using eqn. 5.36 and the period of limit cycle, are written down as follows:

$$(t_{off} + \frac{1}{2}t_f) = (t_{d2} + \frac{1}{2}t_f)\left[\frac{K_t\left[K_R - \frac{1}{2}(t_{d2} + \frac{1}{2}t_f)\right]}{K_R - \frac{1}{2}(t_{d1} + \frac{1}{2}t_r + t_{d2} + \frac{1}{2}t_r)} - 1\right] \qquad \dots (5.61)$$

$$(t_{on} + \frac{1}{2}t_r) = \frac{t_{d1} + \frac{1}{2}t_r + t_{off} + \frac{1}{2}t_f}{K_t - 1} - (t_{d2} + \frac{1}{2}t_f) \qquad \ldots (5.62)$$

$$\text{and } T_p = \frac{K_t}{(K_t - 1)} \left[\frac{(t_{d2} + \frac{1}{2}t_f)K_t \left[K_R - \frac{1}{2}(t_{d2} + \frac{1}{2}t_f) \right]}{K_R - \frac{1}{2}(t_{d1} + \frac{1}{2}t_r + t_{d2} + \frac{1}{2}t_f)} + t_{d1} + \frac{1}{2}t_r - \left(t_{d2} + \frac{1}{2}t_f \right) \right] \qquad \ldots (5.63)$$

The inverse of T_p gives the frequency of oscillations.

Following observations can be made from the above expressions for time segments;

1. The period of limit cycle or frequency of oscillation is a function of K_t *i.e.*, the ratio of U and D.

2. If t_r and t_f are equated to zero, we get exactly the same expressions as in Ref. 2.

3. It is interesting to observe that an equivalent rectangular control torque characteristics can be considered for a trapezoidal control characteristics of Fig. 5.3. The equivalent time segments for rectangular characteristics are given by:

$$t'_{d1} = t_{d1} + \frac{1}{2}t_r \qquad \ldots (5.64)$$

$$t'_{on} = \frac{1}{2}t_r + t_{on} \qquad \ldots (5.65)$$

$$t'_{d2} = t_{d2} + \frac{1}{2}t_f \qquad \ldots (5.66)$$

$$t'_{off} = \frac{1}{2}t_f + t_{off} \qquad \ldots (5.67)$$

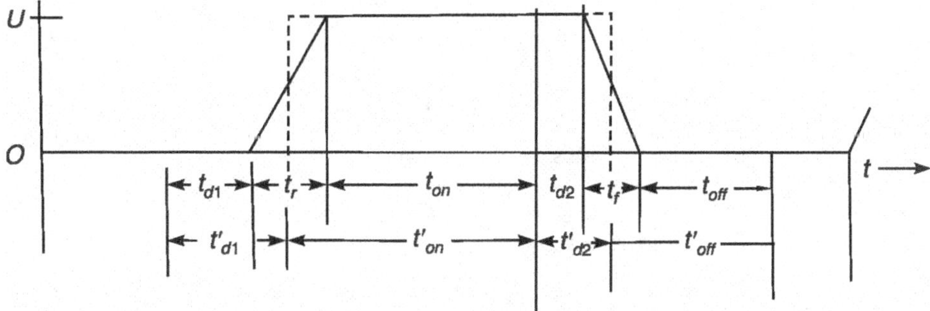

Fig. 5.4 Equivalent Rectangular Control torque Pulse

The vehicle state calculated using rectangular characteristics may be slightly different from those obtained by using trapezoidal characteristics. Appendix 5.1 gives the required expressions for various parameters of limit cycle for rectangular characteristics (or $t_r = t_f = 0$) for easy reference and are taken from Ref. 2.

5.6.3   Determination of the Optimum Width of Dead Zone ($2d_z$)

It can be seen from above analysis that the peak to peak oscillation amplitude of the error function is decided by U, D, K_R and the thrust characteristics and the d_z only serves as the boundary about which it oscillates. The average error can be computed by integrating θ over the entire oscillation and dividing by the period T_p. The average error thus will depend on d_z. If we reduce the value of d_z (Ref. Fig. 5.3) till the minimum of the error function just touches the opposite boundary of the dead zone *i.e.*, $-d_z$, we will get minimum average error without expenditure of any extra control fuel. This value of deadzone is called here as the optimum value of dead zone.

To find the minimum of error function e, first find the value of sign of e at point 6. If e_6 is +ve, the minimum would occur between 5 and 6 and the time can be found from

$$\frac{de}{dt} = \theta + K_R \theta = 0 \qquad \qquad \text{... (5.68)}$$

The required expressions are listed below without showing detailed calculations.

CASE I : $e_6 > 0$, Minimum occurs between 5 and 6

$$t_{e\min} = \left[\frac{\alpha_c - \alpha_d}{n_2} - K_R\right] + \sqrt{\left[\frac{\alpha_c - \alpha_d}{n_2}\right]^2 + K_R^2 - \frac{2\theta_5}{n_2}} \qquad \text{... (5.69)}$$

$$e_{\min} = -d_z \qquad \qquad \text{... (5.70)}$$

$$\therefore d_z = \frac{1}{2}\left[\alpha_c T_1\left(K_R + \frac{1}{2}T_1\right) + \frac{1}{24}\alpha_c t_r^2 - \frac{1}{2}n_2 t_{e\min}^2\left(K_R + \frac{1}{3}t_{e\min}\right) - T_2\left(\theta_1 + K_R\alpha_d + \frac{1}{2}\alpha_d T_2\right)\right] \qquad \text{... (5.71)}$$

where

$$T_1 = \frac{1}{2}t_r + t_{on} + t_{d2} + t_{e\min} \qquad \qquad \text{... (5.72)}$$

$$T_2 = t_{d1} + t_r + t_{on} + t_{d2} + t_{e\min} \qquad \qquad \text{... (5.73)}$$

CASE II. $e_6 < 0$,   minimum occurs between 6 and 7

$$t_{e\min} = -\frac{\theta_6 + K_R\alpha_d}{\alpha_d} \qquad \qquad \text{... (5.74)}$$

$$\therefore \quad d_z = \frac{1}{2}\left[\alpha_c T_3\left[K_R + \frac{1}{2}(T_3 + t_f) + t_{e\min}\right] + \frac{1}{24}\left(t_r^2 - t_f^2\right) - T_4\left(\theta_1 + K_R\alpha_d + \frac{1}{2}\alpha_d T_4\right)\right] \qquad \text{... (5.75)}$$

$$T_3 = \frac{1}{2}t_r + t_{on} + t_{d2} + \frac{1}{2}t_f \qquad \qquad \ldots (5.76)$$

$$T_4 = t_{d1} + t_r + t_{on} + t_{d2} + t_f + t_{e\min} \qquad \qquad \ldots (5.77)$$

In actual practice, the actual value of d_z should be sufficiently higher than this value to avoid activation of opposite side control thrusters due to uncertain values of delays, rise and fall times.

5.6.4 Average Angular Error over a Cycle

The average angular error is given by

$$\text{Average error} = \frac{1}{T_p} \int_0^{T_p} (\theta - \theta_c)\, dt$$

$$= \frac{1}{T_p} \begin{bmatrix} (\theta_1 - \theta_c)t_{d1} + \frac{1}{2}\theta_1 t_{d1}^2 + \frac{1}{6}\alpha_d t_{d1}^3 + (\theta_2 - \theta_c)t_r + \frac{1}{2}\theta_2 t_r^2 + \frac{1}{6}\alpha_d t_r^3 - \frac{1}{24}n_1 t_r^4 \\[2mm] +(\theta_3 - \theta_c)(t_{on} + t_{d2}) + \frac{1}{2}\theta_3(t_{on} + t_{d2})^2 + \frac{1}{6}(\alpha_d - \alpha_c)(t_{on} + t_{d2})^3 \\[2mm] +(\theta_5 - \theta_c)t_f + \frac{1}{2}\theta_5 t_f^2 + \frac{1}{6}(\alpha_d - \alpha_c)t_f^3 + \frac{1}{24}n_2 t_f^4 + (\theta_6 - \theta_c)t_{off} + \frac{1}{2}\theta_6 t_{off}^2 + \frac{1}{6}\alpha_d t_{off}^3 \end{bmatrix}$$

$$\ldots (5.78)$$

5.6.5 Effect of Ramp Input to the System

In launch vehicles, many times the pitch programs are given in terms of constant pitch rates $\dot{\theta}_c$. Indicating the corresponding vehicle states by primes and carrying out similar analysis we get following results:

1. Conditions to be used in the analysis are $\theta_7' = \theta_1'$

    Deadzone boundary:

$$\theta_7' - \theta_{C7}' + K_R\, \dot{\theta}_7' = \theta_1' - \theta_{C1}' + K_R\, \dot{\theta}_1' = d_z' \qquad \qquad \ldots (5.79)$$

$$\theta_7' - \theta_{C7}' = \theta_1' - \theta_{C1}'$$

$$\theta_{C7}' - \theta_{C1}' = \dot{\theta}_C' T_p'$$

$$\theta_7' = \theta_1' + \dot{\theta}_C' T_p'$$

2. Summary of analysis results

$$\theta_1' - \theta_c = \frac{1}{2}\alpha_d\left[t_{off}' + \frac{1}{2}t_f - t_{d1} - \frac{1}{2}t_r\right] + \frac{1}{24}\frac{\left(t_r^2 - t_f^2\right)}{T_p'} \qquad \dots (5.80)$$

Using vehicle state at point 4, we get same expressions for t_{off}', t_{on}' and T_p' as for constant θ_c case.

3. If $e_6 > 0$, error minimum occurs between points 5 and 6

$$t_{e\,min}' = \left[\frac{\alpha_c - \alpha_d}{n_2} - K_R + \sqrt{\left(\frac{\alpha_c - \alpha_d}{n_2}\right)^2 + K_R^2 - 2\frac{(\theta_5' - \theta_c')}{n_2}}\right] \qquad \dots (5.81)$$

$$d_z' = \frac{1}{2}\left[\alpha_c T_1'(K_R + \frac{1}{2}T_1') + \frac{1}{2}\alpha_c t_r^2 - \frac{1}{2}n_2 t_{e\,min}'^2)\left(K_R + \frac{1}{3}t_{e\,min}'\right)\right.$$

$$\left. -T_2'\left[\theta_1' - \theta_c + K_R\alpha_d + \frac{1}{2}\alpha_d T_2'\right]\right] \qquad \dots (5.82)$$

4. If $e_6 < 0$, minimum error occurs between 6 and 7

$$t_{e\,min}' = -\frac{\theta_6' - \theta_c' + K_R\alpha_d}{\alpha_d} \qquad \dots (5.83)$$

$$d_z = \frac{1}{2}\left[\alpha_c T_3'(K_R + \frac{1}{2}(T_3' + t_f) + t_{e\,min}' + \frac{1}{24}(t_r^2 - t_f^2) - T_4'(\theta_1' - \theta_c + K_R\alpha_d + \frac{1}{2}\alpha_d T_4')\right] \qquad \dots (5.84)$$

where T_3' and T_4' have same expressions as Eqs. 5.76 and 5.77 in terms of t_{on}'.

The expression of average error is same as Eq. 5.78 with following replacements:

$(\theta_1 - \theta_c)$ to be replaced by $(\theta_1' - \theta_{c1})$

$(\theta_2 - \theta_c)$ to be replaced by $(\theta_2' - \theta_{C2})$ and so on till

$(\theta_6 - \theta_c)$ to be replaced by $(\theta_6 - \theta_{C6})$

and $\qquad\theta_1$ to be replaced by $(\theta_1' - \theta_C)$

θ_2 to be replaced by $(\theta_2' - \theta_C)$ and so on till θ_6 to be replaced by $(\theta_6' - \theta_C)$

The above analysis leads us to following conclusions:

1. The expressions for t_{on}, t_{off} and T_p remain same for both cases:

(*i*) constant θ_c and (*ii*) ramp θ_c (or $\theta_c = $ constant).

2. Control impulse and duty cycle for both cases remain same

3. The optimum value of deadzone remains same for both cases.

4. The numerical value of θ_1 and hence the rates $\theta_2, \theta_3, ..., \theta_6$ for constant θ_c case are same as the numerical value of $(\theta_1' - \theta_c)$ and hence $(\theta_1' - \theta_c), (\theta_3' - \theta_c'), ..., (\theta_6' - \theta_c)$

   This automatically means that the actual rates $\theta_1, \theta_2'....\theta_6'$ are higher for ramp case by an amount of θ_c or the slope of ramp compared to the rates of constant θ_c case.

5. Using the dead zone equation

$$(\theta_1' - \theta_{c1}) = d_z - K_R\,\theta_1' = d_z - K_R(\theta_1 + \theta_c)$$

$$= d_z - K_R\,\theta_1 - K_R\,\theta_c = (\theta_1 - \theta_c) - K_R\,\theta_c$$

Or
$$(\theta_{C1} - \theta_1') = (\theta_c - \theta_1) + K_R\,\theta_C \qquad\qquad ...\,(5.85)$$

And similarly at other points 2, 3 .....6. This shows that the average tracking error measured as $(\theta_C - \theta)$ in case θ_C as a ramp is more by $K_R\,\theta_C$ than the average error in case of constant θ_C.

It will be interesting to compare this result with Eqn. 3.75 of Chapter 3, which shows that when the pitch program is given by a constant pitch rate θ_C, the average tracking error increases by an amount $K_R\,\theta_C$. This additional error can be reduced if we define the dead zone boundary for control law as :

$$(\theta - \theta_C) + K_R(\theta - \theta_c) = \pm\, d_z \qquad\qquad ...\,(5.86)$$

5.6.6 Special Cases

The analysis given in earlier sections is valid in normal cases where the computed values of t_{on} and t_{off} are positive.

It can be easily shown that

(*i*) for t_{off} to be positive (using Eq. 5.61)

$$K_t > \left[\frac{t_{d2} + t_f}{t_{d2} + \dfrac{1}{2}t_f}\right] \times \frac{A_2}{A_1} \qquad\qquad ...\,(5.87)$$

where
$$A_1 = K_R - \frac{1}{2}(t_{d2} + \frac{1}{2}t_f) \qquad\qquad ...\,(5.88)$$

$$A_2 = K_R - \frac{1}{2}(t_{d1} + \frac{1}{2}t_r + t_{d2} + \frac{1}{2}t_f) \qquad \ldots (5.89)$$

and

(*ii*) for t_{on} to be positive (using Eqs. 5.62 and 5.61)

$$K_t < \frac{A_2(t_{d1} + t_r)}{\frac{1}{2}t_r A_2 - \frac{1}{2}(t_{d1} + \frac{1}{2}t_r)(t_{d2} + \frac{1}{2}t_f)} \qquad \ldots (5.90)$$

Equation 5.87 and 5.90 thus give a desirable range of K_t for satisfactory duty cycle during limit cycle operation.

Consider now following situations:

Case I : Control Torque too High for a Given Disturbance Torque

Let the vehicle have a constant disturbance torque. If now we increase the available control torque level, the value of t_{on} will decrease and after some level, t_{on} will become negative. This means, that the control is switched on when the error crosses the dead zone. After the delay period (t_{d1}), control torque starts rising. The error function starts decreasing and enters the deadzone before the thruster reaches the full level. The thrust continues to rise for a cutoff delay period and then starts decreasing with slope n_2. Two sub cases arise.

Case I-1 : The control thrust does not reach full level even after the cut off delay (t_{d2}) and starts decreasing.

This gives a triangular pulse as shown in Fig. 5.5

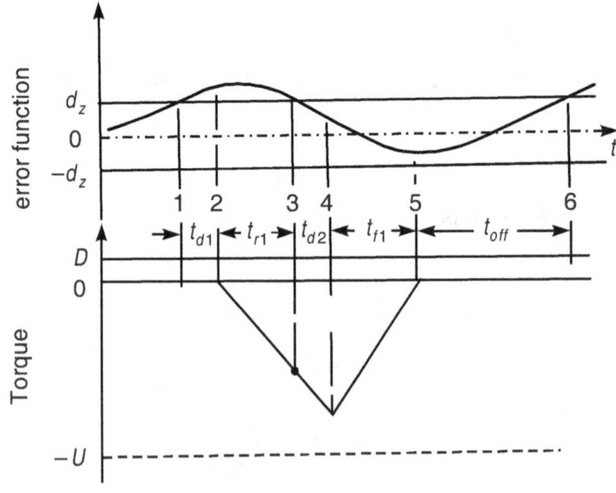

Fig. 5.5

Case I-2: The thrust reaches full level in part of the cut-off delay, remains at full level for the remaining part of delay and then starts falling.

This gives a trapezoidal pulse as shown in Fig. 5.6.

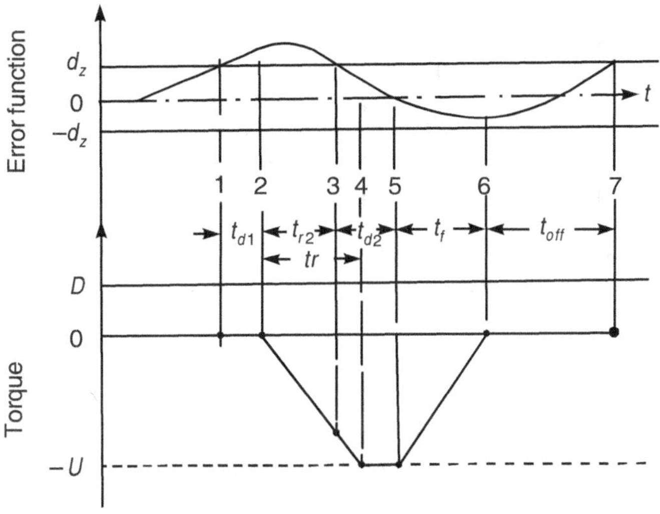

Fig. 5.6

Case II: Control Torque not Adequately Higher than Disturbance Torque

Consider the situation that for a fixed control torque level, the disturbance torque is gradually increasing. The time t_{off} will continuously decrease and at a certain level, t_{off} will become negative. This means that the control will be switched off when error enters the dead zone. The thrust will start falling after cutoff delay (t_{d2}). However, when the thrust falls below the disturbance level, error will again start increasing and cross the dead zone before the thrust falls to zero and switch on command will be issued to thrust again. The thrust is assumed to fall for a duration of on-delay (t_{d1}) and then start rising at the slope n_1, *i.e.*, same slope at which thrust would rise from zero level.

Here again two sub cases can be considered.

Case II-1: The control thrust does not reach zero even after on delay (t_{d1}) after the error crosses the deadzone.

This gives a triangular notch in the control thrust pulse as shown in Fig 5.7

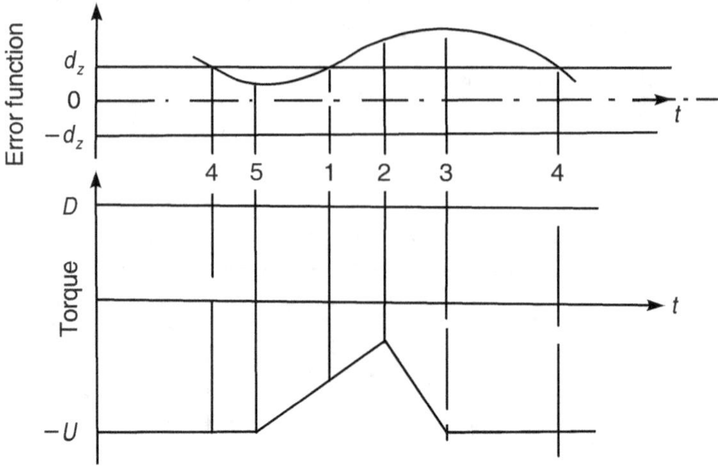

Fig. 5.7

Case II-2: The control thrust reaches zero in part of the on-delay time (t_{d1}) and remains zero during the remaining part of the delay time and then starts rising.

This gives a trapezoidal notch in the control pulse as shown in Fig. 5.8

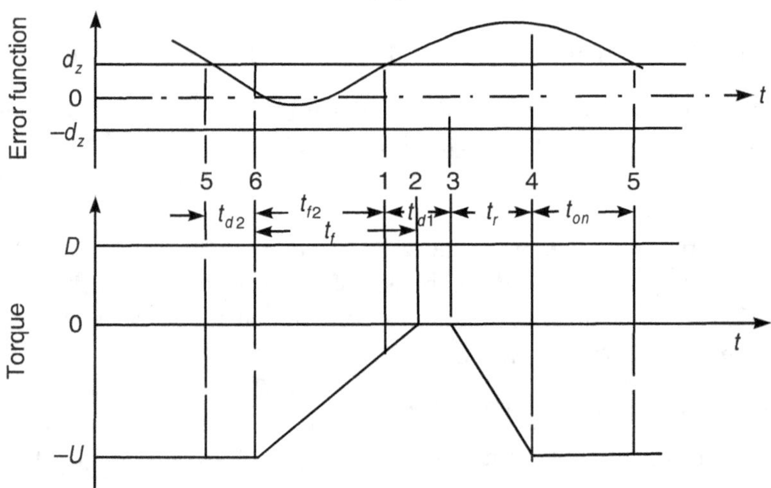

Fig. 5.8

The procedure for calculating various quantities for the special cases is the same as for normal case discussed in previous sections except for the fact that one is required to solve cubic equations to calculate time segments and select the proper root before proceeding further. Appendix 5.2 gives the required expressions for the special cases.

5.7   STABILITY CONDITIONS FOR THE LIMIT CYCLE

For the sake of simplicity of analysis, we consider equivalent rectangular control pulse for the actual trapezoidal control pulse as shown in Fig. 5.3 with the following parameters:

$$t'_{d1} = t_{d1} + \frac{1}{2}t_r = T_R = equivalent\ on\text{-}delay \qquad\qquad \dots (5.91)$$

$$t'_{d2} = t_{d2} + \frac{1}{2}t_f = T_D = equivalent\ cut\text{-}off\ delay \qquad\qquad \dots (5.91)$$

PHASE PLANE DIAGRAM FOR THE SYSTEM

The phase plane diagram for the system is obtained as follows:

$$\theta = \frac{d\theta}{dt} = \alpha_d \qquad and \qquad \frac{d\theta}{dt} = \theta$$

$$\theta\, d\theta = \alpha_d d\theta \qquad\qquad \dots (5.93)$$

or $\qquad\qquad \theta^2 = 2\alpha_d \theta + C_1$ when control is off and

$$\theta_2 = 2(\alpha_d - \alpha_c)\theta + C_2 \ \text{when control is ON}$$

$$= -2(\alpha_c - \alpha_d)\theta + C_2 \ \text{when control is ON.} \qquad\qquad \dots (5.94)$$

These are the parabolic trajectories in phase plane. The dead zone boundaries are given by:

$$-\theta_C + \theta + K_R\,\theta = \pm d_z$$

This gives two straight lines for $+d_z$ and $-d_z$ with a slope of $= -\dfrac{1}{K_R}$

Fig. 5.9 shows the required phase plane diagram for trajectories and dead zone for the limit cycle under consideration.

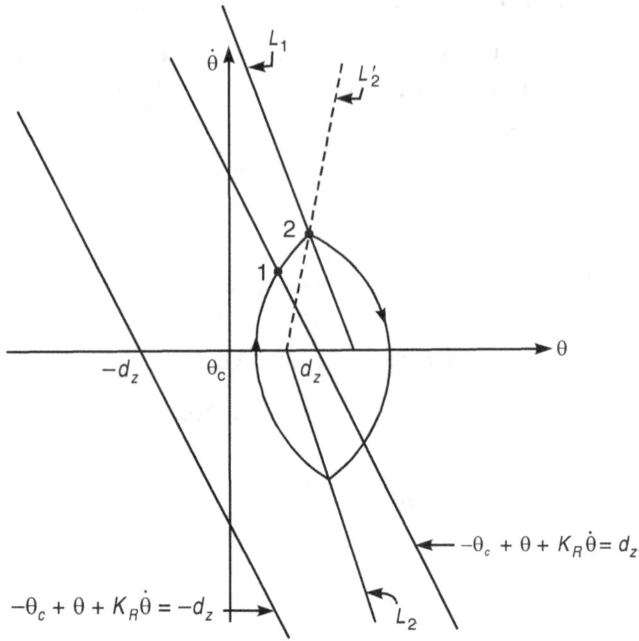

Fig. 5.9 Limit cycle Phase plane diagram

To obtain the equation for switching line L_1

$$\theta = \theta_2 = \theta_1 + \alpha_D T_R \qquad \qquad \text{... (5.95)}$$

$$\theta = \theta_2 = \theta_1 + \theta_1 T_R + \frac{1}{2}\alpha_D T_R^2 \qquad \qquad \text{... (5.96)}$$

Eliminating θ_1 and θ_1 from Eqs. 5.95 and 5.96 and dead zone boundary

$$-\theta_C + \theta + (K_R - T_R)\theta = d_z + \alpha_d T_R (K_R - \frac{1}{2}T_R) \qquad \qquad \text{... (5.97)}$$

Similarly, equation for L_2 is obtained as

$$-\theta_C + \theta + (K_R - T_D)\,\theta = d_z - (\alpha_C - \alpha_d)T_D(K_R - \frac{1}{2}T_D) \qquad \qquad \text{... (5.98)}$$

If we take the mirror image (L_2') of L_2 about the θ axis, it will intersect L_1 in Point 2. For limit cycle to exist, we have

$$0 < \theta_2 (= -\theta_4) < \infty$$

where θ_2 is the rate at intersection of L_1 and L_2'.

we get,

$$\theta_2 = \frac{\alpha_d T_R \left[K_R - \frac{1}{2} T_R \right] + (\alpha_c - \alpha_d) T_D \left(K_R - \frac{1}{2} T_D \right)}{2 K_R - T_R - T_D} \qquad \dots (5.99)$$

Let
$$K_R = r(T_R + T_D)/2$$

Then it can be shown that

$$\theta_2 = \frac{(r-1)T_R^2 + (K_t - 1)(r-1)T_D^2 + r T_R T_D}{(r-1)(T_R + T_D)}$$

For θ_2 to be positive, we must have

(1) $r > 1$ or $K_R > (T_R + T_D)/2$

$$\text{OR} \qquad (K_R > \frac{1}{2}\left[td_1 + \frac{1}{2}t_r + t_{d2} + \frac{1}{2}t_f \right]) \qquad \dots (5.100)$$

(2) $\qquad\qquad K_t > 1$ or $U > D$ $\qquad\qquad\qquad \dots (5.101)$

OR (control torque > disturbance torque)

These are the sufficiency conditions for the stable limit cycle.

If $r < 1$, then numerator of θ_2 expression on RHS of (5.99) must also be negative. This is satisfied if

$$r < \frac{T_R^2 + (K_t - 1)T_D^2}{[T_R^2 + (K_t - 1)T_D^2 + K_t T_R T_D]} \qquad \dots (5.102)$$

For such a condition, there may occur a limit cycle. However, it can be shown by drawing actual sketches of parabolas that in all such cases any disturbance to limit cycle leads to diverging oscillations showing that the limit cycle is not a stable one. On the other hand, if the conditions 5.100 and 5.101 are satisfied one can show that any disturbance to the limit cycle leads to an oscillation which converges back to the same limit cycle.

By drawing sketches for various cases one can find that

(*i*) A system without delays and $K_R = 0$, gives pure oscillator and maintains initial amplitude of oscillation.

(*ii*) A system without delays and $K_R > 0$ gives converging oscillation and final maximum rate during limit cycle will only be decided by the shortest on-time a RCS system can give.

If magnitude of slope of the deadzone boundary is smaller than the slope of parabola with control ON, the vehicle state will slide down the deadzone boundary instead of having converging oscillations (see Fig. 5.10).

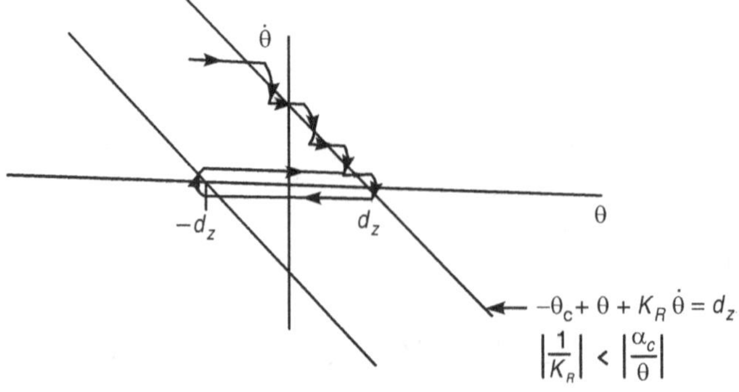

Fig. 5.10

(*iii*) Delays introduce a break in the switching lines for positive θ and negative θ and also changes the slope of switching lines from that of deadzone boundary.

It gives a stable limit cycle provided the conditions given by Eq. 5.100 and 5.101 are satisfied. Any disturbance given to the limit cycle by changing the vehicle state converges back to the same limit cycle.

(*iv*) If delays are present but $K_R = 0$ (*i.e.* condition 5.100 not met) one can easily see by drawing sketches that the system leads to diverging oscillations.

5.8   LIMIT CYCLE ANALYSIS FOR ZERO DISTURBANCE

5.8.1   Three Phases of Controlled Motion

The state of a vehicle with a big initial error and with zero disturbance can be brought down to a final steady state oscillation in two ways.

The first approach which is a very inefficient approach is studied in detail in Ref. 4 and is shown diagrammatically in Fig. 5.11.

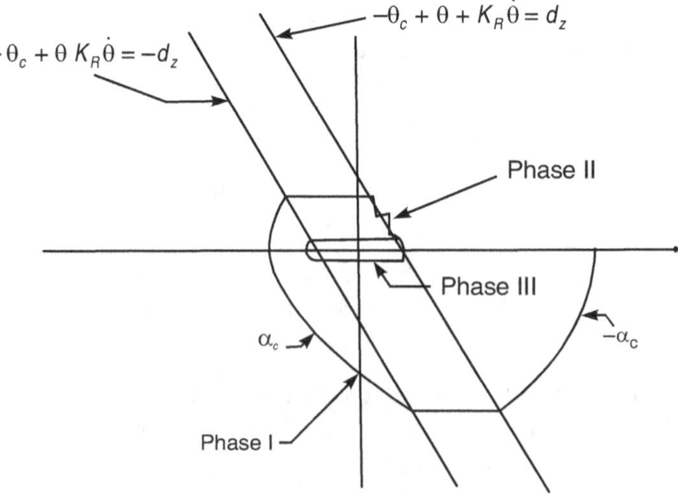

Fig. 5.11

The total transition is considered in three phases.

Phase I. This consists of control phase with decaying amplitude and with an approximate slope of envelope given by

$$\frac{d\theta_{env}}{dt} = -\alpha_c(K_R - T_D) - \frac{K_R\alpha_c^2 T_D + \frac{\alpha_c^2}{2}T_D^2}{\alpha_c(K_R - T_D) - \sqrt{\alpha_c^2 K_R^2 + 2\alpha_c\theta_0}}$$

$$= -\alpha_c K_R \text{ if } K_R \gg T_R, T_D \qquad \text{... (5.103)}$$

or $$\theta_{env} = \theta_0 - \alpha_c K_R t \qquad \text{... (5.104)}$$

The phase I exists if $\theta_0 > \alpha_c K_R^2$ ... (5.105)

Phase I continues till the start of Phase II which occurs when slope of the trajectory and the switching line matches or when

$$\frac{d\theta}{d\theta} = \pm\frac{\alpha_c}{\theta} = -\frac{1}{K_R}$$

Ignoring the value of d_z, we get $\theta + K_R\theta = 0$

This gives the envelope value of θ for Phase I to end, *i.e.,*

$$\theta_1 = \alpha_c K_R^2 \qquad \text{... (5.106)}$$

and duration of Phase I is given by

$$T_1 = \frac{\theta_0}{\alpha_c K_R} - K_R \qquad \text{... (5.107)}$$

Phase II. This phase continues till the steady state limit cycle peak rate is reached. At this value of the rate, switching-on the control leads to the change in sign of the rate and system gets into steady oscillation.

Using the switching boundary equation

$$\theta = -\frac{\theta \pm d_z}{K_R}$$

$$\therefore \quad \frac{d\theta}{\theta \pm d_z} = -\frac{dt}{K_R}$$

or $$T_{II} = K_R \ln\left(\frac{\theta_1 \pm d_z}{\theta_{II} \pm d_z}\right) = K_R \ln\left(\frac{\theta_I}{\theta_{II}}\right) \qquad (5.108)$$

Phase III : Steady state oscillation:

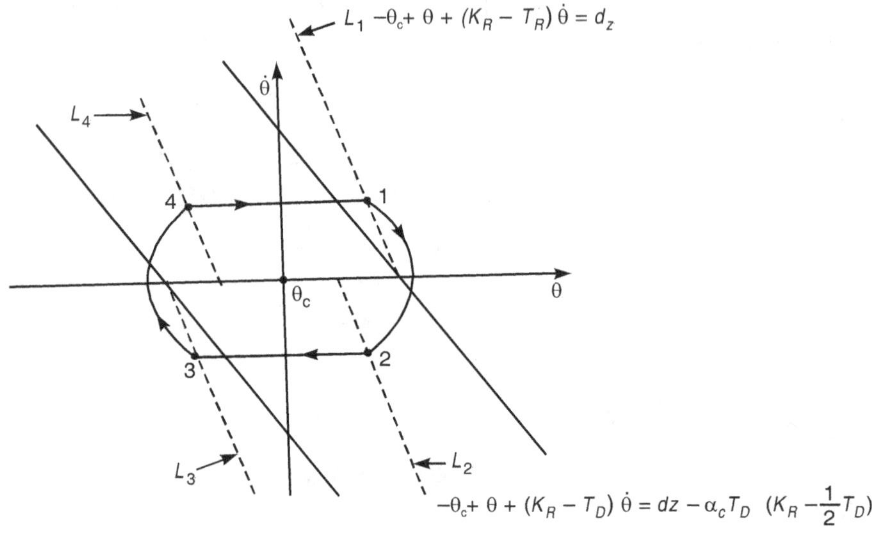

Fig. 5.12 Limit Cycle for Zero Disturbance

From Eq. 5.97, it is seen that L_1 switch line intersects θ axis at $\left[d_z + \alpha_d T_R (K_R - \frac{1}{2}T_R) \right]$ which is equal to d_z for the present case since $\alpha_d = 0$.

The switch line L_2 intersects θ-axis at $\left[d_z - \alpha_c T_D (K_R - \frac{1}{2}T_D) \right]$ and similarly for L_3 and L_4.

The eq. 5.99 gives the maximum rate during the limit cycle as ($\alpha_d = 0$)

$$(i) \qquad \theta_1 = \frac{\alpha_c T_D (K_R - \frac{1}{2}T_D)}{2K_R - T_R - T_D} \qquad \qquad \ldots (5.109)$$

$$-\theta_c + \theta_1 = -\frac{\alpha_c T_D (K_R - \frac{1}{2}T_D)}{2K_R - T_R - T_D}(K_R - T_R) + d_z$$

$$(ii)\ \text{Amplitude } (-\theta_c + \theta_1)_{max} = (-\theta_c + \theta_1) + \frac{\theta_1^2}{2\alpha_c} \qquad \ldots (5.110)$$

$$(iii)\ \text{Total on-time during the cycle, } t_{on} = 4\frac{\theta_1}{\alpha_c} \qquad \ldots (5.111)$$

(*iv*) Total off time during the cycle, $t_{off} = 4\dfrac{\theta_1}{\theta^l}$

$$\hspace{12cm} \dots (5.112)$$

(*v*) Period of limit cycle, $T_p = 4\left[\dfrac{\theta_1}{\alpha_c} + \dfrac{\theta}{\theta}\right]$

$$\hspace{12cm} \dots (5.113)$$

(*vi*) Percent on time $= \dfrac{100\, t_{on}}{Tp} = \dfrac{400\, \theta_1/\alpha_c}{Tp}$

$$\hspace{12cm} \dots (5.114)$$

The above expressions for limit cycle parameters are applicable, provided the switching off condition point 2 and 4 fall within the deadzone. If we narrow down the dead zone progressively, a stage will come for which the switch off condition will occur after the trajectory goes out of deadzone. The opposite control motor will be switched on but during part of the on delay time T_R, state will follow parabolic trajectory and afterwards constant rate trajectory. The limiting conditions for switching to occur inside the dead zone are (Ref. 4 for details):

(1) Minimum dead zone for switch off points 2 and 4 to occur within the dead zone are given by

$$d_{z\,\min} = \theta_1(K_R - \tfrac{1}{2}T_R) = \dfrac{\alpha_c T_D\left[K_R - \dfrac{1}{2}T_D\right]\left[K_R - \dfrac{1}{2}T_R\right]}{\left[2K_R - \dfrac{T_R + T_D}{2}\right]} \qquad \dots (5.115)$$

(2) Maximum amplitude for which switch point occurs inside the deadzone

$$\theta_{\max} = \theta_1 + \dfrac{\theta_1^2}{2\alpha_c} \qquad \dots (5.116)$$

The limit cycle parameters in case of switch off taking place outside the dead zone are given by:

$$\theta_1 = \dfrac{2\alpha_c T_d\left[K_R - \dfrac{1}{2}T_D\right] - 2\delta}{2K_R - T_R - T_D - \dfrac{2\delta}{\alpha_c K_R}} \qquad \dots (5.117)$$

$$\theta = \dfrac{\left(K_R \alpha_c T_D - \dfrac{\alpha_c}{2}T_D^2\right)\left(T_R - T_D + \dfrac{2\delta}{\alpha_c K_R}\right)}{2K_R - T_R - T_D - \dfrac{2\delta}{\alpha_c K_R}} \qquad \dots (5.118)$$

The $\theta_{\max}$, percentage on time, period and impulse per cycle can be obtained similar to case of switching inside the deadzone.

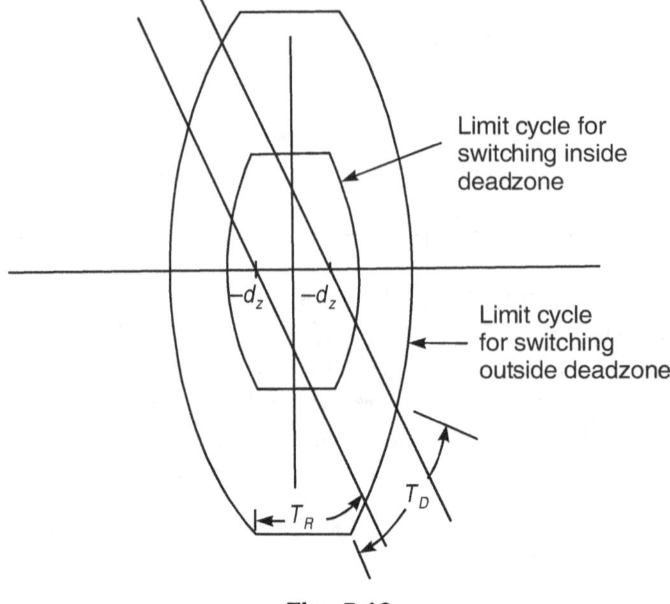

Fig. 5.13

5.8.2   Sensitivity of Impulse Consumption to Control Parameters

It will be of interest to see how the control impulse consumption varies with control force level when the disturbance is zero. For launch vehicles during coast phase the accuracy is not very critical, since it has no implication on mission performance. (This is not true in case of satellites where pointing accuracy of antennas or sensors will be an important consideration). Here we consider that the deadzone is adequately large so that the control switch off takes place within the dead zone.

The control impulse consumed per second is given by

$$I_s = \frac{On\ time \times force}{period} \qquad \qquad \dots (5.119)$$

Ignoring T_R and T_D in comparison to K_R we get

$$I_s = \frac{\dfrac{4\theta_1}{\alpha_c} \times \dfrac{I\alpha_c}{l_c}}{4\left[\dfrac{\theta_1}{\alpha_c} + \dfrac{\theta}{\theta_1}\right]}$$

$$= \frac{\dfrac{I}{l_c}\,\alpha_c T_D\,\dfrac{T_D}{2} * \alpha_c \dfrac{T_D}{2}}{\alpha_c \dfrac{T_D}{2} \cdot \dfrac{T_D}{2} + d_z - K_R \alpha_c \dfrac{T_D}{2}}$$

$$= \frac{\dfrac{I}{l_c}\,\alpha_c^2 \left(\dfrac{T_D}{2}\right)^2}{d_z - \alpha_c \dfrac{T_D}{2}\left[K_R - \dfrac{T_D}{2}\right]} = \frac{\dfrac{l_c}{I}\,F_c^2 \left(\dfrac{T_D}{2}\right)^2}{d_z - \alpha_c \dfrac{T_D}{2}\left[K_R - \dfrac{T_D}{2}\right]} \qquad \text{... (5.120)}$$

The term $\alpha_c \dfrac{T_D}{2}\left[K_R - \dfrac{T_D}{2}\right]$ is much smaller than d_z. Hence we see that the impulse consumption per sec. varies directly as (1) the square of control force, (2) square of cut off delay and (3) inversely as deadzone. This indicates that one should use small control force, small cut off delays and large deadzone during coast phase to minimize fuel consumption.

5.9 EFFICIENT METHODS TO BRING DOWN LARGE INITIAL ANGULAR ERROR AND RATE OR THE FAULT TOLERANT CONTROL LOGIC

We have seen in earlier section that the normal control logic brings down the large initial angle and rate in a very inefficient way such as Phase I, Phase II and finally the limit cycle oscillation. The situation will be worse if the rate gyro is saturated due to large rate. This need not occur in a normal functioning of the control system but can occur in case of some temporary failures such as one of the actuator controlling aerodynamic surfaces malfunctions and drives the surface to the maximum limit. This has occurred in one of the flights for roll control just before first stage separation. The case is illustrated with following data:

Rate gyro saturation limit = 30°/sec

$$K_R = 0.5$$

$$d_z = \pm 1°$$

Consider the situation where initial rate $\approx 60°$ per sec. and the vehicle is spinning when the RCS is switched on. The angle sensor senses −ve angle when the angle exceeds 180°. Due to this, the error function changes sign and opposite control motor will be switched on which will try to increase the roll rate instead of reducing it. The switching boundary will also get modified due to rate saturation as shown in Fig 5.14 (Ref. 6, 7).

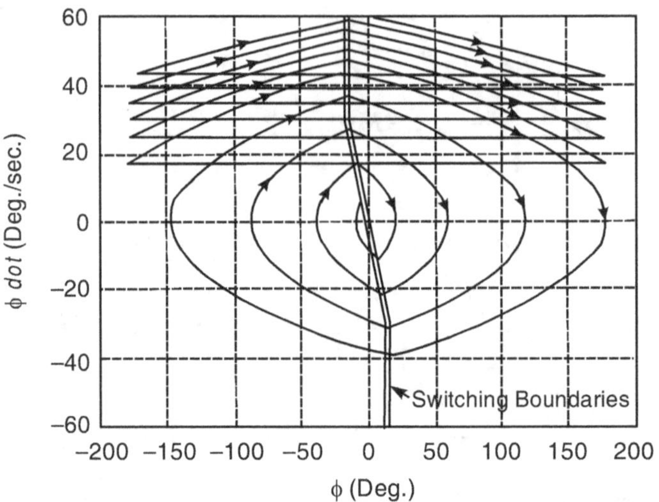

Fig. 5.14

The control logic will function as follows:

For rate $\dot{\phi} > 30$, $\dot{\phi} = 30$

$$K_R \dot{\phi} = 0.5 x 30 = 15$$

$\therefore$
$$e = \phi + K_R \dot{\phi}$$
$$= \phi + 15 > 1°, \alpha = -\alpha_c \qquad \text{... (5.121)}$$
$$< -1°, \alpha = +\alpha_c$$
$$|e| < 1°, \quad \alpha = 0$$

$\therefore$
$$\alpha = -\alpha_c \text{ during } -14 < \phi < 180° \qquad \text{... (5.122)}$$
$$= +\alpha_c \text{ during } -180 < \phi < -16°$$

Thus, in a revolution, the RCS will function in a controlling mode over an angle of 194° and as a disturbance mode over an angle of 164° or as a net control over only 30° per revolution. This will make capture phase very long and control fuel will get wasted unnecessarily and pitch and yaw also will not work satisfactorily due to high spin rate.

Ref. 7 discusses four different logics which can bring down the rates more efficiently. Some of the simple logics are discussed below briefly.

1. CONTROL LAW 1

One simple logic would be to activate control only based on angular rate if the rate is high, and follow the normal logic when rate is brought down below a certain level. The control law is given below:

$$\text{if } \left|\theta\right| < \theta_d \quad e = -\theta_c + \theta + K_R\,\dot\theta$$

$$\text{if } \left|\theta\right| > \theta_d \quad e = \theta \text{ and } d_z = 0 \qquad \qquad \text{... (5.123)}$$

$$\text{Then} \quad \alpha = -\alpha_c \qquad \text{if} \qquad e > d_z \qquad \qquad \text{... (5.124)}$$

$$= +\alpha_c \qquad \text{if} \qquad e < -d_z$$

$$= 0 \qquad \text{if} \qquad \left|e\right| < d_z$$

The phase plane diagram would look like Fig. 5.15.

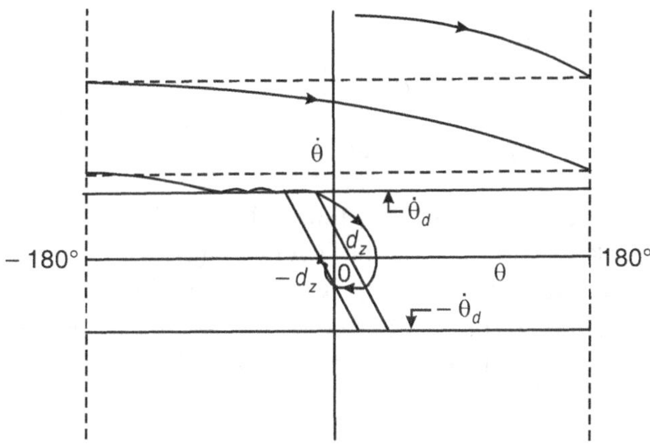

Fig. 5.15

2. CONTROL LAW 2

This control logic uses a limiter in the angular error channel. The block diagram is shown below in Fig. 5.16 and corresponding phase plane diagram is shown in Fig. 5.17.

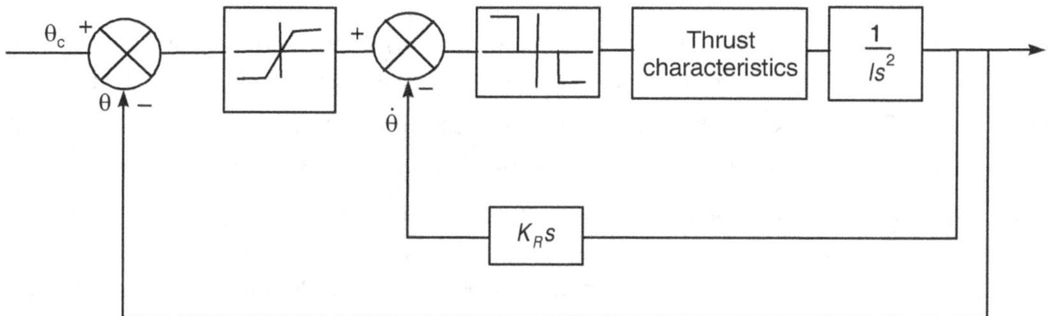

Fig. 5.16

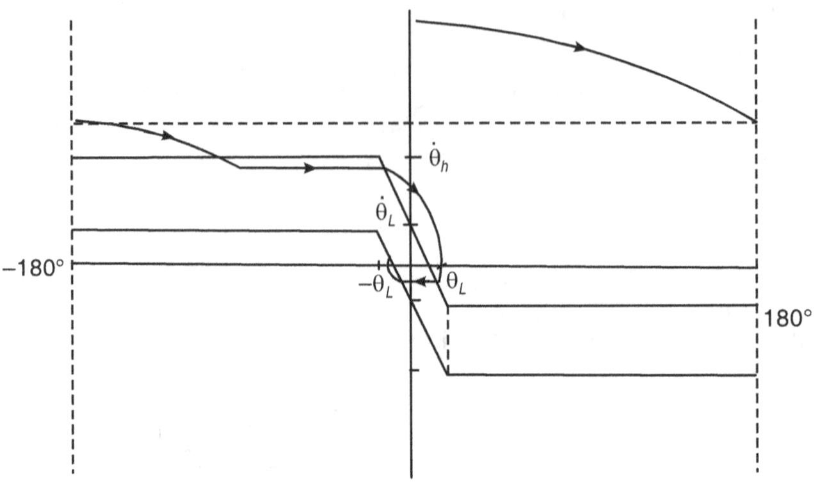

Fig. 5.17

This logic is simple to implement and also fuel efficient. The limiter value on the angular error and rate feed back gain K_R decide the level of angular rate above which the system works to bring down the rate. When the vehicle state enters the rate dead band, the control is switched off and the angular error gets corrected in coasting mode without expenditure of control fuel till it enters the normal dead zone boundary. The rate dead band is given by

$$\dot{\theta}_h - \dot{\theta}_L = Rate\ dead\ band = \frac{2dz}{K_R} \qquad \qquad ...\ (5.125)$$

The rate at which the system will coast depends on the cut off delay and fall time and will lie between

$$\dot{\theta}_h = \frac{d_z - \theta_L}{K_R} \quad and \quad \dot{\theta}_L = \frac{-d_z - \theta_L}{K_R} \qquad \qquad ...\ (5.126)$$

where θ_L is the limit on $(-\theta_c + \theta)$. For positive $\dot{\theta}_L$ and $\dot{\theta}_h$ (2nd quadrant) the limit value θ_L is negative quantity. The limiter value must be adequately higher than the deadzone to ensure that $\dot{\theta}_L > 0$. If the θ_L is less than the d_z, $\dot{\theta}_L$ will become negative and system state is likely to drift away while still remaining in rate dead band and control will fail to correct the error.

This control logic is simple to implement and efficient from the fuel expenditure considerations since a large angle can be corrected without expenditure of fuel. However, if for some reason it is desired to change θ_c at a rate faster than θ obtained with the thruster cut off delay or $(\dot{\theta}_L + \alpha_c T_D)$, this logic will give trouble since θ_c will increase faster than θ and the error $(-\theta_c + \theta)$ will grow but will get limited due to limiter and the control system will ignore any further increase in the error and one will find that the control system does not act, inspite of the fact that there is a large attitude error. Hence, in such case, one should select the limiter value θ_L such that $\theta_L \geq \theta_c$.

5.10 REACTION CONTROL DURING ATMOSPHERIC PHASE

As stated earlier, the reaction control is generally not preferred during high dynamic pressure region. However, one may be required to use it during low dynamic pressure region. It is easier to set the design parameters using trajectory simulation. However, as discussed in earlier cases, the analysis of limit cycle gives greater insight into the functioning of the control system and hence, it is discussed here briefly. Readers are advised to refer JM Mendel (Refs. 9, 10) for further details.

Assuming $\theta \approx \alpha$, the vehicle dynamics is given by

$$\theta \pm M_\alpha \theta = u \qquad \qquad \dots (5.127)$$

where $u = 0$, $+\alpha_c$ or $-\alpha_c$ and upper sign is for stable vehicle and lower sign is for unstable vehicle. Solving these equations and bringing it in the form of phase plane equations, one gets

Case I : $u = 0$

$$\theta^2 \pm M_\alpha \theta^2 = \theta_0^2 \pm M_\alpha \theta_0^2 \qquad \qquad \dots (5.128)$$

Case II: $u = \alpha_c$

$$\theta^2 \pm M_\alpha [\theta \quad \theta_L]^2 = \theta_0^2 \pm M_\alpha (\theta_0 \quad \theta_L)^2 \qquad \qquad \dots (5.129)$$

Case III: $u = -\alpha_c$

$$\theta^2 \pm M_\alpha [\theta \pm \theta_L]^2 = \theta_0^2 \pm M_\alpha (\theta_0 \pm \theta_L)^2 \qquad \qquad \dots (5.130)$$

where $\theta_L = \dfrac{\alpha_c}{M_\alpha}$ = maximum angle of attack which can be balanced by control torque

$$\dots (5.131)$$

M_α is a positive quantity and upper signs are for stable vehicle and lower signs are for unstable vehicle.

Following observations can be made:

1. Stable vehicle follows elliptical trajectory (Fig. 5.18)

2. Unstable vehicle follows an hyperbolic trajectory with asymptotes given by (Fig 5.19)

$$\theta = \pm \sqrt{M_\alpha} \, (\theta \pm \theta_L) \qquad \qquad \dots (5.132)$$

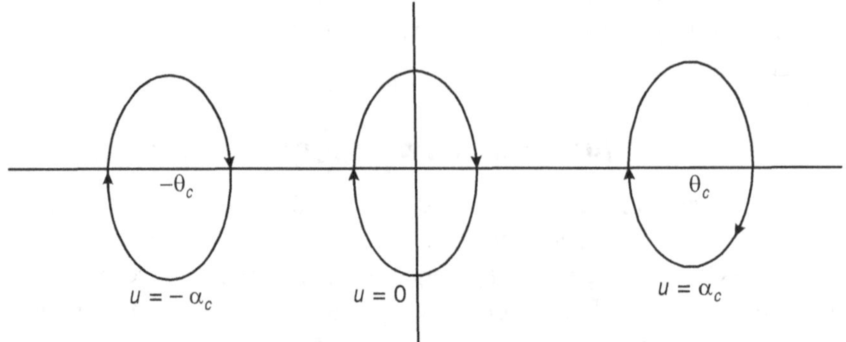

Fig. 5.18 Stable vehicle

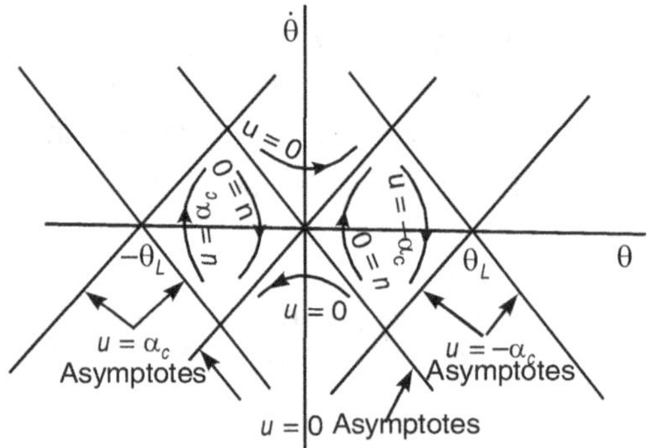

Fig. 5.19 Unstable vehicle

The limit cycle characteristics for this case also can be obtained on similar lines as in Section 5.7. However, obtaining of vehicle delayed state after T_R or T_D is slightly more complicated than in Section 5.7 due to aerodynamic term M_α and state transition matrix (STM) approach is adopted for convenience where the state equations are given by

$$\begin{bmatrix} \dot{\theta} \\ \ddot{\theta} \end{bmatrix} = \begin{bmatrix} 0 & 1 \\ M_\alpha & 0 \end{bmatrix} \begin{bmatrix} \theta \\ \dot{\theta} \end{bmatrix} + \begin{bmatrix} 0 \\ 1 \end{bmatrix} u \qquad \qquad \text{... (5.133)}$$

and the state transition matrix for time(T) is given by

$$\phi = \begin{bmatrix} \phi_{11} & \phi_{12} \\ \phi_{21} & \phi_{22} \end{bmatrix} \qquad \qquad \text{... (5.134)}$$

TABLE 5.1

STM	Stable Vehicle	Unstable vehicle
$\phi_{11}(T)$	$\cos\left(\sqrt{M_\alpha}\,T\right)$	$\cosh\left(\sqrt{M_\alpha}\,T\right)$
$\phi_{12}(T)$	$\sin\left(\sqrt{M_\alpha}\,T\right)/\sqrt{M_\alpha}$	$\sinh\left(\sqrt{M_\alpha}\,T\right)/\sqrt{M_\alpha}$
$\phi_{21}(T)$	$-M_\alpha\cdot\phi_{12}(T)$	$M_\alpha\cdot\phi_{12}(T)$
$\phi_{22}(T)$	$\phi_{11}(T)$	$\phi_{11}(T)$

As in case of Section 5.4 to 5.6, it is assumed that the dead zone is designed in such a way that the control force is required to act only on one side. For example, for an unstable vehicle and θ_c positive value, vehicle will tend to increase θ beyond θ_c and control will bring it back (Fig. 5.20b). For a stable case and positive θ_c, vehicle will tend to reduce θ to zero and control will increase it when the state goes out of dead zone (Fig. 5.20a).

Fig. 5.20c shows the limit cycle for an unstable vehicle with $\theta_c = 0$. In this case, control acts on both sides of deadzone and the vehicle state follows a hyperbolic curve when control is off (or $\alpha_c = 0$).

Consider now the vehicle initial state to be on the normal or (undelayed) dead zone boundary. The actual or delayed switch line is then obtained by eliminating θ_0 and $\dot\theta_0$ from following equations.

$$\theta(T) = \theta_{11}(T)\theta_0 + \theta_{12}(T)\,\dot\theta_0 \pm [1-\theta_{11}(T)]\theta_L.u \qquad \text{... (5.135)}$$

$$\dot\theta(T) = \theta_{21}(T)\theta_0 + \theta_{22}(T)\dot\theta_0 + \theta_{12}(T)\theta_L.u$$

and using

$$-\theta_c + \theta_0 + K_R\,\dot\theta_0 = \pm d_z$$

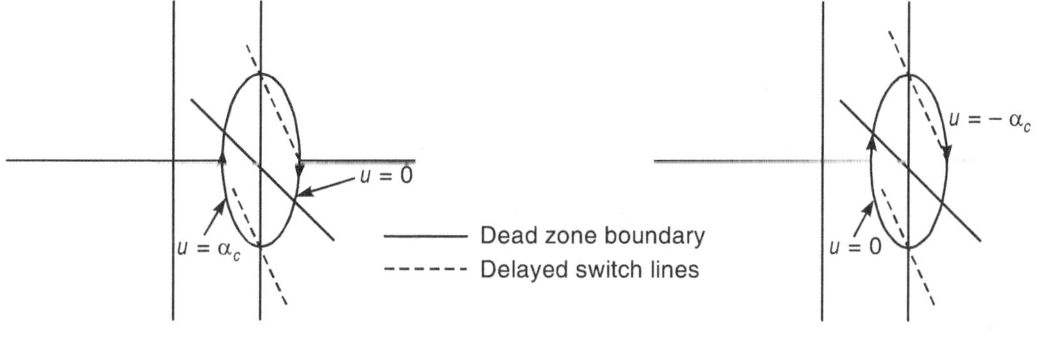

(a) stable vehicle, $\theta_c > 0$ (b) unstable vehicle, $\theta_c > 0$

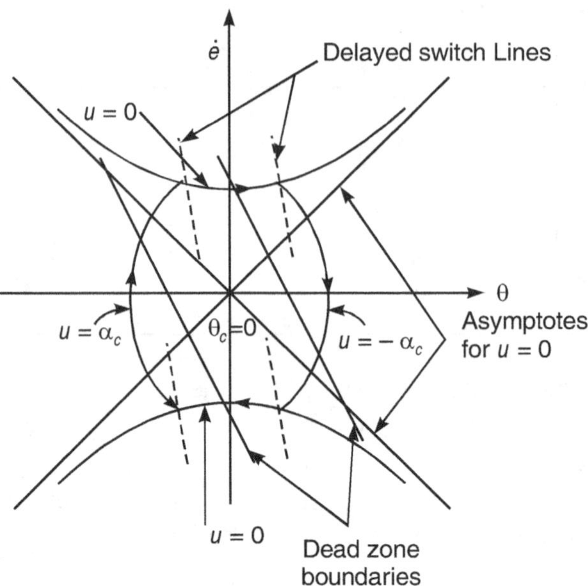

(c) unstable vehicle, $\theta_c = 0$

Fig. 5.20

By following procedure similar to Section 5.7, the switching state θ_s and θ_s are given below. Referring to Fig. 5.20(*a*) for stable configuration:

1. Delayed switch line for control off command:

$$\text{Delay time} = T_D, \quad \theta_L = \frac{\alpha_c}{M_\alpha}$$

$$\theta - \frac{K_1}{K_2}\theta = \left(\frac{\theta_c + d_z}{K_2}\right)D + a_1 + a_2\frac{K_1}{K_2} \qquad \dots (5.136)$$

2. Delayed switch line for control on-command:

$$\text{Delay time} = T_R, \quad \theta_L = 0$$

$$\theta - \frac{K_1'}{K_2'}\theta = \left(\frac{d_z + \theta_c}{K_2'}\right)D \qquad \dots (5.137)$$

3. Switching state

$$\theta_s = \left(\frac{d_z + \theta_c}{K_2'}\right)D + \frac{K_1'}{K_2'}\left\{(d_z + \theta_c)D\left[\frac{1}{K_2'} - \frac{1}{K_2}\right]\left[\frac{K_1}{K_2} + \frac{K_1'}{K_2'}\right] - \frac{\left(a_1 + a_2\dfrac{K_1}{K_2}\right)}{\left(\dfrac{K_1}{K_2} + \dfrac{K_1'}{K_2'}\right)}\right\} \quad \dots (5.138)$$

$$\theta_s = (d_z + \theta_c) + \left[\frac{1}{K_2'} - \frac{1}{K_2}\right]\left[\frac{K_1}{K_2} + \frac{K_1'}{K_2'}\right] - \frac{\left[a_1 + a_2\dfrac{K_1}{K_2}\right]}{\left[\dfrac{K_1}{K_2} + \dfrac{K_1'}{K_2'}\right]} \quad \dots (5.139)$$

$$\text{where} \quad D = \phi_{11}\phi_{22} - \phi_{12}\phi_{21} \qquad \dots (5.140)$$

$$a_1 = (\phi_{11} - 1)\theta_L \qquad \dots (5.141)$$

$$a_2 = \phi_{12}\theta_L \qquad \dots (5.142)$$

$$K_1 = K_R\phi_{11} - \phi_{12} \quad ; \quad K_1' = K_R\phi_{11}' - \phi_{12}' \qquad \dots (5.143)$$

$$K_2 = K_R\phi_{21} - \phi_{22} \quad ; \quad K_{21}' = K_R\phi_{21}' - \phi_{22}' \qquad \dots (5.144)$$

4. For $T_R = T_D$, the equations become simple ($K_1 = K_1'$ and $K_2 = K_2'$)

$$\theta_s = \left(\frac{d_z + \theta_c}{K_2'}\right)D - \left(\frac{a_1 K_2 + a_2 K_1}{2K_2}\right) \qquad \dots (5.145)$$

$$\theta_s = -\left(\frac{a_1 K_2 + a_2 K_1}{2K_1}\right) \qquad \dots (5.146)$$

5. For unstable configuration, the equations can be obtained by using $\theta_L = -\dfrac{\alpha_c}{M_\alpha}$

By actual integration of the equations, Ref. 5.11 has shown that the on time and period of limit cycle are given as in Table 5.2.

TABLE 5.2

Parameter	Stable vehicle	Unstable vehicle
1. On time (t_{on})	$\dfrac{1}{\sqrt{M_\alpha}}\sin^{-1}\left[\dfrac{2\sqrt{M_\alpha}\,g_s\,\theta_s}{M_\alpha g_s^2 + \theta_s^{\,2}}\right]$	$\dfrac{1}{\sqrt{M_\alpha}}\,l_n\left[\dfrac{\sqrt{M_\alpha}\,g_s + \theta_s}{\sqrt{M_\alpha}\,g_s - \theta_s}\right]$
2. Period (T_p)	$\dfrac{1}{\sqrt{M_\alpha}}\left[\sin^{-1}\left(\dfrac{2\sqrt{M_\alpha}\,g_s\,\theta_s}{M_\alpha g_s^2 + \theta_s^2}\right)\right.$ $\left. + \sin^{-1}\left(\dfrac{2\sqrt{M_\alpha}\,\theta_s\,\theta_s}{M_\alpha \theta_s^2 + \theta_s^2}\right)\right]$	$\dfrac{1}{\sqrt{M_\alpha}}\,l_n\left[\left(\dfrac{\sqrt{M_\alpha}\,\theta_s + \theta_s}{\sqrt{M_\alpha}\,\theta_s - \theta_s}\right)\left(\dfrac{\sqrt{M_\alpha}\,g_s + \theta_s}{\sqrt{M_\alpha}\,g_s - \theta_s}\right)\right]$

where $g_s = \theta_L - \theta_s$

3. Peak to peak amplitude

$$\left[-\theta_L + \sqrt{(\theta_s - \theta_L)^2 \pm \frac{1}{M_\alpha}\theta_s^2} \pm \sqrt{\theta_s^2 + \frac{\theta_s^2}{M_\alpha}}\,\right]$$

(upper sign for stable dynamics, lower sign for unstable dynamics)

4. Average error $\quad -\theta_c + \theta_L\left(\dfrac{t_{on}}{T_p}\right)$

5. Control impulse per second $\quad \dfrac{t_{on} \times F_c}{T_p}$

5.11   DESIGN OF AN ON-OFF REACTION CONTROL SYSTEM

The design is illustrated here for a vehicle operating out of atmosphere.

Table 5.3 gives variation of parameters required for the design where

F_d – Equivalent disturbance force at control location assuming the thrust misalignment acts in pitch plane.

α_c – Angular acceleration due to control force.

α_d – Angular acceleration due to disturbance force.

TABLE 5.3

t	F_d	α_d rad/s²	α_c rad/s²	K_t
0	59.58	0.03432	0.14112	4.111
10	104.86	0.06782	0.15843	2.336
20	140.924	0.10568	0.18391	1.741
30	177.38	0.16141	0.22324	1.383
37	206.29	0.22375	0.26537	1.186
40	195.51	0.23082	0.28925	1.253
42	173.46	0.21633	0.30547	1.412
44	63.70	0.08399	0.32244	3.839
46	19.894	0.02732	0.33718	12.341

The thruster delay characteristics are given by:

$$td_1 = 0.025 \qquad t_r = 0.095 \qquad t_{d2} = 0.015 \qquad t_f = 0.135$$

1. Fig. 5.21 gives variation of F_d with time. The area under the curve = 6404 Nsec.

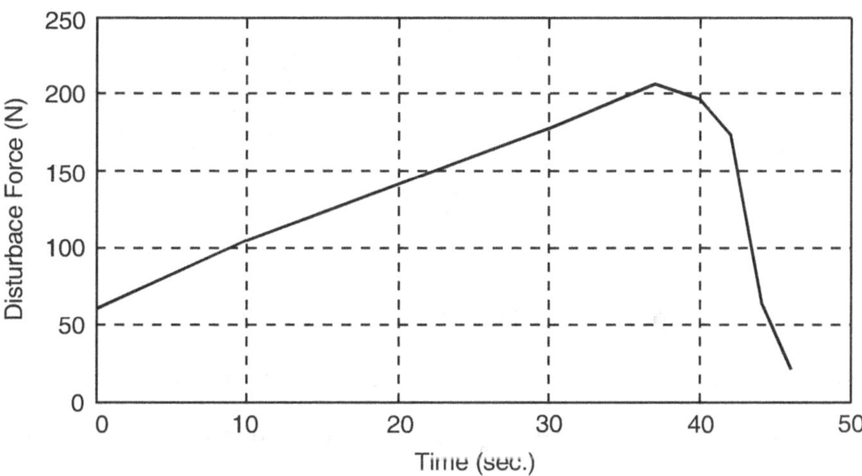

Fig. 5.21 Disturbance Force Vs. Time

This would be control impulse for pitch control system provided the dead zone is designed such that the oscillation occurs about only one dead zone boundary.

2. Since actual thrust misalignment can occur in any plane, say at an angle η with pitch plane.

   Then pitch plane impulse $= 6404\cos\eta$

   Yaw plan impulse $= 6404\sin\eta$

   Total impulse $= 6404\,(\cos\eta + \sin\eta)$

   It can be found that the maximum total impulse requirement will occur if $\eta = 45°$. Therefore, the total control impulse required for pitch and yaw control is given by:

   Total impulse for pitch and yaw $= 6404\sqrt{2} = 9056$ N sec

3. Appendix 5.3 gives a program to compute limit cycle characteristics for a given set of parameters. Select a value of K_R (which satisfies condition given by Eq. 5.100) and obtain limit cycle characteristics for various values of K_R. Fig. 5.22 gives a sample characteristics at one instant.

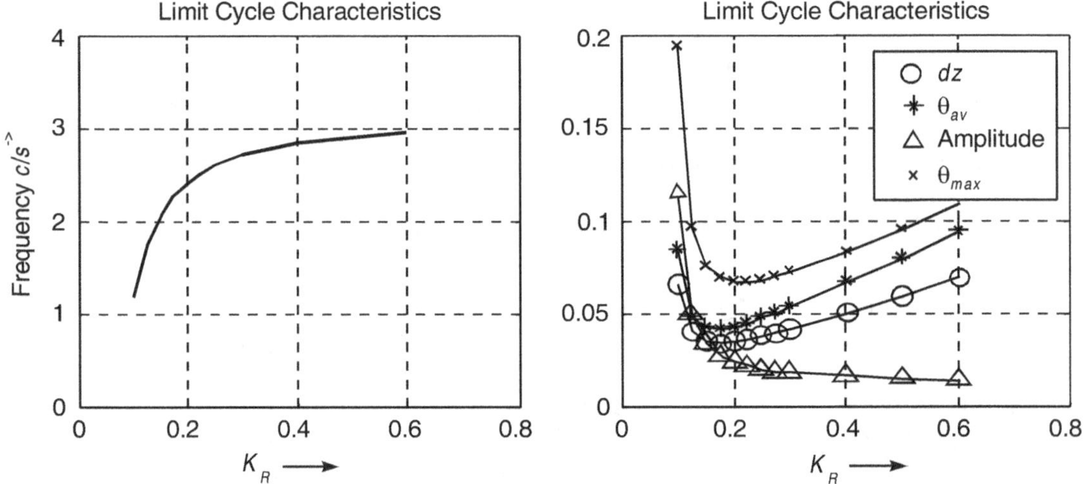

Fig. 5. 22 Limit Cycle Characteristics Vs. K_R

It is seen that the frequency and amplitude is nearly constant for $K_R > 0.2$.

The average error $\theta_{average}$ and θ_{max} (*i.e.*, maximum of the angular error) are nearly minimum for $K_R = 0.2$.

Fig. 5.23 gives transient response for initial error for $K_R = 0.2$.

The study should be repeated at various instants and the K_R can be fixed to suit for the entire flight.

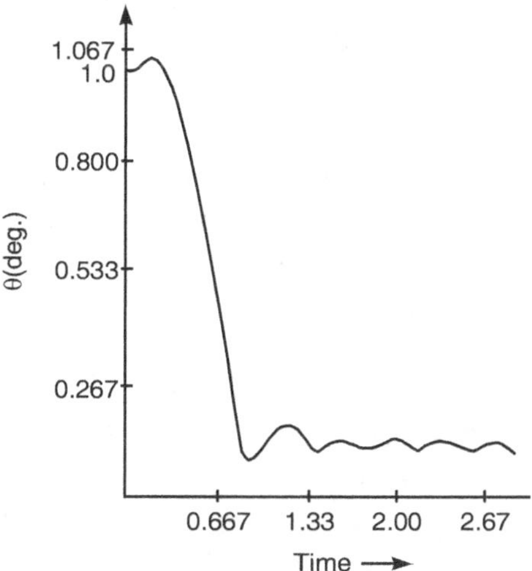

Fig. 5.23

4. The variation of the optimum dead zone is plotted in Fig. 5.24. It is seen that if deadzone is set as $d_z = \pm\,0.2°$, it would satisfy the condition of oscillation about only one boundary for entire burning phase.

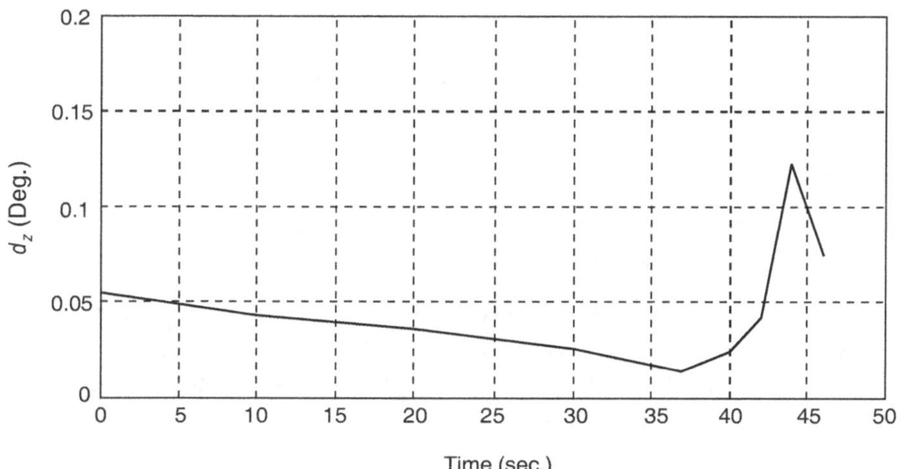

Fig. 5.24 Variation of Optimum Semi Dead Zone with Time

5. Fig. 5.24 indicates that the required deadzone becomes high when K_t becomes high.

In actual practice, thrust misalignment in a given plane may vary from zero to maximum. Hence, one should use sufficiently higher deadzone such that the control impulse for zero disturbance case does not become very high due to control acting on both sides.

1. The control delays are also many times found to vary from the original specifications. Hence, it is better to fix K_R on a higher side than the optimum value obtained from the characteristics (Fig. 5.22) so that one does not have to update the design whenever delays are found to be increased.

2. Fig. 5.25 gives the comparison of the limit cycle obtained using analytical equations with the results of 6DOF trajectory simulation for a given set of parameters. It is seen that the comparison is excellent and the analytical results can be very useful for the design of the reaction control system.

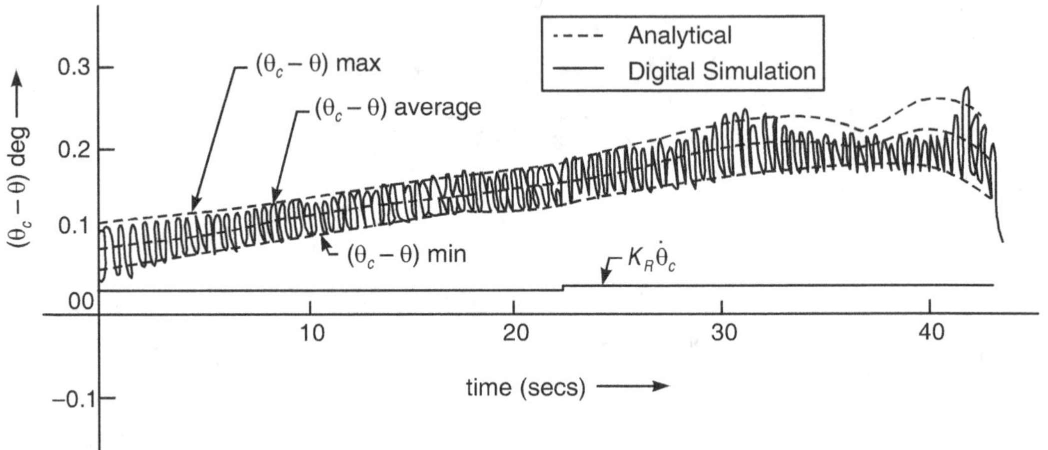

Fig. 5.25 Attitude Error Profile

5.12   CONCLUSIONS

This chapter discusses the on off control system in sufficient details since the literature on this topic is very little and not frequently used.

The limit cycle has been discussed for constant disturbance case, zero disturbance case and for atmospheric phase for both stable and unstable vehicle. General tendency of the designers is to fix the design parameters using simulation. However, the simulation does not give the required insight into the functioning of the limit cycle.

This chapter brings out following important conclusions:

1. For a constant disturbance case, the control torque impulse is equal to the disturbance torque impulse provided the deadzone is selected higher than the optimum deadzone.

2. For a zero disturbance case, the control impulse consumption per second varies directly as the square of control force, square of cut off delay and inversely with the deadzone.

3. The limit cycle characteristics obtained using analytical results give an excellent match with 6DOF simulation results.

REFERENCES

1. **Edgar C. Lineberry Jr., Edwin C Foudriate:** Applications of describing function analysis to the study of an on-off Reaction control system, NASA, TND 654, JAN. 1961.

2. **N.V. Kadam:** Analytical expressions for preliminary estimation of control impulse and prediction of steady state performance of a roll stabilization system, Tech Memo No. SSTC/CG1/TM/06/73.

3. **J. M. Mendel:** A steady state approach to the analysis and synthesis of controllers for a reaction jet controlled vehicle, AIAA Guidance, Control and Flight mechanics Conference Princeton, New Jersey, Aug. 18-20, 1969.

4. **Lawrence W. Taylor, Jr:** An analytical approach to the design of an automatic discontinuous control system, NASA TN D 630, Apr. 1961.

5. **N.V. Kadam:** A new method of determining the limit cycle characteristics and the design of an ON-OFF reaction control system, VSSC.TR.00059:77.

6. **N.V. Kadam:** A new method of determining the limit cycle characteristics and the design of an ON-OFF reaction control system, VSCC. T.R. 00059:7, J of Aeronautical Society of India Vol 35, No.1 Feb. 1983.

7. **N.V. Kadam:** On the modification of control logic to avoid wrong motor actuation due to sign change of sensed angle at 180°, VSSC/CGD/CSA/06/1982.

8. **N.V. Kadam:** Practical design of flight control systems—Some problems and their solutions, Defence Science Journal - Special Issue on Guidance & Control for missiles, Vol. 55, No. 3, Jul. 2005, pp. 211-221.

9. **J.M. Mendel:** On-off limit cycle controllers for reaction Jet controlled systems, IEEE transactions on automatic control Vol. AC-15, No. 3, June 1970, pp. 285-299.

10. **J.M. Mendel:** Performance cost functions for a Reaction Jet controlled system during an on-off limit Cycle, IEEE Transactions on automatic control Vol. AC-13, No. 4, Aug. 1968, pp. 362-368.

11. **J.M. Mendel:** Optimal Bang Bang Limit Cycle performance for a non-linear control system, DAC-60676 June 1967, Douglas, Missile & Space Division.

APPENDIX 5.1 | **LIMIT CYCLE CHARACTERISTICS FOR EQUIVALENT RECTANGULAR CONTROL PULSE**

Following expressions are taken from the author's report given in Ref. 5.

Let

$$T_R = t_{d1} + \frac{1}{2}t_r = t'_{d1} \qquad \text{... A.5.1.1}$$

$$T_D = t_{d2} + \frac{1}{2}t_f = t'_{d2} \qquad \text{... A.5.1.2}$$

Following the similar steps of analysis as given in Section 5, the expressions for various characteristics are given below:

$$t_{off} = T_D \left[\frac{K_t(K_R - \frac{1}{2}T_D)}{K_R - \frac{1}{2}(T_R + T_D)} - 1 \right] \qquad \text{... A.5.1.3}$$

$$t_{on} = \frac{T_R + t_{off}}{K_t - 1} - T_D \qquad \text{... A.5.1.4}$$

$$Tp = \frac{K_t}{K_t - 1} \left[\frac{K_t T_D(K_R - \frac{1}{2}T_D)}{K_R - \frac{1}{2}(T_R + T_D)} + T_R - T_D \right] \qquad \text{... A.5.1.5}$$

$$d_z = \frac{1}{2}\alpha_D t_{off} \left[K_R - \frac{1}{2}T_R \right] \qquad \text{... A.5.1.6}$$

$$\theta_1 = \frac{1}{2}\alpha_D \left[t_{off} - T_R \right] \qquad \text{... A.5.1.7}$$

$$\theta_{max} = \theta_1 + \alpha_D T_R = \frac{1}{2}\alpha_D(t_{off} + T_R) \qquad \text{... A.5.1.8}$$

$$\theta_1 = d_z - \frac{1}{2}K_R\alpha_D \left[t_{off} - T_R \right] \qquad \text{... A.5.1.9}$$

$$\theta_{max} = d_z + \frac{1}{2}\alpha_D \left[t_{off}T_R - K_R(t_{off} - T_D) + \frac{(t_{off} + T_R)^2}{4(K_t - 1)} \right]$$

$$\text{... A.5.1.10}$$

| **APPENDIX 5.2** | **LIMIT CYCLE CHARACTERISTICS FOR SPECIAL CASES** |

A. EQUATIONS FOR THE SPECIAL CASE I-1

This case occurs when control torque is very large for a given disturbance torque. Use of equations for the normal case will give $t_{on} < 0$. The illustrative control pulse with various time segments is given in Fig. A-5.2A.

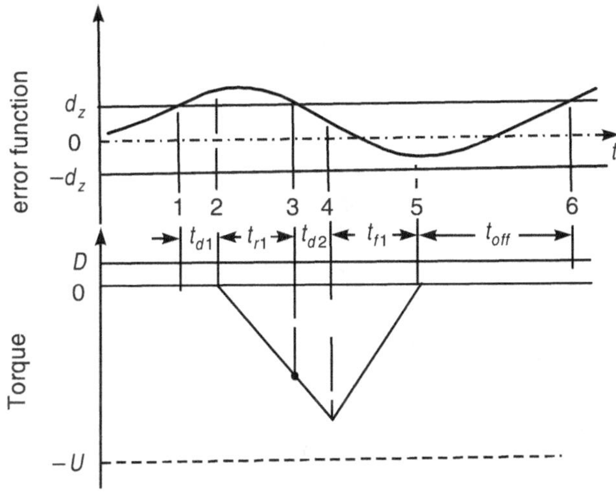

Fig. A-5.2A

This case is distinguished by the condition

$$0 \le t_{r1} \le t_r - t_{d2} \qquad \qquad ...\,(A.1)$$

The useful expressions for various quantities are listed below without giving the algebraic details.

Consider θ_1 and $\dot{\theta}_1$ as the initial state of the system. Then

$$\theta_2 = \theta_1 + \alpha_d t_{d1} \qquad \qquad ...\,(A.2)$$

$$\theta_2 = \theta_1 + \theta_1 t_{d1} + \frac{1}{2}\alpha_d t_{d1}^2 \qquad \qquad ...\,(A.3)$$

$$\theta_3 = \theta_2 + \alpha_d t_{r1} - \frac{1}{2}n_1 t_{r1}^2 \qquad \qquad ...\,(A.4)$$

$$\theta_3 = \theta_2 + \theta_2 t_{r1} + \frac{1}{2}\alpha_d t_{r1}^2 - \frac{1}{6}n_1 t_{r1}^3 \qquad \qquad ...\,(A.5)$$

$$\theta_4 = \theta_2 + \alpha_d(t_{r1} + t_{d2}) - \frac{1}{2}n_1(t_{r1} + t_{d2})^2 \qquad \text{... (A.6)}$$

$$\theta_4 = \theta_2 + \theta_2(t_{r1} + t_{d2}) + \frac{1}{2}\alpha_d(t_{r1} + t_{d2})^2 - \frac{1}{6}n_1(t_{r1} + t_{d2})^3 \qquad \text{...(A.7)}$$

$$\theta_5 = \theta_4 + [\alpha_d - n_1(t_{r1} + t_{d2})t_{f1} + \frac{1}{2}n_2 t_{f1}^2 \qquad \text{... (A.8)}$$

$$\theta_5 = \theta_4 + \theta_4 t_{f1} + \frac{1}{2}[\alpha_d - n_1(t_{r1} + t_{d2})]t_{f1}^2 + \frac{1}{6}n_2 t_{f1}^3 \qquad \text{... (A.9)}$$

$$\theta_6 = \theta_5 + \alpha_d\, t_{off} \qquad \text{... (A.10)}$$

$$\theta_6 = \theta_5 + \theta_5 t_{off} + \frac{1}{2}\alpha_d\, t_{off}^2 \qquad \text{... (A.11)}$$

$$n_1(t_{r1} + t_{d2}) = n_2 t_{f1} \qquad \text{... (A.12)}$$

$$\alpha_d T_p = \frac{1}{2}n_1(t_{r1} + t_{d2})(t_{r1} + t_{d2} + t_{f1})$$

$$= \frac{1}{2}\frac{n_1}{n_2}(n_1 + n_2)(t_{r1} + t_{d2})^2$$

$$= \text{Area of control pulse}$$

where
$$T_p = t_{d1} + t_{r1} + t_{d2} + t_{f1} + t_{off} \qquad \text{... (A.13)}$$

$$\theta_1 = \frac{1}{2}\alpha_d\left[t_{off} + \frac{1}{3}t_{f1} - (t_{d1} + \frac{t_{r1} + t_{d2}}{3})\right] \qquad \text{... (A.14)}$$

$$\theta_{max} = \theta_2 + \frac{1}{2}\frac{\alpha_d^2}{n_1} \qquad \text{... (A.15)}$$

$$\theta_1 = \theta_c + \delta - K_R \theta_1 \qquad \text{... (A.16)}$$

For maximum value of θ, check the sign of θ_4

If
$$\theta_4 < 0,$$

$$t_{\theta max} = \frac{\alpha_d + \sqrt{\alpha_d^2 + 2n_1\theta_2}}{n_1} \qquad \text{... (A.17)}$$

$$\theta_{max} = \theta_2 + \theta_2 t_{\theta max} + \frac{1}{2}\alpha_d t_{\theta max}^2 - \frac{1}{6}n_1 t_{\theta max}^3 \qquad \text{... (A.18)}$$

If $\theta_4 > 0$

$$t_{\theta max} = \frac{(n_2 t_{f1} - \alpha_d) \pm \sqrt{(n_2 t_{f1} - \alpha_d)^2 - 2n_2\,\theta_4}}{n_2} \qquad \text{... (A.19)}$$

select $t_{\theta max}$ such that $0 < t_{\theta max} \le t_{f1}$

$$\theta_{max} = \theta_4 + \theta_4 t_{\theta max} + \frac{1}{2}(\alpha_d - n_2 t_{f1}) t_{\theta max}^2 + \frac{1}{6} n^2 t_{\theta max}^3 \qquad \text{... (A.20)}$$

For minimum value of θ, check the sign of θ_5 :

If $\theta_5 > 0$

$$t_{\theta min} = \frac{(n_2 t_{f1} - \alpha_d) + \sqrt{(n_2 t_{f1} - \alpha_d)^2 - 2n_2\,\theta_4}}{n_2} \qquad \text{... (A.21)}$$

$$\theta_{min} = \theta_4 + \theta_4 t_{\theta min} + \frac{1}{2}(\alpha_d - n_2 t_{f1}) t_{\theta min}^2 + \frac{1}{6} n^2 t_{\theta min}^3 \qquad \text{... (A.22)}$$

If $\theta_5 < 0$

$$t_{\theta min} = -\frac{\theta_5}{\alpha_d} \qquad \text{... (A.23)}$$

$$\theta_{min} = \theta_5 - \frac{1}{2}\frac{\theta_5^2}{\alpha_d} \qquad \text{... (A.24)}$$

To get the value of t_{r1}, solve the following cubic equation:

$$A t_{r1}^3 + B t_{r1}^2 + C t_{r1} + D = 0$$

where

$$A = \frac{n_1(3n_1 + n_2)}{12 n_2} \qquad \text{... (A.26)}$$

$$B = \left\{ \frac{1}{4}\frac{n_1(n_1 + n_2)}{n_2}(t_{d1} + 2t_{d2}) - \left(\frac{n_2 + 2n_1}{6 n_2}\right)\alpha_d - \frac{1}{2}n_1 K_R \right\} \qquad \text{... (A.27)}$$

$$C = \left\{ \frac{1}{4}\frac{n_1(n_1 + n_2)}{n_2}(t_{d2}^2 + 2t_{d1}t_{d2}) - \alpha_d\left(\frac{n_1 + 2n_2}{3 n_2}\right)(t_{d1} + t_{d2}) + K_R \alpha_d \right\} \qquad \text{... (A.28)}$$

$$D = \left\{ \frac{1}{4} \frac{n_1(n_1 + n_2)}{n_2} t_{d1} t_{d2}^2 - \frac{1}{2} \alpha_d t_{d1}^2 - \left(\frac{n_1 + 2n_2}{3n_2} \right) \alpha_d t_{d1} t_{d2} + K_R \alpha_d t_{d1} \right\} \quad \ldots \text{(A.29)}$$

Select that value of t_{r1} which is real, positive and which satisfies the condition $0 \le t_{r1} \le t_r - t_{d2}$.

If none of the roots of Equation (A.25) satisfies these criteria, Case I-1 does not occur. Hence, one should go for Case I-2.

Expressions for optimum deadzone are given below:

Check the sign of

$$\frac{dE}{dt}(5) = \theta_5 + K_R \alpha_d \quad \ldots \text{(A.30)}$$

If $\dfrac{dE}{dt}(5) > 0$

$$t_{E\min} = \left(\frac{n_2 t_{f1} - \alpha_d}{n_2} \right) - K_R + \sqrt{ \left(\frac{n_2 t_{f1} - \alpha_d}{n_2} \right)^2 + K_R^2 - \frac{2\theta_4}{n_2} } \quad \ldots \text{(A.31)}$$

$$\delta = \frac{1}{2} \left\{ \frac{1}{2} n_1 T_1^2 (K_R + \frac{1}{3} T_1) - \frac{1}{2} (n_1 + n_2) t_{E\min}^2 (K_R + \frac{1}{3} t_{E\min} - T_2 (\theta_1 + K_R \alpha_d + \frac{1}{2} \alpha_d T_2) \right\}$$

$$\ldots \text{(A.32)}$$

where
$$T_1 = t_{r1} + t_{d2} + t_{E\min} \quad \ldots \text{(A.33)}$$
$$T_4 = t_{d1} + tr_1 + t_{d2} + t_{E\min} \quad \ldots \text{(A.34)}$$

If $\dfrac{dE}{dt}(5) < 0$

$$t_{E\min} = -\frac{\theta_5 + K_R \alpha_d}{\alpha_d} \quad \ldots \text{(A.35)}$$

$$\delta = \frac{1}{2} \left\{ \frac{1}{2} n_1 T_3^2 (K_R + \frac{1}{3} T_3 + t_{E\min}) - \frac{1}{2} (n_1 + n_2) t_{f1}^2 (K_R + \frac{1}{3} t_{f1} + t_{E\min}) \right.$$

$$\left. - T_4 (\theta_1 + K_R \alpha_d + \frac{1}{2} \alpha_d T_4) \right\} \quad \ldots \text{(A.36)}$$

where

$$T_3 = t_{r1} + t_{d2} + t_{f1} \qquad \dots \text{(A.37)}$$

$$T_4 = t_{d1} + t_{r1} + t_{d2} + t_{f1} + t_{E\,min} \qquad \dots \text{(A.38)}$$

B. EQUATIONS FOR THE SPECIAL CASE I-2

This case also occurs when the use of equations for the normal case gives $t_{on} < 0$

The illustrative control pulse with various time segments is given in Fig. A-5.2B.

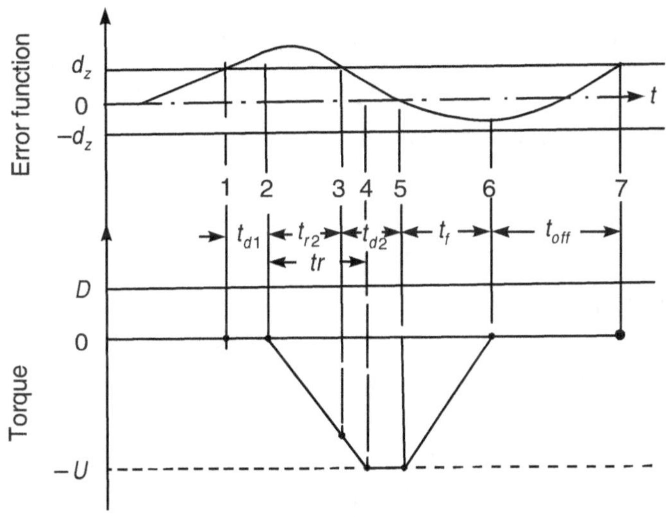

Fig. A-5.2B

This case is distinguished by the condition

$$t_r - t_{d2} \le t_{r2} \le t_r \qquad \dots \text{(B.1)}$$

The useful expressions for various quantities are listed below without giving the algebraic details.

Consider θ_1 and $\dot{\theta}_1$ as the initial state of the system. Then

$$\dot{\theta}_2 = \dot{\theta}_1 + \alpha_d t_{d1} \qquad \dots \text{(B.2)}$$

$$\theta_2 = \theta_1 + \dot{\theta}_1 t_{d1} + \frac{1}{2}\alpha_d d_1^2 \qquad \dots \text{(B.3)}$$

$$\theta_3 = \theta_2 + \alpha_d t_{r2} - \frac{1}{2}n_1 t_{r2}^2 \qquad \dots \text{(B.4)}$$

$$\theta_3 = \theta_2 + \theta_2 t_{r2} + \frac{1}{2}\alpha_d t_{r2}^2 - \frac{1}{6} n_1 t_{r2}^3 \qquad \ldots (B.5)$$

$$\theta_4 = \theta_2 + \alpha_d t_r - \frac{1}{2} n_1 t_r^2 \qquad \ldots (B.6)$$

$$\theta_4 = \theta_2 + \theta_2 t_r + \frac{1}{2}\alpha_d t_r^2 - \frac{1}{6} n_1 t_r^3 \qquad \ldots (B.7)$$

$$\theta_5 = \theta_4 + (\alpha_d - \alpha_c)(t_{r2} + t_{d2} - t_r) \qquad \ldots (B.8)$$

$$\theta_5 = \theta_4 + \theta_4 (t_{r2} + t_{d2} - t_r) + \frac{1}{2}(\alpha_d - \alpha_c)(t_{r2} + t_{d2} - t_r)^2 \qquad \ldots (B.9)$$

$$\theta_6 = \theta + \alpha_d t_f - \frac{1}{2}\alpha_c t_f \qquad \ldots (B.10)$$

$$\theta_6 = \theta_5 + \theta_5 t_f + \frac{1}{2}(\alpha_d - \alpha_c)t_f^2 + \frac{1}{6} n_2 t_f^3 \qquad \ldots (B.11)$$

$$\theta_7 = \theta_6 + \alpha_d t_{off} \qquad \ldots (B.12)$$

$$\theta_7 = \theta_6 t_{off} + \frac{1}{2}\alpha_d t_{off}^2 \qquad \ldots (B.13)$$

$$T_p = t_{d1} + t_{r2} + t_{d2} + t_f + t_{off} \qquad \ldots (B.14)$$

$$\alpha_d T_p = \alpha_c\left(t_{r2} + t_{d2} + \frac{1}{2}t_f - \frac{1}{2}t_r\right) \qquad \ldots (B.15)$$

$$= \text{Area of control pulse}$$

$$\theta_1 = \frac{1}{2}\alpha_d\left[t_{off} + \frac{1}{2}t_f - \left(t_{d1} + \frac{1}{2}t_r\right)\right] + \frac{1}{24}\frac{\alpha_c(t_r^2 - t_f^2)}{T_p} \qquad \ldots (B.16)$$

$$\theta_{max} = \theta_2 + \frac{1}{2}\frac{\alpha_d^2}{n_1} \qquad \ldots (B.17)$$

$$\theta_1 = \theta_c + \delta - K_R \theta_1 \qquad \ldots (B.18)$$

For maximum value of θ, check the sign of θ_4.

If $\theta_4 < 0$,

$$t_{\theta max} = \frac{\alpha_d + \sqrt{\alpha_d^2 + 2n_1 \theta_2}}{n_1} \qquad \ldots (B.19)$$

$$\theta_{max} = \theta_2 + \theta_2 t_{\theta\,max} + \frac{1}{2}\alpha_d t_{\theta\,max}^2 - \frac{1}{6}n_1 t_{\theta\,max}^3 \qquad \text{... (B.20)}$$

If $\theta_4 > 0$,

$$t_{\theta\,max} = \frac{\theta_4}{(\alpha_c - \alpha_d)} \qquad \text{... (B.21)}$$

$$\theta_{max} = \theta_4 + \theta_4 t_{\theta\,max} + \frac{1}{2}(\alpha_d - \alpha_c)\, t_{\theta\,max}^2 \qquad \text{... (B.22)}$$

For minimum value of θ, check the sign of θ_6 :

If $\theta_6 > 0$

$$t_{\theta\,min} = \frac{(\alpha_c - \alpha_d) + \sqrt{(\alpha_c - \alpha_d)^2 - 2n_2\,\theta_5}}{n_2} \qquad \text{... (B.23)}$$

$$\theta_{min} = \theta_5 + \theta_5 t_{\theta\,min} + \frac{1}{2}(\alpha_d - \alpha_c)\, t_{\theta\,min}^2 + \frac{1}{6}n_2 t_{\theta\,min}^3 \qquad \text{... (B.24)}$$

If $\theta_6 < 0$

$$t_{\theta\,min} = -\frac{\theta_6}{\alpha_d} \qquad \text{... (B.25)}$$

$$\theta_{min} = \theta_6 - \frac{1}{2}\frac{\theta_6^2}{\alpha_d} \qquad \text{... (B.26)}$$

To get the value of t_{r2}, solve the following cubic equation:

$$A t_{r2}^3 + B t_{r2}^2 + C t_{r2} + D = 0 \qquad \text{... (B.27)}$$

where $\qquad\qquad A = 1 \qquad\qquad\qquad\qquad\qquad\qquad\qquad\qquad\qquad\qquad$ ... (B.28)

$$B = 3(K_R - t_r) \qquad \text{... (B.29)}$$

$$C = -6\left\{\frac{\alpha_d}{n_1}\left[K_R - \frac{1}{2}\left(t_{d1} + \frac{1}{2}t_r + t_{d2} + \frac{1}{2}tf\right)\right] + \frac{1}{2}t_r\left(t_{d2} + \frac{1}{2}t_f + t_{d1} - \frac{1}{2}t_r\right)\right\} \qquad \text{... (B.30)}$$

$$D = -6 t_{d1}\left\{\frac{\alpha_d}{n_1}\left[K_R - \frac{1}{2}\left(t_{d1} + \frac{1}{2}t_r + t_{d2} + \frac{1}{2}t_f\right)\right] + \frac{1}{2}t_r\left(t_{d2} + \frac{1}{2}t_f - \frac{1}{2}t_r\right)\right\} \qquad \text{... (B.31)}$$

Select that value of t_{r2} which is real, positive and which satisfies the condition

$t_r - t_{d2} \le t_{r2} \le t_r$

Case I-2 does not occur if none of the roots of (B.27) satisfies the conditions.

To find expressions for deadzone, check the sign of

$$\frac{\partial E}{\partial t}(6) = \theta_6 + K_R \alpha_d \qquad \qquad \dots \text{(B.32)}$$

If $\dfrac{\partial E}{\partial t}(6) > 0$,

$$t_{E\min} = \frac{\alpha_c - \alpha_d}{n_2} - K_R + \sqrt{\left(\frac{\alpha_c - \alpha_d}{n_2}\right)^2 + K_R^2 - \frac{2\theta_5}{n_2}} \qquad \dots \text{(B.33)}$$

$$\delta = \frac{1}{2}\left\{\alpha_c T_1\left(K_R + \frac{1}{2}T_1\right) + \frac{1}{24}\alpha_c t_r^2 - \frac{1}{2}n_2 t_{E\min}^2\left(K_R + \frac{1}{3}t_{E\min}\right) - T_2\left(\theta_1 + K_R \alpha_d + \frac{1}{2}\alpha_d T_2\right)\right\}$$

$$\dots \text{(B.34)}$$

where

$$T_1 = t_{r2} + t_{d2} - \frac{1}{2}t_r + t_{E\min} \qquad \qquad \dots \text{(B.35)}$$

$$T_2 = t_{d1} + t_{r2} + t_{d2} + t_{E\min} \qquad \qquad \dots \text{(B.36)}$$

If $\dfrac{\partial E}{\partial t}(6) = 0$, $t_{E\min} = \dfrac{\theta_6 + K_R \alpha_d}{\alpha_d}$ $\dots \text{(B.37)}$

$$\delta = \frac{1}{2}\left\{\alpha_c T_3\left[K_R + \frac{1}{2}(T_3 + t_f) + t_{E\min}\right] + \frac{1}{24}\alpha_c(t_r^2 - t_f^2) - T_4\left(\theta_1 + K_R \alpha_d + \frac{1}{2}\alpha_d T_4\right)\right\}$$

$$\dots \text{(B.38)}$$

where

$$T_3 = t_{r2} + t_{d2} - \frac{1}{2}t_r + \frac{1}{2}t_f \qquad \qquad \dots \text{(B.39)}$$

$$T_4 = t_{d1} + t_{r2} + t_{d2} + t_f + t_{E\min} \qquad \qquad \dots \text{(B.40)}$$

It may be noted here that as the control pulse has the shape similar to the normal case, most of the expressions for various quantities are also similar to those for the normal case.

C. EQUATIONS FOR THE SPECIAL CASE II-1

This case occurs when the disturbance torque is very high for a given control torque. Use of equations for the normal case will give $t_{off} < 0$. The illustrative control torque vs. time curve is given in Fig. A-5.2C.

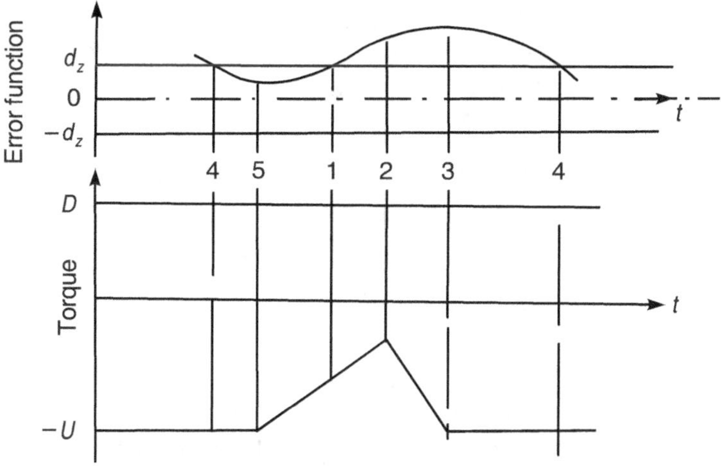

Fig. A-5.2C

This case is distinguished by the condition

$$0 \le t_{f1} \le t_f - t_{d1} \qquad \text{... (C.1)}$$

The useful expressions for various quantities are listed below without giving the algebraic details:

Consider θ_4 and $\dot{\theta}_4$ as the starting state of the system so that we get symmetric results with Case I-1. Then

$$\dot{\theta}_5 = \dot{\theta}_4 - (\alpha_c - \alpha_d)t_{d2} \qquad \text{... (C.2)}$$

$$\theta_5 = \theta_4 + \dot{\theta}_4 t_{d2} - \frac{1}{2}(\alpha_c - \alpha_d)t_{d2}^2 \qquad \text{... (C.3)}$$

$$\dot{\theta}_1 = \dot{\theta}_5 - (\alpha_c - \alpha_d)\,t_{f1} + \frac{1}{2}n_2 t_{f1}^2 \qquad \text{... (C.4)}$$

$$\theta_1 = \theta_5 + \dot{\theta}_5 t_{f1} - \frac{1}{2}(\alpha_c - \alpha_d)t_{f1}^2 + \frac{1}{6}n_2 t_{f1}^3 \qquad \text{... (C.5)}$$

$$\dot{\theta}_2 = \dot{\theta}_5 - (\alpha_c - \alpha_d)(t_{f1} + t_{d1}) + \frac{1}{2}n_2(t_{f1} + t_{d1})^2 \qquad \text{... (C.6)}$$

$$\theta_2 = \theta_5 + \dot{\theta}_5(t_{f1} + t_{d1}) - \frac{1}{2}(\alpha_c - \alpha_d)(t_{f1} + t_{d1})^2 + \frac{1}{6}n_2(t_{f1} + t_{d1})^3 \qquad \text{... (C.7)}$$

$$\dot{\theta}_3 = \dot{\theta}_2 - (\alpha_c - \alpha_d)\,t_{r1} + n_2(t_{f1} + t_{d1})\,t_{r1} - \frac{1}{2}n_1 t_{r1}^2 \qquad \text{... (C.8)}$$

$$\theta_3 = \theta_2 + \dot{\theta}_2 t_{r1} - \frac{1}{2}(\alpha_c - \alpha_d)\,t_{r1}^2 + \frac{1}{2}n_2(t_{f1} + t_{d1})\,t_{r1}^2 - \frac{1}{6}n_1 t_{r1}^3 \qquad \text{... (C.9)}$$

$$\theta'_4 = \theta_3 - (\alpha_c - \alpha_d)t_{on} \qquad \dots (C.10)$$

$$\theta'_4 = \theta_3 + \theta_3 t_{on} - \frac{1}{2}(\alpha_c - \alpha_d)t_{on}^2 \qquad \dots (C.11)$$

check that $\theta'_4 = \theta_4$ and $\theta'_4 = \theta_4$

$$n_2(t_{f1} + t_{d1}) = n_1 t_{r1} \qquad \dots (C.12)$$

$$T_p = t_{d1} + t_{r1} + t_{on} + t_{d2} + t_{f1}$$

$$\alpha_d T_p = \alpha_c T_p - \frac{1}{2} n_1 t_{r1}(t_{f1} + t_{d1} + t_{r1}) \qquad \dots (C.13)$$

$$= \text{Area under the control torque vs. time curve}$$

$$\theta_4 = -(\alpha_c - \alpha_d)\left[t_{on} + \frac{1}{3}t_{r1} - \left(t_{d2} + \frac{t_{f1} + t_{d1}}{3} \right) \right] \qquad \dots (C.14)$$

$$t_{\dot\theta max} = \frac{n_2(t_{f1} + t_{d1}) - (\alpha_c - \alpha_d)}{n_1} \qquad \dots (C.15)$$

$$\theta_{max} = \theta_2 + \frac{1}{2}n_1\left[n_2(t_{f1} + t_{d1}) - (\alpha_c - \alpha_d) \right]^2$$

$$\theta_1 = \theta_c + \delta - K_R\,\theta_1 \qquad \dots (C.16)$$

For maximum value of θ, check the sign of θ_3

If $\theta_3 < 0$
$$t_{\theta max} = \frac{n_1 t_{r1} - (\alpha_c - \alpha_d) + \sqrt{[n_1 t_{r1} - (\alpha_c - \alpha_d)]^2 + 2n_1\,\theta_2}}{n_1} \qquad \dots (C.17)$$

$$\theta_{max} = \theta_2 + \theta_2 t_{\theta max} + \frac{1}{2}\left[n_1 t_{r1} - (\alpha_c - \alpha_d) \right]t_{\theta max}^2 - \frac{1}{6}n_1 t_{\theta max}^3 \qquad \dots (C.18)$$

If $\theta_3 > 0$

$$t_{\theta max} = \frac{\theta_3}{(\alpha_c - \alpha_d)} \qquad \dots (C.19)$$

$$\theta_{max} = \theta_3 + \frac{1}{2}\frac{\theta_3}{(\alpha_c - \alpha_d)} \qquad \dots (C.20)$$

for minimum value of θ, check the sign of θ_2

If $\theta_2 > 0$

$$t_{\theta\min} = \frac{\alpha_c - \alpha_d}{n_2} + \sqrt{\left(\frac{\alpha_c - \alpha_d}{n_2}\right)^2 - 2\frac{\theta_5}{n_2}} \qquad \ldots \text{(C.21)}$$

$$\theta_{\min} = \theta_5 + \theta_5 t_{\theta\min} + \frac{1}{2}(\alpha_d - \alpha_c)t_{\theta\min}^2 + \frac{1}{6}n_2 t_{\theta\min}^3 \qquad \ldots \text{(C.22)}$$

If $\theta_2 < 0$

$$t_{\theta\min} = \frac{n_1 t_{r1} - (\alpha_c - \alpha_d) + \sqrt{[n_1 t_{r1} - (\alpha_c - \alpha_d)]^2 + 2n_1\theta_2}}{n_1} \qquad \ldots \text{(C.23)}$$

$$\theta_{\min} = \theta_2 + \theta_2 t_{\theta\min} + \frac{1}{2}[n_1 t_{r1} - (\alpha_d - \alpha_c)]t_{\theta\min}^2 - \frac{1}{6}n_1 t_{\theta\min}^3 \qquad \ldots \text{(C.24)}$$

To get the value of t_{f1}, solve the following cubic equation

$$A\, t_{f1}^3 + B\, t_{f1}^2 + C\, t_{f1} + D = 0 \qquad \ldots \text{(C.25)}$$

where
$$A = \frac{n_2(n_1 + 3n_2)}{12 n_1} \qquad \ldots \text{(C.26)}$$

$$B = \left\{ \frac{1}{4}\frac{n_2}{n_1}(n_1 + n_2)(t_{d2} + 2t_{d1}) - (\alpha_c - \alpha_d)\left(\frac{n_1 + 2n_2}{6n_1}\right) - \frac{1}{2}n_2 K_R \right\} \qquad \ldots \text{(C.27)}$$

$$C = \left\{ \frac{1}{4}\frac{n_2(n_1 + n_2)}{n_1}(t_{d1}^2 + 2t_{d1}t_{d2}) - (\alpha_c - \alpha_d)\left(\frac{2n_1 + n_2}{3n_1}\right)(t_{d1} + t_{d2}) + K_R(\alpha_c - \alpha_d) \right\}$$

$$\ldots \text{(C.28)}$$

$$D = \left\{ \frac{1}{4}\frac{n_2(n_1 + n_2)}{n_1}(t_{d1}^2 t_{d2}) - (\alpha_c - \alpha_d)\left(\frac{2n_1 + n_2}{3n_1}\right)t_{d1}t_{d2} - \frac{1}{2}(\alpha_c - \alpha_d)t_{d2}^2 + (\alpha_c - \alpha_d)K_R t_{d2} \right.$$

$$\ldots \text{(C.29)}$$

(Note the symmetry of eqns. (C.25) to (C.29) with eqns. (A.25 to A.29)

Select that value of t_{f1} which is real, positive and which satisfies the condition

$$0 \le t_{f1} \le t_f - t_{d1}$$

This case does not occur if none of the roots of Eqn. (C.25) satisfies these conditions. Hence, one should go to Case II-2.

To find the expressions for optimum dead zone, it may be noted that the error must reach minimum between the points (5) and (1). Then

$$t_{E\min} = \frac{\alpha_c - \alpha_d}{n_2} - K_R + \sqrt{\left(\frac{\alpha_c - \alpha_d}{n_2}\right)^2 + K_R^2 - 2\frac{\theta_5}{n_2}} \qquad \dots \text{(C.30)}$$

$$\delta = \frac{1}{2}(t_{d2} + t_{E\min})\left[(\alpha_c - \alpha_d)\left\{K_R + \frac{1}{2}(t_{d2} + t_{E\min})\right\} - \theta_4\right] - \frac{1}{4}n_2 t_{E\min}^2\left[K_R + \frac{1}{3}t_{E\min}\right] \qquad \dots \text{(C.31)}$$

D. EQUATIONS FOR THE SPECIAL CASE II-2

This case also occurs when the disturbance torque is very high for a given control torque. Use of equations for the normal case gives $t_{off} < 0$. The illustrative control torque vs. time curve is given in Fig. A-5.2D.

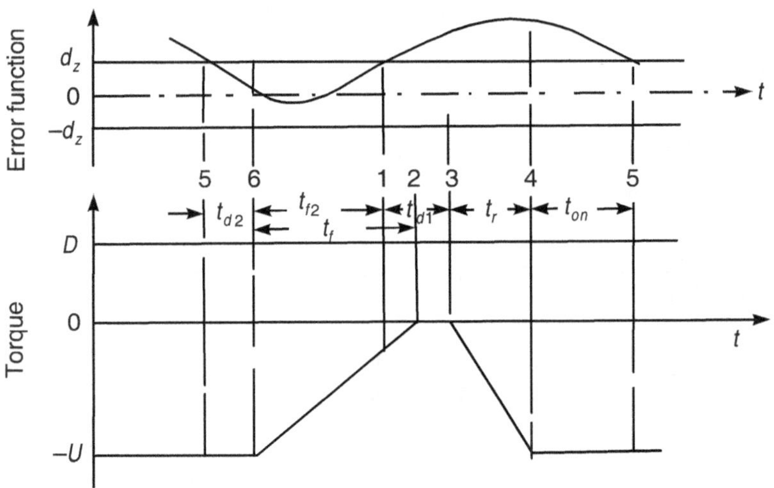

Fig. A-5.2D

This case is distinguished by the condition:

$$t_f - t_{d1} \le t_{f2} \le t_f \qquad \dots \text{(D.1)}$$

The useful expressions for various quantities are listed below without giving the algebraic details:

Consider θ_5 and θ_5 as the starting state of the system so that we get symmetric results with Case I-2. Then,

$$\theta_6 = \theta_5 - (\alpha_c - \alpha_d)t_{d2} \qquad \text{... (D.2)}$$

$$\theta_6 = \theta_5 + \theta_5 t_{d2} - \frac{1}{2}(\alpha_c - \alpha_d)t_{d2}^2 \qquad \text{... (D.3)}$$

$$\theta_1 = \theta_6 - (\alpha_c - \alpha_d)t_{f2} + \frac{1}{2}n_2 t_{f2}^2 \qquad \text{... (D.4)}$$

$$\theta_1 = \theta_6 + \theta_6 t_{f2} - \frac{1}{2}(\alpha_c - \alpha_d)t_{f2}^2 + \frac{1}{6}n_2 t_{f2}^3 \qquad \text{... (D.5)}$$

$$\theta_2 = \theta_6 - (\alpha_c - \alpha_d)t_f + \frac{1}{2}n_2 t_f^2 \qquad \text{... (D.6)}$$

$$\theta_2 = \theta_6 + \theta_6 t_{f2} - \frac{1}{2}(\alpha_c - \alpha_d)t_{f2}^2 + \frac{1}{6}n_2 t_{f2}^3 \qquad \text{... (D.7)}$$

$$\theta_3 = \theta_2 + \alpha_d(t_{d1} + t_{f2} - t_f) \qquad \text{... (D.8)}$$

$$\theta_3 = \theta_2 + \theta_2(t_{d1} + t_{f2} - t_f) + \frac{1}{2}\alpha_d(t_{d1} + t_{f2} - t_f)^2 \qquad \text{... (D.9)}$$

$$\theta_4 = \theta_3 + \alpha_d t_r - \frac{1}{2}n_1 t_r^2 \qquad \text{... (D.10)}$$

$$\theta_4 = \theta_3 + \theta_3 t_r + \frac{1}{2}\alpha_d t_r^2 - \frac{1}{6}n_1 t_r^3 \qquad \text{... (D.11)}$$

$$\theta'_5 = \theta_4 - (\alpha_c - \alpha_d)t_{on} \qquad \text{... (D.12)}$$

$$\theta'_5 = \theta_4 + \theta_4 t_{on} - \frac{1}{2}(\alpha_c - \alpha_d)t_{on}^2 \qquad \text{... (D.13)}$$

check that $\quad \theta'_5 = \theta_5$ and $\theta'_5 = \theta_5$

$$T_p = t_{d1} + t_r + t_{on} + t_{d2} + t_{f2} \qquad \text{.. (D.14)}$$

$$\alpha_d T_p = \alpha_c\left(t_{d2} + \frac{1}{2}t_f + \frac{1}{2}t_r + t_{on}\right) \qquad \text{... (D.15)}$$

$$= \text{Area under the control torque vs. time curve.}$$

$$0_5 - -\frac{1}{2}\alpha_c \frac{\left(t_{d1} + t_{f2} + \frac{1}{2}t_r - \frac{1}{2}t_f\right)}{T_p}\left[t_{on} + \frac{1}{2}t_r - \left(t_{d2} + \frac{1}{2}t_f\right)\right] + \frac{\alpha_c(t_r^2 - t_f^2)}{24 T_p} \qquad \text{... (D.16)}$$

$$\theta_{max} = \theta_3 + \frac{1}{2}\frac{\alpha_d^2}{n_1} \qquad \text{... (D.17)}$$

$$\theta_1 = \theta_c + \delta - K_R \theta_1 \qquad \text{... (D.18)}$$

For maximum value of θ, check the sign of θ_4

$$\text{If } \theta_4 < 0, \qquad t_{\theta\max} = \frac{\alpha_d + \sqrt{\alpha_d^2 + 2n_1\theta_3}}{n_1} \qquad \text{... (D.19)}$$

$$\theta_{\max} = \theta_3 + \theta_3 t_{\theta\max} + \frac{1}{2}\alpha_d t_{\theta\max}^2 - \frac{1}{6}n_1 t_{\theta\max}^3 \qquad \text{... (D.20)}$$

$$\text{If } \theta_4 > 0, \qquad t_{\theta\max} = \frac{\theta_4}{(\alpha_c - \alpha_d)} \qquad \text{... (D.21)}$$

$$\theta_{\max} = \theta_4 + \theta_4 t_{\theta\max} - \frac{1}{2}(\alpha_c - \alpha_d) t_{\theta\max}^2 \qquad \text{... (D.22)}$$

For minimum value of θ, check the sign of θ_2

$$\text{If } \theta_2 > 0, \qquad t_{\theta\min} = \frac{(\alpha_c - \alpha_d) + \sqrt{(\alpha_c - \alpha_d)^2 - 2n_2\theta_6}}{n_2} \qquad \text{... (D.23)}$$

$$\theta_{\min} = \theta_6 + \theta_6 t_{\theta\min} - \frac{1}{2}(\alpha_c - \alpha_d) t_{\theta\min}^2 + \frac{1}{6}n_2 t_{\theta\min}^3 \qquad \text{... (D.24)}$$

$$\text{If } \theta_2 < 0, \qquad t_{\theta\min} = \frac{-\theta_2}{\alpha_d} \qquad \text{... (D.25)}$$

(This is true only if $t_{\theta\min} \leq (t_{f2} + t_{d1} - t_f)$. If this is not satisfied one must calculate the correct value).

$$\theta_{\min} = \theta_2 + \theta_2 t_{\theta\min} + \frac{1}{2}\alpha_d t_{\theta\min}^2 \qquad \text{... (D.26)}$$

$$= \theta_2 - \frac{1}{2}\frac{\theta_2^2}{\alpha_d}$$

To get the value of t_{f2}, solve the following cubic equation:

$$A t_{f2}^3 + B t_{f2}^2 + C t_{f2} + D = 0 \qquad \text{... (D.27)}$$

$$\text{where} \qquad A = 1 \qquad \text{... (D.28)}$$

$$B = 3\,(K_R - t_f) \qquad \text{... (D.29)}$$

$$C = -6\left\{\frac{(\alpha_c - \alpha_d)}{n_2}\left[K_R - \frac{1}{2}(t_{d1} + \frac{1}{2}t_r + t_{d2} + \frac{1}{2}t_f)\right] + \frac{1}{2}t_f(t_{d1} + \frac{1}{2}t_r + t_{d2} - \frac{1}{2}t_f)\right\}$$

$$... (D.30)$$

$$D = -6t_{d2}\left\{\frac{(\alpha_c - \alpha_d)}{n_2}\left[K_R - \frac{1}{2}(t_{d1} + \frac{1}{2}t_r + t_{d2} + \frac{1}{2}t_f)\right] + \frac{1}{2}t_f(t_{d1} + \frac{1}{2}t_r - \frac{1}{2}t_f)\right\}$$

$$... (D.31)$$

(Note the symmetry of Eqns. (D.27) to (D.30) with equations (B.27) to (B.30)).

Select the value of t_{f2}, which is real, positive and which satisfies the condition

$$t_f - t_{d1} \leq t_{f2} \leq t_f$$

Case II-2 does not occur if none of the roots of Eqns. (D.27) satisfies these conditions.

To find expression for dead zone it may be noted that the error function must reach minimum between the points (6) and (1).

Then $\quad t_{E\min} = \frac{(\alpha_c - \alpha_d)}{n_2} - K_R + \sqrt{\left(\frac{\alpha_c - \alpha_d}{n_2}\right)^2 + K_R^2 - 2\frac{\theta_6}{n_2}}$ $\qquad$ ... (D.32)

$$\delta = \frac{1}{2}(t_{d2} + t_{E\min})\left[(\alpha_c - \alpha_d)\left\{K_R + \frac{1}{2}(t_{d2} + t_{E\min})\right\} - \theta_5\right] - \frac{1}{4}n_2 t_{E\min}^2\left[K_R + \frac{1}{3}t_{E\min}\right]$$

$$... (D.33)$$

APPENDIX 5.3	**COMPUTER PROGRAM FOR THE ANALYSIS OF ON-OFF REACTION CONTROL SYSTEM**

This Appendix gives a computer program for analyzing the limit cycle for the case of constant disturbance due to thrust misalignment as discussed in Sections 5.4 to 5.7 and Appendix 5.2. The variables used are similar to those used in the text and need no explanation. The limit cycle characteristics can be calculated at different flight instants and designer needs to fix the final parameters using his judgement. The limit cycle characteristics can once again be recomputed to predict the control system performance for the chosen parameters.

Following are some of the variables needing explanations:

ALPC   – α_c

ALPD   – α_d

RLC   – Control moment arm

RLT   – Disturbance moment arm

ETA   – Thrust misalignment angle

RKK   – Array of rate feedback gains (RK) for which limit cycle is to be calculated

CONF   – Control force

DSTF   – Disturbance force.

```fortran
C         Program for the analysis of ON-OFF Reaction Control system
          DIMENSION TME(40),RIYY(40),RKK(40),CCG(40),THRST(20)
C         DIMENSION AA(4),BB(4),CC(4),DISTF(40)
          DIMENSION A(4),RR(3),RI(3), C(4)
          DIMENSION B(4),D(4),DD(4)
C         COMPLEX Z(10)

c         ALPC = Angular Acceleration due to control torque
c         ALPD = Angular Acceleration due to disturbance torque
c         RLC = Control moment arm
c         RLT = Disturbance moment arm
c         ETA = Thrust Misalignment angle
c         RKK = Array of rate feedback gains(KR) for which limit
c               cycle is to be calculated
c         CONF = Control force
c         DSTF = Disturbace force
c         RKT = Ratio of Control torque to Disturbace torque

          open(1,file='Input.dat')
          open (2,file='INPUT_KR1.dat')
          open(112,file='output.dat')

          READ(1,*)IN,IK,ITS,IKD,IKS
          READ(1,*)TD1,TR,TD2,TF,ALPC,ALPD

          do I = 1,IK
          read(2,*)RKK(I)
          end do

C         write(*,*)RKK

C         READ 1,(TME(I),I=1,IN)
C         READ 1,(RKK(I),I=1,IK)

C         IF(ITS.EQ.1) GO TO 3181
C         READ 1,(THRST(I),I=1,IN)
C         READ 1,(RIYY(I),I=1,IN)
C         READ 1,(CCG(I),I=1,IN)
C         READ 1,TDI,TR,TD2,TF,CONF,RLC,RLT,ETA
C         PRINT 11
C         PRINT 3,(TME(I),I=1,IN)
C         PRINT 3,(THRST(I),I=1,IN)
C         PRINT 3,(RIYY(I),I=1,IN)
C         PRINT 3,(CCG(I),I=1,IN)
C         PRINT 3,(RKK(I),I=1,IK)
C         PRINT 3,TDI,TR,TD2,TF,CONF,RLC,RLT,ETA
```

```
C3181   CONTINUE
        RTOD=45./ATAN(1.0)
C       DO 100 II=1,IN
C       PRINT 11
C       TIME=TME(II)
C       PRINT 4,TIME
C       IF(ITS.EQ.1) GO TO 3182
C       RIY=RIYY(II)
C       CG=CCG(II)
C       DSTF=THRST(II)*ETA/RTOD
C       ALPC=(CG-RLC)*CONF/RIY
C       ALPD=(CG-RLT)*DSTF/RIY
C       PRINT 2
C       PRINT 3,TD1,TR,TD2,TF
C       GO TO 3183
C3182   CONTINUE
C       READ 1,ALPD,ALPC
C       PRINT 3,AIPD,ALPC
C3183   CONTINUE
        IF(IKS.EQ.1.OR.ITS.EQ.1) PRINT 3187
        RKT=ALPC/ALPD
        AN1=ALPC/TR
        AN2=ALPC/TF
        DO 100 JJ=1,IK
        RK=RKK(JJ)

        A1=RK-0.5*(TD2+0.5*TF)
        A2=RK-0.5*(TD1+0.5*TR+TD2+0.5*TF)
        RKTMX=(TD1+TR)*A2/(0.5*TR*A2-0.5*(TD1+0.5*TR)*(TD2+0.5*TF))
        RKTMN=(TD2+TF)*A2/((TD2+0.5*TF)*A1)
        TOFF=(TD2+0.5*TF)*(RKT*A1/A2-1.0)-0.5*TF


C       write(*,*) RKTMX,RKTMN,TOFF

        IF (TOFF.LT.0.) GO TO 800

        TON=(TOFF+0.5*TF+TD1+0.5*TR)/(RKT-1.)-(TD2+0.5*(TF+TR))

        IF (TON.LT.0.) GO TO 400

        TP=TD1+TR+TON+TD2+TF+TOFF
        THDT1=0.5*ALPD*(TOFF+0.5*TF-(TD1+0.5*TR))
       &+ALPC*(TR*TR-TF*TF)/(24.*TP)
        THDT2=THDT1+ALPD*TD1
        THDT3=THDT2+(ALPD-0.5*ALPC)*TR
```

```
      THDT4=THDT3+(ALPD-ALPC)*TON
      THDT5=THDT4+(ALPD-ALPC)*TD2
      THDT6=THDT5+(ALPD-0.5*ALPC)*TF
      THDT7=THDT6+ALPD*TOFF

      TTDMX=TR/RKT
      THDMX=THDT2+0.5*ALPD*TR/RKT
      EDT6=THDT6+RK*ALPD
      IF (EDT6.LT.0.) GO TO 10
      TMIN1=((ALPC-ALPD)/AN2-RK)+SQRT(((ALPC-ALPD)/
     &AN2)**2+RK*RK-2.0*THDT5/AN2)
      T1=0.5*TR+TON+TD2+TMIN1
      T2=TD1+TR+TON+TD2+TMIN1
      DZ1=0.5*(ALPC*T1*(RK+0.5*T1)+ALPC*TR*TR/24.
     &-0.5*AN2*TMIN1*TMIN1*(TMIN1/3.+RK)
     &-T2*(THDT1+RK*ALPD+0.5*ALPD*T2))
      DZ=DZ1
      TMIN2=0
      DZ2=0.
      GOTO 20
10    TMIN2=-(THDT6+RK*ALPD)/ALPD
      T3=0.5*TR+TON+TD2+0.5*TF
      T4=TD1+TR+TON+TD2+TF+TMIN2
      DZ2=0.5*(ALPC*T3*(0.5*T3+0.5*TF+TMIN2+RK)
      1+ALPC*(TR*TR-TF*TF)/24. -T4*(THDT1+ALPD*(RK+0.5*T4)))
      DZ=DZ2
      TMIN1=0
      DZ1=0.
20    CONTINUE
c     If Limit cycle is to be calculated for a specified Dead-
c     Zone, Uncoment
c     the next line with appropriate Dead-Zone value
c     DZ=0.5/RTOD
      TH1=DZ-RK*THDT1
      TH2=TH1+THDT1*TD1+0.5*ALPD*TD1*TD1
      TH3=TH2+THDT2*TR+(0.5*ALPD-ALPC/6.0)*TR*TR
      TH4=TH3+THDT3*TON+0.5*(ALPD-ALPC)*TON*TON
      TH5=TH4+THDT4*TD2+0.5*(ALPD-ALPC)*TD2*TD2
      TH6=TH5+THDT5*TF+0.5*(ALPD-ALPC)*TF*TF+ALPC*TF*TF/6.0
      TH7=TH6+THDT6*TOFF+0.5*ALPD*TOFF*TOFF

c     Calculation of Average Theta Error
      THS1=THS1*TD1+TH2*TR+TH3*(TON+TD2)+TH5*TF+TH6*TOFF
      THS2=0.5*(THDT1*TD1*TD1+THDT2*TR*TR+THDT3*(TON+TD2)**2
     &+THDT5*TF*TF+THDT6*TOFF*TOFF)
      THS3=(ALPD/6.0)*(TD1**3+TR**3+(TON+TD2)**3+TF**3+TOFF**3)
```

```
      &-(ALPC/6.0)*((TON+TD2)**3+TF**3)
       THS4=(AN2*TF**4-AN1*TR**4)/24.0

       AVTHE=(THS1+THS2+THS3+THS4)/TP


C*****************Maximum and Minimum Error Computation

       VLTON=TD1+TR+TON
       IF(THDT3.GT.0.) GO TO 30
       TMAX1=ALPD+SQRT(ALPD*ALPD+2*AN1*THDT2))/AN1
       THMAX1=TH2+THDT2*TMAX1+0.5*ALPD*TMAX1*TMAX1-AN1*TMAXI**3/6.
       THMAX=THMAX1
       TMAX2=0
       THMAX2=0.0
       GOTO 40
30     TMAX2=THDT3/(ALPC-ALPD)
       THMAX2=TH3+0.5*THDT3*THDT3/(ALPC-ALPD)
       THMAX=THMAX2
       TMAX1=0
       THMAX1=0
40     CONTINUE
       IF (THDT6.LE.0.) GO TO 45
       TMN1=(ALPC-ALPD+SQRT((ALPC-ALPD)**2-2*AN2*THDT5))/AN2
       THMIN1=TH5+THDT5*TMN1+0.5*(ALPD-
      &ALPC)*TMN1*TMN1+AN2*TMN1*TMN1/6
       THMIN=THMIN1
       TMN2=0
       THMIN2=0
       GOTO 50
45     TMN2=-THDT6/ALPD
       THMIN2=TH6+THDT6*TMN2+0.5*ALPD*TMN2*TMN2
       THMIN=THMIN2
       TMN1=0
       THMIN1=0
50     CONTINUE
       GOTO 105


C******************************************

400    ICASE=11
       C(4)=-6.*ALPD*TD1/AN1*(0.5*RKT*(TD2+0.5*(TF-TR))+A2)
       C(3)=-6.*ALPD/AN1*(0.5*RKT*(TD2+TD1+0.5*(TF-TR))+A2)
       C(2)=3.0*(RK-TR)
       C(1)=1.0

       CALL PROOT (3,C,RR,RI,IER)
```

```
C         DO 1218 I = 1,3
C         RR(I) = REAL(Z(I))
C         RI(I) = AIMAG(Z(I))
C1218     CONTINUE
          DOWN=TR-TD2
          UP=TR
          CALL SELECT(T,DOWN,UP,RR,RI,MCASE)

C         PRINT 1000,ICASE,MCASE,T
1000      FORMAT (1X, "CASE-"I2,"="I2, 2X, "T="F5.3)
          IF (MCASE.GT.0) GO TO 412
          GOTO 500
412       TR2=T
C         PRINT 415,TR2
415       FORMAT (1X,"TR2"G14.6)

413       TOFF=(RKT-1)*(TR2+TD2+0.5*(TF-TR))-TD1-0.5*TR-0.5*TF
          TON=0.0
          TP=TD1+TR2+TD2+TF+TOFF

          THDT1=0.5*ALPD*(TOFF+0.5*TF-TD1-0.5*TR)+ALPC*
     &(TR*TR-TF*TF)/(24.*TP)

          THDT2=THDT1+ALPD*TD1
          THDT3=THDT2+ALPD*TR2-0.5*AN1*TR2*TR2
          THDT4=THDT2+ALPD*TR-0.5*AN1*TR*TR
          THDT5=THDT4-(ALPC-ALPD)*(TR2+TD2-TR)
          THDT6=THDT5+ALPD*TF-0.5*ALPC*TF
          THDT7=THDT6+ALPD*TOFF

          TTDMX=TR/RKT
          THDMX=THDT2+0.5*ALPD*TR/RKT
          EDT6=THDT6+RK*ALPD
          IF(EDT6.LT.0)GO TO 431
          TMIN1=((ALPC-ALPD)/AN2-RK)+SQRT(((ALPC-ALPD)/AN2)**2+RK*RK
     &-2.0*THDT5/AN2)
          T1=TR2+TD2-0.5*TR+TMIN1
          T2=TD1+TR2+TD2+TMIN1
          DZ1=0.5*(ALPC*T1*(RK+0.5*T1)+ALPC*TR*TR/24-
     &0.5*AN2*TM1N1*TM1N1*
     &(RK+TMIN1/3)-T2*(THDT1+RK*ALPD+0.5*ALPD*T2))
          DZ=DZ1
          TMIN2=0
          DZ2=0
          GO TO 433
```

```
431        TMIN2=-(THDT6+RK*ALPD)/ALPD
           T3=TR2+TD2+0.5*TF-0.5*TR
           T4=TD1+TR2+TD2+TF+TMIN2
           DZ2=0.5*(ALPC*T3*(RK+0.5*(T3+TF)+TMIN2)+ALPC*(TR*TR-
          &TF*TF)/24-T4*(THDT1+RK*ALPD+0.5*ALPD*T4))
           DZ=DZ2
           TMIN1=0
           DZ1=0
433        CONTINUE

c          If Limit cycle is to be calculated for a specified Dead-
c          Zone) Uncoment
c          the next line with appropriate Dead - Zone value

c          DZ=0.5/RTOD

           TH1=DZ-RK*THDT1
           TH2=TH1+THDT1*TD1+0.5*ALPD*TD1*TD1
           TH3=TH2+THDT2*TR2+0.5*ALPD*TR2*TR2-AN1*TR2**3/6
           TH4=TH2+THDT2*TR2+0.5*ALPD*TR*TR-AN1*TR2**3/6
           TH5=TH4+THDT4*(TR2+TD2-TR)-0.5*(ALPC-ALPD)*(TR2+TD2-TR)**2
           TH6=TH5+THDT5*TF+0.5*(ALPD-ALPC)*TF*TF+ALPC*TF*TF/6
           TH7=TH6+THDT6*TOFF+0.5*ALPD*TOFF*TOFF

C          Calculation of Avarage Theta Error
           THS1=TH1*TD1+TH2*TR+TH4*(TR2+TD2-TR)+TH5*TF+TH6*TOFF
           THS2=0.5*(THDT1*TD1*TD1+THDT2*TR*TR+THDT4*(TR2+TD2-TR)**2+
          &THDT5*TF*TF+THDT6*TOFF*TOFF)
           THS3=(ALPD/6)*(TD1**3+TR**3+(TR2+TD2-
          &TR)**3+TF**3+TOFF**3)-(ALPC/6.)*((TR2+TD2-TR)**3+TF**3)
           THS4=(AN2*TF**4-AN1*TR**3)/24
           AVTHE=(THS1+THS2+THS3+THS4)/TP

C*******************************************

           VLTON=TD1+TR2
           IF (THDT4.GT.O.O)GO TO 434
           TMAX1=(ALPD+SQRT(ALPD*ALPD+2. *AN1*THDT2))/AN1
           TMAX=TMAX1
           THMAX1=TH2+THDT2*TMAX+0.5*ALPD*TMAX*TMAX-AN1*TMAX**3/6
           THMAX=THMAX1
           TMAX2=0
           THMAX2=0
           GO TO 437
434        TMAX2=THDT4/(ALPC-ALPD)
           THMAX2=TH4+THDT4*TMAX2-0.5*(ALPC-ALPD)*TMAX2*TMAX2
```

```
            TMAX=TMAX2
            THMAX=THMAX2
            TMAX1=0.0
            THMAX1=0.0
437         IF (THDT6.LE.0) GO TO 435
            TMN1= (ALPC-ALPD+SQRT((ALPC-ALPD)**2-2*AN2*THDT5))/AN2
            THMIN1=TH5+THDT5*TMN1+0.5*(ALPD-ALPC)*TMN1*TMN1
      &+AN2*TMN1**3/6
            THMIN=THMIN1
            TMN2=0
            THMIN2=0
            GOTO 105
435         TMN2=-THDT6/ALPD
            THMIN2=TH6+THDT6*TMN2+0.5*ALPD*TMN2*TMN2
            TMN1=0.0
            THMIN=THMIN2
            THMIM1=0.0
            GOTO 105

500         ICASE=12
            TRTF=TR/(TR+TF)
            A(4)=0.25*ALPC/TR*(TR+TF)*TD1*TD2*TD2/TR-0.5*ALPD*TD1*TD1-
      &(TF+2.0*TR)*TD1*TD2/3.0/TR*ALPD+RK*ALPD*TD1
            A(3)=0.25*ALPC*(TR+TF)*TD2*(TD2+2.0*TD1)/(TR*TR)-
      &(TF+2.0*TR)*(TD1+TD2)/(3.0*TR)*ALPD+RK*ALPD
            A(2)=0.25*ALPC*(TR+TF)*(TD1+2.0*TD2)/TR/TR-(TR+2.0*TF)*
      &ALPD/6.0/TR-0.5*RK*AN1

            A(1)=ALPC*(TR+3.0*TF)/(12.0*TR*TR)

            CALL PROOT (3,A,RR,RI,IER)
C           CALL ZRPOLY(A,3,Z,IER)
C           DO 1216 I=1,3
C           RR(I)=REAL(Z(I))
C           RI(I)=AIMAG(Z(I))

C1216       CONTINUE
            DOWN=0.0
            UP=TR-TD2
            CALL SELECT(T,DOWN,UP,RR,RI,MCASE)
            PRINT 1000,ICASE,MCASE,T
            IF (MCASE.GT.0) GO TO 512
            GOTO 800
512         TR1=T
513         TRD=TR1+TD2
            PRINT 515,TR1
```

```
515        FORMAT (1X, "TR1" G14.6)
           TF1=AN1*TRD/AN2
           TOFF=0.5*AN1*TRD*(TRD+TF1)/ALPD-(TD1+TRD+TF1)
           TON=0
           TP=TD1+TRD+TF1+TOFF

           THDT1=0.5*ALPD*(TOFF-TD1+(TF1-TRD)/3.0)
           THDT2=THDT1+ALPD*TD1
           THDT3=THDT2+ALPD*TR1-0.5*AN1*TR1*TR1
           THDT4=THDT2+ALPD*TRD-0.5*AN1*TRD*TRD
           THDT5=THDT4+(ALPD-AN1*TRD)*TF1+0.5*AN2*TF1*TF1
           THDT6=THDT5+ALPD*TOFF
           THDT7=0.0

           TTDMX=ALPD/AN1
           THDMX=THDT2+0.5*ALPD*ALPD/AN1
           EDT5=THDT5+RK*ALPD
           IF(EDT5.LT.0.0) GO TO 531
           S1=(AN1*TRD-ALPD)/AN2
           TEMIN1=S1-RK+SQRT(S1*S1+RK*RK-2.0*THDT4/AN2)
           TT1=TRD+TEMIN1
           TT2=TD1+TT1
           DZ1=0.5*(0.5*AN1*TT1*TT1*(RK+TT1/3.0)-
          &0.5*(AN1+AN2)*TEMIN1
          &*TEMIN1*(RK+TEMIN1/3.0)-TT2*(THDT1+RK*ALPD+0.5*ALPD*TT2))
           TMIN1=TEMIN1
           TMIN2=0.0
           DZ=DZ1
           DZ2=0.0
           GO TO 533
531        TEMIN2=-(THDT5+RK*ALPD)/ALPD
           TT3=TRD+TF1
           TT4=TD1+TT3+TEMIN2
           DZ2=0.5*(0.5*AN1*TT3*TT3*(RK+TEMIN2+TT3/3.0)-
          &0.5*(AN1+AN2)*TF1*TF1*(RK+TEMIN2+TF1/3.0)
          &-TT4*(THDT1+RK*ALPD+0.5*ALPD*TT4))
           DZ=DZ2
           TMIN1=0.0
           TMIN2=TEMIN2
533        CONTINUE
C          If Limit cycle is to be calculated for a specified Dead-
           Zone, Uncoment the next line with appropriate Dead - Zone
           value
c          DZ=0.5/RTOD

           TH1=DZ-RK*THDT1
```

```
      TH2=THI+THDT1*TD1+0.5*ALPD*TD1*TD1
      TH3=TH2+THDT2*TR1+0.5*ALPD*TR1*TR1-AN1*TR1**3.0/6.0
      TH4=TH2+THDT2*TRD+0.5*ALPD*TRD*TRD-AN1*TRD**3.0/6.0
      TH5=TH4+THDT4*TF1+0.5*(ALPD-AN1*TRD)*TF1*TF1+AN2*TF1**3.0/
     &6.0
      TH6=TH5+THDT5*TOFF+0.5*ALPD*TOFF*TOFF
      TH7=0.0

c     Calculation of Average Theta Error

      THS1=TH1*TD1+TH2*(TR1+TD2)+TH4*TF1+TH5*TOFF
      THS2=0.5*(THDT1*TD1*TD1+THDT2*(TR1+TR2)**2+THDT4*TF1*TF1
     &+THDT5*TOFF*TOFF)
      THS3=(ALPD/6)*(TD1**3+(TR1+TD2)**3+TF1**3+TOFF**3)
     &-AN1*(TR1+TD2)*TF1**3/6
      THS4=(AN2*TF1**4-AN1*(TR1+TD2)**4)/24
      AVTHE=(THS1+THS2+THS3+THS4)/TP
      VLTON=TD1+TR1
      IF(THDT4.GT.0.) GO TO 541
      TMAX1=(ALPD+SQRT(ALPD*ALPD+2.*AN1*THDT2))/AN1
      THMAX1=TH2+THDT2*TMAX1+0.5*ALPD*TMAX1*TMAX1-
     &AN1*TMAX1**3/6
      THMAX=THMAX1
      TMAX2=0
      THMAX2=0
      GOTO 521
541   TMO=(TF1-ALPD/AN2)**2-2.*THDT4/AN2
      IF (TMO.GE.0.) GO TO 542
      PRINT 543
543   FORMAT (1X, "TMAX IS IMAGINARY")
      GOTO 521
542   TM1=TF1-ALPD/AN2+SQRT(TMO)
      TM2=TF1 - ALPD/AN2-SQRT(TMO)
      IF (TM1.GT.TF1) GO TO 547
      IF (TM1.NE.TM2) GO TO 545
      GO TO 547
545   PRINT 551
551   FORMAT (1X, "TWO VALUES FOR THMAX")
      THMAX=0
      GOTO 521
547   TMAX2=TM2
      THMAX2=TH4+(THDT4+0.5*(ALPD-
     &AN2*TF1)*TMAX2+AN2*TMAX2*TMAX2/6)*
     &TMAX2
      TMAX=TMAX2
      THMAX=THMAX2
```

```
           TMAX1=0
           THMAX1=0
521        IF(THDT5.LE.0.0) GO TO 525
           TMN1=(AN1*TRD-ALPD+SQRT((AN1*TRD-ALPD)**2-2*AN2*THDT4))/
          &AN2
           THMIN1=TH4+THDT4*TMN1+0.5*(ALPD-
          &AN1*TRD)*TMN1*TMN1+AN2*TMN1**3/6
           THMIN=THMIN1
           TMN2=0
           THMIN2=0
           GOTO 527
525        TMN2=-THDT5/ALPD
           THMIN2=TH5-0.5*THDT5*THDT5/ALPD
           THMIN=THMIN2
           TMN1=0
           THMIN1=0
527        CONTINUE
           GOTO 105
800        ICASE=21

           B(4)=-6.*TD2*(A2*(ALPC-ALPD)/AN2+0.5*TF*(TD1+0.5*(TR-TF)))
           B(3)=-6.*(A2*(ALPC-ALPD)/AN2+0.5*TF*(TD1+TD2+0.5*(TR-TF)))
           B(2)=3.*(RK-TF)
           B(1)=1.0

           CALL PROOT(3.B.RR.RI.IER)
C          WRITE(*,*)RR,RI
C          DO 1217 I=1,3
C          RR(I)=REAL(Z(I))
C          RI(I)=AIMAG(Z(I))
C1217      CONTINUE
           DOWN=TF-TD1
           UP=TF
           CALL SELECT (T,DOWN,UP,RR,RI,MCASE)
C          PRINT 1000,ICASE,MCASE,I
           IF(MCASE.GT.0) GOTO 812
           GOTO 900
812        TF2=T
           PRINT 815,TF2
815        FORMAT (1X,"TF2" G14.6)
           TON=(TD1+TD2+TF2+TR-RKT*(TD2+0.5*(TF+TR)))/(RKT-1)
           TP=TD1+TD2+TF2+TR+TON
           TOFF=0
           THDT5=-0.5*ALPC*(TD1+TF2+0.5*(TR-TF))*(TON-TD2+0.5*(TR-
          &TF))/TP+ALPC*(TR*TR-TF*TF)/(24.*TP)
           THDT6=THDT5-(ALPC-ALPD)*TD2
```

```fortran
      THDT1=THDT6-(ALPC-ALPD)*TR2+0.5*AN2*TF2*TF2
      THDT2=THDT6-(ALPC-ALPD)*TF+0.5*AN2*TF*TF
      THDT3=THDT2+ALPD*(TD1+TF2-TF)
      THDT4=THDT3+ALPD*TR-0.5*AN1*TR*TR
      THDTD5=THDT4-(ALPC-ALPD)*TON
      THDT7=THDTD5

      TTDMX=ALPD/AN1
      THDMX=THDT3+0.5*ALPD*ALPD/AN1
      THDMN=THDT6-0.5*(ALPC-ALPD)**2/AN2
      TMIN1=(ALPC-ALPD)/AN2-RK+SQRT(((ALPC-ALPD)/AN2)**2+RK*RK-
     &2*THDT6/AN2)
      TEMIN=TMIN1
      T5=TD2+TEMIN
      DZ1=0.5*T5*((ALPC-ALPD)*(RK+0.5*T5)-THDT5)-
     &0.25*AN2*TEMIN*TEMIN*(RK+TEMIN/3)
      DZ=DZ1
      DZ2=0
      TMIN2=0.0

C     If Limint cycle is to be calculated for a specified Dead-
C     Zone, Uncoment
c     the next line with appropriate Dead - Zone value

c     DZ=0.5/RTOD

      TH5=DZ-RK*THDT5
      TH6=TH5+THDT5*TD2-0.5*(ALPC-ALPD)*TD2*TD2
      TH1=TH6+THDT6*TF2-0.5*(ALPC-ALPD)*TF2*TF2+AN2*TF2**3/6
      TH2=TH6+THDT6*TF-0.5*(ALPC-ALPD)*TF*TF+AN2*TF**3/6
      TH3=TH2+THDT2*(TD1+TF2-TF)+0.5*ALPD*(TD1+TF2-TF)**2
      TH4=TH3+THDT3*TR+0.5*ALPD*TR*TR-AN1*TR**3/6
      THD5=TH4+THDT4*TON-0.5*(ALPC-ALPD)*TON*TON
      TH7=THD5

c     Calculation of Average Theta Error

      THS1=TH6*TF+TH2*(TD1+TF2-TF)+TH3*TR+TH4*(TON+TD2)
      THS2=0.5*(THDT6*TF*TF+THDT2*(TD1+TF2-
     &TF)**2+THDT3*TR*TR+THDT4*(TON+TD2)**2)
      THS3=(ALPD/6.0)*(TF**3+(TD1+TF2-TF)**3+TR**3+(TON+TD2)**3)
     &-(ALPC/6.0)*(TF**3+(TON+TD2)**3)
               THS4=(AN2*TF**4-AN1*TR**4)/84
      AVTHE=(THS1+THS2+THS3+THS4)/TP

      VLTON=TD1+TR+TON
```

```
         IF(THDT4.GT.0.0)GO TO 834
         TMAX1=(ALPD+SQRT(ALPD*ALPD+2*AN1*THDT3))/AN1
         THMAX1=TH3*(THDT3+0.5*ALPD*TMAX1-AN1*TMAX1*TMAX1/6)*TMAX1
         THMAX=THMAX1
         TMAX2=0
         THMAX2=0
         GOTO 837
834      TMAX2=THDT4/(ALPC-ALPD)
         THMAX2=TH4+0.5*THDT4*THDT4/(ALPC-ALPD)
         THMAX=THMAX2
         TMAX1=0
         THMAX1=0
837      IF(THDT2.LE.0) GO TO 835
         TMN1=(ALPC-ALPD+SQRT((ALPC-ALPD)**2-2*AN2*THDT6))/AN2
         THMIN1=TH6+(THDT6-0.5*(ALPC-ALPD)*TMN1+AN2*TMN1*TMN1/
        &6)*TMN1
         THMIN=THMIN1
         TMN2=0
         THMIN2=0
         GO TO 105
835      TMN2=-THDT2/ALPD
         THMIN2=TH2+(THDT2+0.5*ALPD*TMN2)*TMN2
         THMIN=THMIN2
         TMN1=0
         THMIN1=0

         GOTO 105

900      1CASE=22
         D(4)=0.25*AN2*(AN1+AN2)*TD1*TD1*TD2/AN1-(ALPC-ALPD)
        &*(2*AN1+AN2)*TD1*TD2/(3*AN1)
        &-0.5*(ALPC-ALPD)*TD2*TD2+(ALPC-ALPD)*RK*TD2
         D(3)=0.25*AN2*(AN1+AN2)*(TD1+2.0*TD2)*TD1/AN1+(ALPC-
        &ALPD)*(RK-(TD1+TD2)*(2*AN1+AN2)/(3*AN1))
         D(2)=0.25*AN2*(AN1+AN2)*(TD2+2*TD1)/AN1-(ALPC-ALPD)
        &*(AN1+2*AN2)/(6*AN1)-0.5*AN2*RK
         D(1)=AN2*(AN1+3*AN2)/(AN1*12)

         CALL PROOT (3,D,RR,RI,IE)
         DOWN=0.0
         UP=TF-TD1
         CALL SELECT(T,DOWN,UP,RR,RI,MCASE)
C        PRINT 1000,ICASE,MCASE,T
         IF(MCASE.GT.0)GO TO 912
         GO TO 100
```

```
912      TF1=T
         PRINT 915,TF1
915      FORMAT(1X,"TF1" G14.6)
         TR1=AN2*(TF1+TD1)/AN1
         TON=0.5*AN1*TR1*(TF1+TD1+TR1)/(ALPC-ALPD)-
         (TD2+TF1+TD1+TR1)
         TOFF=0
         TP=TD2+TF1+TD1+TR1+TON

         THDT4=-0.5*(ALPC-ALPD)*(TON+TR1/3-(TD2+(TF1+TD1)/3))
         THDT5=THDT4-(ALPC-ALPD)*TD2
         THDT1=THDT5-(ALPC-ALPD)*TF1+0.5*AN2*TF1*TF1
         THDT2=THDT5-(ALPC-ALPD)*(TF1+TD1)+0.5*AN2*(TF1+TD1)**2
         THDT3=THDT2-(ALPC-ALPD)*TF1+AN2*(TF1+TD1)*TR1-
        &0.5*AN1*TR1*TR1
          THDTD4=THDT3-(ALPC-ALPD)*TON
          THDT6=THDTD4
          THDT7=0.0

          TTDMX=(AN2*(TF1+TD1)-(ALPC-ALPD)/AN1
          THDMX=THDT2-TTDMX*(ALPC-ALPD-AN2*(TF1+TD1)+0.5*AN1*TTDMX)
          THDMN=THDT5-0.5*(ALPC-ALPD)**2/AN2
          TMIN1=(ALPC-ALPD)/AN2-RK+SQRT(((ALPC-ALPD)/AN2)**2+RK*RK
         &-2*THDT5/AN2)
          T6=TD2+TMIN1
          DZ1=0.5 *T6*((ALPC-ALPD)+(RK+0.5*T6)-THDT4)-0.25*AN2*
         &TMIN1*TMIN1*(RK+TMIN1/3)
          DZ=DZ1

c        If Limit cycle is to be calculated for a specified Dead-
c        Zone, Uncoment

c        the next line with appropriate Dead- Zone value

c        DZ=0.5/RTOD

         TH4=DZ-RK*THDT4
         TH5=TH4+(THDT4-0.5*(ALPC-ALPD)*TD2)*TD2
         TH1=TH5+(THDT5-0.5*(ALPC-ALPD)*TF1+AN2*TF1*TF1/6)*TF1
         TH2=TH5+(THDT5-0.5*(ALPC-ALPD)*(TF1+TD1)+AN2*(TF1+TD1)**2/
        &6)*(TF1+TD1)
         TH3=TH2+(THDT2-0.5*(ALPC-ALPD)*TR1+0.5*AN2*(TF1+TD1)*TR1
        &-AN1*TR1*TR1/6)*TR1
          THD4=TH3+THDT3*TON-0.5*(ALPC-ALPD)*TON*TON
          TH6=THD4
          TH7=0.0
```

```
c         Calculation of Average Theta Error

          THS1=TH5*(TF1+TD1)+TH2*TR1+TH3*(TON+TD2)
          THS2=0.5*(THDT5*(TF1+TD1)**2+THDT2*TR1*TR1+
         &THDT3*(TON+TD2)**2)
          THS3=(ALPD/6.0)*((TF1+TD1)**3+TR1**3+(TON+TD2)**3)-(ALPC/
         &6.0)*((TF1+TD1)**3+TR1**3+(TON+TD2)**3
         &+AN2*(TF1+TD1)*TR1**3/6
          THS4=(AN2*(TF1+TD1)**4-AN1*TR1**4)/24
          AVTHE=(THS1+THS2+THS3+THS4)/TP

          VLTON=TD1+TR1+TON
          IF(THDT3.GT.0) GO TO 934
          TMAX1=(AN2*(TF1+TD1)-(ALPC-ALPD)+SQRT((ALPC-ALPD-
         &AN2*(TF1+TD1))**2+2*AN1*THDT2))/AN1
          THMAX1=TH2+(THDT2+0.5*(AN2*(TF1+TD1)-(ALPC-ALPD))*TMAX1
         &-AN1*TMAX1*TMAX1/6)*TMAX1
          THMAX=THMAX1
          TMAX2=0
          THMAX2=0
          GOTO 937
934       TMAX2=THDT3/(ALPC-ALPD)
          THMAX2=TH3+0.5*THDT3*THDT3/(ALPC-ALPD)
          THMAX=THMAX2
          TMAX1=0
          THMAX1=0

937       TMN1=(ALPC-ALPD+SQRT((ALPC-ALPD)**2-2.0*AN2*THDT5))/AN2
          THMIN1=TH5+(THDT5-0.5*(ALPC-ALPD)*TMN1+AN2*TMN1*TMN1/
         &6)*TMN1
          THMIN=THMIN1
          TMN2=0.0
          THMIN2=0
          IF(TMN1.LE.(TF1+TD1))GO TO 105
          PRINT 941
941       FORMAT (1X,"WARNING FOR THMIN")
105       CONTINUE

          E1=(TH1+RK*THDT1)*RTOD
          E2=(TH2+RK*THDT2)*RTOD
          E3=(TH3+RK*THDT3)*RTOD
          E4=(TH4+RK*THDT4)*RTOD
          E5=(TH5+RK*THDT5)*RTOD
          E6=(TH6+RK*THDT6)*RTOD
          E7=(TH7+RK*THDT7)*RTOD
```

```
      TH1=TH1*RTOD
      TH2=TH2*RTOD
      TH3=TH3*RTOD
      TH4=TH4*RTOD
      TH5=TH5*RTOD
      TH6=TH6*RTOD
      TH7=TH7*RTOD

      THDT1=THDT1*RTOD
      THDT2=THDT2*RTOD
      THDT3=THDT3*RTOD
      THDT4=THDT4*RTOD
      THDT5=THDT5*RTOD
      THDT6=THDT6*RTOD
      THDT7=THDT7*RTOD

      THMAX1=THMAX1*RTOD
      THMAX2=THMAX2*RTOD
      DZ1=DZ1*RTOD
      DZ2=DZ2*RTOD
      DZ=DZ*RTOD
      THMAX=THMAX*RTOD
      THDMX=THDMX*RTOD
      THMIN1=THMIN1*RTOD
      THMIN2=THMIN2*RTOD
      THMIN=THMIN*RTOD

      RADFRQ=6.28318/TP
      CYFRQ=1./TP
      AMP=(THMAX-THMIN)/2.0
      AVTHE=AVTHE*RTOD
      DUTY=1.0/RKT

      WRITE(112,88)RK,DZ
C     WRITE(12222,88)RK,DZ,DETY,CYFRQ,AVTHE,AMP,THMAX,THMIN
88    format(1x,10(F14.10,1x))

      IF(ITS.EQ.1) GO TO 3184
      IF(IKD.EQ.0) GO TO 702
C     PRINT 2
C     PRINT 701,RK,RKT,RKTMN,RKTMX
C     PRINT 5,TH1,TH2,TH3,TH4,TH5,TH6,TH7
C     PRINT 6,THDT1,THDT2,THDT3,THDT4,THDT5,THDT6,THDT7
C     PRINT 7,E1,E2,E3,E4,E5,E6,E7
C     PRINT 9,TMAX1,THMAX1,TMAX2,THMAX2,TMIN1,DZ1,TMIN2,DZ2
C     PRINT 9,TMN1,THMIN1,TMN2,THMIN2
```

```
702       CONTINUE
          IF(IKS.EQ.1) GO TO 3184
c         PRINT 8,  TON,TOFF,TP,DZ.THMAX,THDMX,THMIN
c         PRINT 12, RADFRQ,CYFRQ,DUTY,AMP,AVTHE,VLTON
          GO TO 3185
3184      CONTINUE
          DETY=1.0/DZ
C         PRINT 3186,RK,DZ,DETY,CYFRQ,AVTHE,AMP,THMAX,THMIN

3185      CONTINUE
3186      FORMAT(26X,F6.3,7F13.6)
3187      FORMAT(29X,"KA",  6X,  "DZ",9X,  "1/DZ",9X,"FRQ",11X,
         &"TH-AV",7X,"AMPLITUDE",9X,"TH-MAX",8X,"TH-MIN")
c1        FORMAT(8F10.0)
c2        FORMAT(1X,100(1H-))
c3        FORMAT(5X,10(F10.4,2X))
c4        FORMAT(15X,"TIME" F10.3)
c5        FORMAT(2X,"THETA",5X,7(E10.3,3X))
c6        FORMAT(2X, "THDOT",5X,7(E10.3,3X))
c7        FORMAT(2X,"ERROR",5X,7(E10.3,3X))
c8        FORMAT(2X,"TON" E10.3X, "TOFF" E10.3,3X,"TP" E10.3X,
c        &"DZ" E10.3,3X,"THMAX" E10.3,3X."THDMX" E10.3,3X,
c        &"THMIN"E10.3)
c9        FORMAT(2X,8(E10.3,3X))
c11       FORMAT(1X,100(1H*))
c12       FORMAT(2X,"RADFRQ" E10.3,3X,"CYFRQ" E10.3,3X, "DUTY",
c        &E10.3,3X,"AMP" E10.3,3X,"AVTHE" E10.3,3X,"VLTON" E10.3)
c701      FORMAT(20X,"KR" F6.3,2X,"KT" F6.3,2X,"KTMIN" F6.3,2X,
c        &"KTMAX" F6.3)
100       CONTINUE

101       STOP
          END
          SUBROUTINE SELECT(T,DOWN,UP,RR,RI,MCASE)
          DIMENSION RR(3),RI(3)

          IREAL=0
          DO 1401 I=1,3
          IF(RI(I).NE.0.0) GO TO 1401
          IR=I
          IREAL=IREAL+1
1401      CONTINUE
          IF(IREAL.EQ.3) GO TO 1403
          T=RR(IR)
          IF(T.GE.DOWN.AND.T.LE.UP) GO TO 1413
          GO TO 1500
```

```
1403      DO 1411 I=1,3
          IL=I
          IF(RR(I).GE.DOWN.AND.RR(I).LE.UP) GO TO 1412
1411      CONTINUE
          GO TO 1500
1412      T=RR(IL)
1413      CONTINUE
          MCASE=1
          GO TO 1415
1500      MCASE=0
1415      RETURN
          END


          SUBROUTINE PROOT(N,A,U,V,IR)
C         IMPLICIT DOUBLE PRECISION(A-H, O-Z)
C         THIS SUBROUTINE USES A MODIFIED BARTSTOW METHOD
C         TO FIND THE ROOTS OF A POLYNOMIAL
c         Ref for this Subroutine is "Computer programs for
c         Computational Assistance
c         in the Study of Linear Control Theory" -- By James.L Melsa
          &Stephen K.Jones
```

CHAPTER 6

Design Validation and Flight Trial Experiences

6.1 INTRODUCTION

Confirmation of satisfactory performance of the control system design in flight by conducting a flight trial is a costly affair. Further, in case of missiles, the performance needs to be confirmed for different trajectories covering the entire zone of operation which will require large number of flight trials. Hence, it needs utmost care in ensuring the satisfactory performance of the control system so that the number of flight trials can be minimized. It may be recalled that we have made number of simplifications in the mathematical model to enable use of standard linear system design techniques. Hence, it is essential to test the design by incorporating all the complex features of the mathematical model and ensure its satisfactory performance.

Design is validated in following steps:

6.2 DESIGN VALIDATION

6.2.1 Stability Margins

Stability margins are calculated using linear model incorporating the dynamics of all the elements inside the control loop.

The margins are computed for various perturbations in the parameters. Generally, it is done using nominal, upper bound and lower bound of data perturbation.

The nominal data should satisfy the desired specifications on the margins. For the perturbed data, the margins may be degraded. One may accept degradation, say, upto 4db gain margin and 25° phase margin. However, the acceptance needs to be based on satisfactory results of 6 degrees of freedom trajectory simulation. The flexibility mode shape and frequency data also needs to be varied as nominal, upper bound and lower bound data. Experience has shown that the mode shape data many times was found to be outside the predicted bounds. Hence, the design must have adequate cushion so that the degradation is graceful and does not change scenario from stable to unstable performance.

6.2.2 Step Response Studies

The design is done in a frequency domain. When the step response study is conducted, the performance can be correlated to the stability margins. The separate study in time domain and frequency domain enables one to detect and avoid some programming and modelling errors and gives confidence when the time domain response is as per expectation from the stability margins.

6.2.3 Six Degree of Freedom Trajectory Simulation

This simulation enables one to incorporate all the features of subsystem models including coupling between the pitch, yaw and roll motion, nonlinearities and external disturbances such as wind, wind gust and wind shears and disturbances due to misalignments. It enables one to incorporate nonlinearities such as dead zone, saturation, backlash etc. and various disturbances such as thrust misalignments, *CG* offsets.

The detailed simulation enables one not only to validate the adequacy of the design stability margins but also adequacy of control force levels, stored control impulse, maximum control surface deflection levels, maximum deflection rate ($\dot{\delta}_{max}$) and acceleration and the hinge moments. It also enables to test the performance of the system in various operational zones and varying propulsion performance.

6.2.4 Trajectory Simulation Incorporating Flexibility and Propellant Sloshing

This enables testing the stability of flexible modes and slosh modes. Many times, slosh modes may be found to be having less margins. Its effect on the vehicle can be seen only in this study. The simulation may show continuous oscillations of slosh angle with a very low decay rate. However, the vehicle body rate may not show any significant effect. In such case, these oscillations may be treated as acceptable instead of adding additional hardware such as baffles or tank partitioning to improve damping which will involve weight penalty and time. The effect of additional loads due to slosh motion on tank walls needs to be considered in the design of structure.

6.2.5 Hardware-in-loop Simulation[1,2]

Inspite of adequate care in modeling, there will always be some hardware features which are not incorporated in the mathematical models exactly. Hardware-in-loop simulation is used to avoid these deficiencies in the model. In this simulation, the trajectory simulation needs to be in real time. The angular motions of the missile are simulated using angular motion simulator. The sensors such as inertial measurement system and rate gyros are put in the angular motion simulator. The sensors are subjected to body angular rates, p, q, r by giving appropriate Euler angle rates to the three gimbals of the motion simulator. Translational motion, however, cannot be simulated physically and these variables are obtained from 6DOF simulation program. The sensor outputs are then fed to onboard computer which has flight software in it and gives out the actuator commands. The commands are then given to physical actuators and their feedback is used in trajectory simulation program as the actuator deflection.

The hardware-in-loop simulation results are then compared with 6DOF trajectory simulation results and any differences need to be accounted for.

6.3   SOFTWARE VALIDATION

This validation is carried out specifically for the flight software resident in onboard computer (OBC) to ensure:

(*i*) Conformance with the design

(*ii*) Adequate safety features on logics

(*iii*) Avoidance of mathematical singularities

(*iv*) Adequate time buffers in computational cycles.

6.4   END TO END SIGN CHECKS

The loop design and simulation programs follow certain sign conventions based on the designers conceptualization or some standard conventions. When the various packages are integrated in the missile or launch vehicles, many times practical constraints come into picture and the packages need to be integrated in a particular way. In that case, some of the sensors may be put with their sensitive axis opposite to the assumed convention or the positive command to actuator may deflect it in opposite direction to the assumed conventions. These aspects in addition to inadvertent wrong wiring connections can lead to disastrous consequences unless those are properly checked and errors are removed.

End-to-end sign checks are carried out in following steps:

(1) Draw a sketch indicating the actual mountings of the sensors showing their sensitive axes and all the actuators and RCS motors (Fig. 6.1)

(2) Make a table indicating the appropriate change in control effector (control surface orientation or movement of gimbaled engine or RCS motor) when a positive and negative motion is given about the sensitive axis of each sensor.

For the sake of convenience of testing, this table can be split into two parts.

(*i*) Table which indicates the sign of output command of onboard computer (OBC) for positive and negative motion about the sensitive axis of each sensor.

(*ii*) Control effector (control surface deflection or movement of gimbaled engine or operating RCS motor) motion for the positive and negative commands from the OBC.

(3) Physical checks to be carried out for positive and negative motion about each sensor axis and recording the onboard computer commands and physical movement of control effector for positive and negative OBC commands to each control effector.

The results of these tests must be as per the table giving the expected response.

A simple case is taken for illustration.

Fig. 6.1 shows the mounting of sensor (INS) and engine actuators and fin actuators. It may be seen that the fin actuators are mounted such that for the same input signal, the two fins move in different directions.

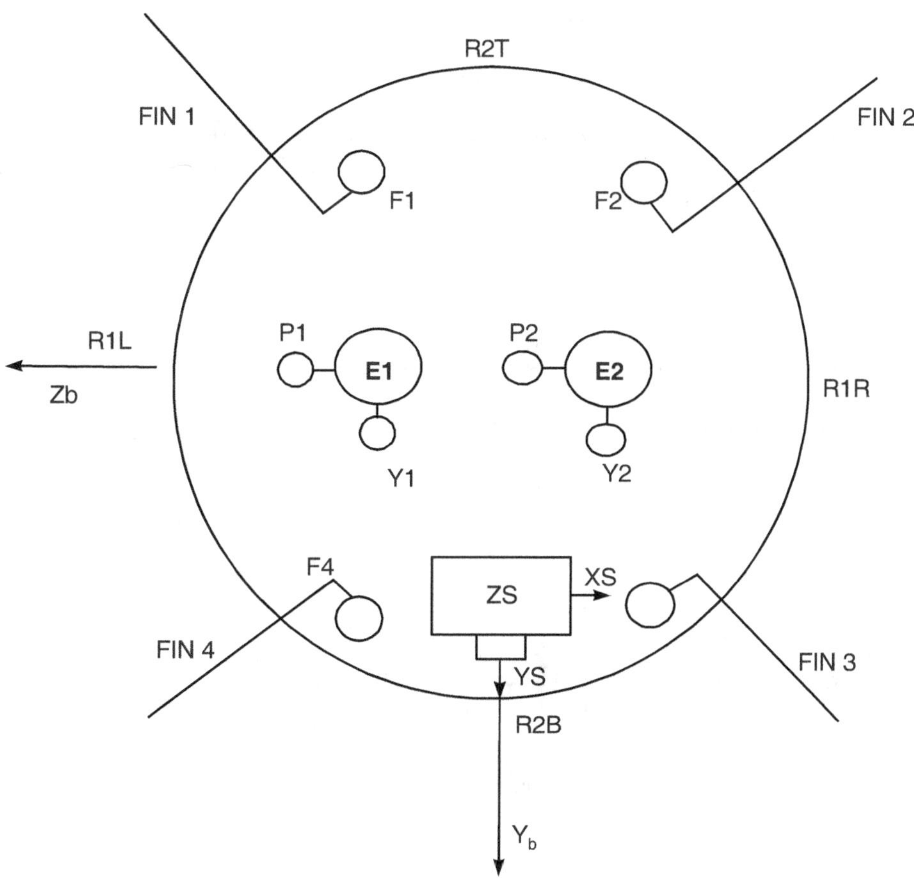

Fig. 6.1 Schematic Diagram of sensor and actuator mounting

Fig. 6.2 shows the physical mounting of actuators and the piston motion for positive and negative input commands.

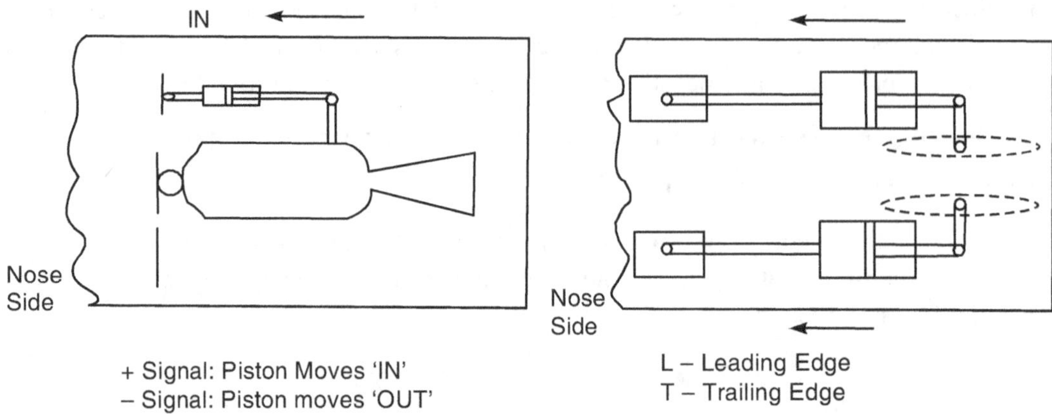

Fig. 6.2

Fig. 6.3 shows the direction of control force when the control fin moves in a particular direction and direction of control force when a particular SITVC port is opened by actuator command. Similar diagrams can be drawn for various control effectors.

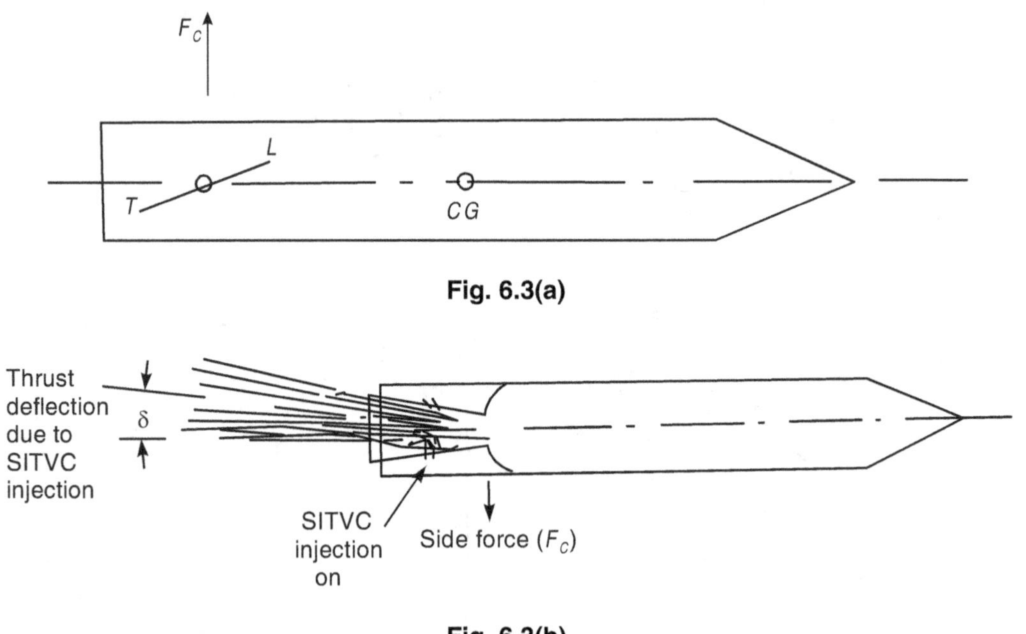

Fig. 6.3(a)

Fig. 6.3(b)

Table 6.1 gives the physical motion of the fins and engines when a particular command is issued from onboard computer.

TABLE 6.1: Control Effector Response to OBC Commands

Control surface	Actuator input	Trailing edge moves to	Actuator input	Trailing edge moves to
F1	+ ve	R2T	– ve	R1L
F2	+ ve	R2T	– ve	R1R
F3	+ ve	R2B	– ve	R1R
F4	+ ve	R2B	– ve	R1L

Engine actuator	Actuator command	Nozzle end moves to	Actuator command	Nozzle end moves to
P1	+ ve	R1L	– ve	R1R
P2	+ ve	R1L	– ve	R1R
Y1	+ ve	R2B	– ve	R2T
Y2	+ ve	R2B	– ve	R2T

Table 6.2 gives the OBC commands to actuators when a physical motion is given about each axis of INS.

TABLE 6.2: OBC Response and Actuator Response to Motion About INS Axis

INS Motion	OBC Command	Exit end of engine moves to
+ @ X	Y1 – ve	E1 to R2T
	Y2 + ve	E1 to R2B
– @ X	Y1 + ve	E1 to R2B
	Y2 – ve	E2 to R2T
+ @ Y	P1 + ve	E1 and E2 to R1L
	P2 + ve	
– @ Y	P1 – ve	E1 and E2 to R1R
	P2 – ve	
+ @ Z	Y1 – ve	E1 and E2 to R2T
	Y2 – ve	
–ve @ Z	Y1 + ve	E1 and E2 to R2B
	Y2 + ve	

(Contd ...)

TABLE 6.2 (Continued...)

INS Motion	OBC Command	Leading edge of fin moves to
+ @ X	F1 − ve	R2T
	F2 − ve	R2T
	F3 + ve	R1L
	F4 + ve	R1R
−@ X	F1 + ve	R1L
	F2 + ve	R1R
	F3 − ve	R2B
	F4 − ve	R2B
+ @ Y	F1 − ve	R2T
	F2 + ve	R2T
	F3 + ve	R1L
	F4 − ve	R1R
− @ Y	F1 + ve	R1L
	F2 − ve	R2T
	F3 − ve	R2B
	F4 + ve	R1L
+ @ Z	F1 + ve	R1L
	F2 − ve	R2B
	F3 + ve	R1R
	F4 − ve	R2T
− ve @ Z	F1 − ve	R2T
	F2 + ve	R1R
	F3 − ve	R2B
	F4 + ve	R1L

For simplicity of testing the signs, a pitch program is stored in computer which asks missile to move vertically upwards. After starting the program, the sensor is given the required motion and the OBC output is monitored. To avoid very high commands and big jerks, the actuators are disconnected in this phase of testing. The OBC commands are monitored for motion about each axis of INS and tabulated. In second phase of test, the actuators are connected and given power supply and their physical motion is monitored for each OBC commands.

Similar tests are conducted by giving appropriate physical motion to each sensor (rate gyros etc.) and monitoring the OBC response and response of each actuator or RCS motors to different OBC commands. A provision needs to be made for connecting external pneumatic low pressure air for testing purposes to know exactly which RCS motor is getting operated for a particular OBC commands.

The end-to-end proper polarity can thus be checked to ensure proper connections during missile integration.

6.5 FLIGHT TRIAL EXPERIENCES

6.5.1 From Design to Real Life Scenario

We have discussed the control system design in the previous chapters in sufficient details. In most of the literature, the design of control system restricts itself to design of control systems gains and design of compensators and filters. During actual implementation of the control design in onboard computer, the designer has to address many other issues to get satisfactory performance of the system. Some of these issues, if not properly addressed, can make the design totally ineffective in some abnormal scenario. Resolution of errors in proper body axes frame, sensor characteristics, actuator characteristics, mounting of sensors and actuators, limiting the operational zones of various signals are some of these factors. A design needs to incorporate some features wherever possible to make the system robust in unforeseen circumstances. Many times, the awareness about the need of such features comes only after encountering catastrophic failures during the flight trials. These robustness features when seen individually, appear minor in nature and never get adequately publicized. Their significance, many times is not appreciated by the new designers who mainly concentrate on the gain schedule and compensator design. Sometimes it is observed that, published literature also indicates such provisions. However, its importance is not realized and is quickly forgotten till there is another mishap. We have attempted to highlight number of robustness features at appropriate places in this book. Some of these are the lessons learnt during flight trials[3]. We discuss below briefly some of these problems which should serve as a caution during the design implementation phase.

6.5.2 Roll Oscillation due to Tail-Wag-Dog Effect

This particular case relates to a missile having liquid propellant two engine configurations. The two engines are mounted in pitch plane and can be moved in pitch and yaw planes. Both the engines are moved simultaneously for pitch and yaw control and differentially for achieving roll control. The simplified dynamic equation for roll motion is as follows:

$$I_{xx}\,\ddot{\phi} = 2T_E L_a \delta + 2M_R L_R L_a \delta \qquad \ldots (6.1)$$

The initial design had stability margin of

      Gain Margin   = 8.6db

      Phase Margin   = 37°

      Peak after TWD frequency = –8.5db

The margin calculations assumed second order actuator.

The flight performance showed 14 Hz oscillations in the roll channel. The flight was a failure. The recovered flight hardware showed damage at the base of engine lug due to fatigue failure. Obviously high stresses were induced due to roll oscillations. The simulation using δ and $\dot{\delta}$ (extracted from 2nd order actuator dynamics) did not reproduce these oscillations.

The step response test on actuator showed approximately second order dynamics. However, the hydraulic pressure on both sides of piston showed low damped pressure oscillations at about 14 Hz. It was felt that the δ obtained from second order actuator transfer function may not be indicative of the actual δ experienced by the engine. Hence the actual δ was derived by putting an accelerometer on the exit end of the nozzle and dividing by the length L from the gimbal position (See Fig. 6.4). The flight roll oscillations were reproduced in hardware-in-loop test when the derived δ as above was incorporated in the simulation.

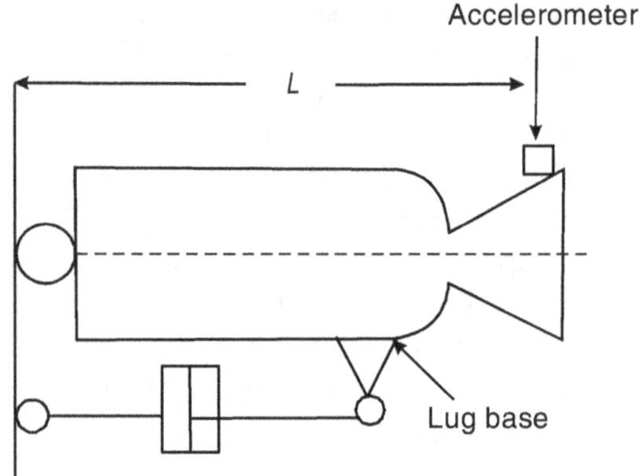

Fig. 6.4 Gimballed Engine Configuration

Fig. 6.5 shows the gain phase plot as per the original design.

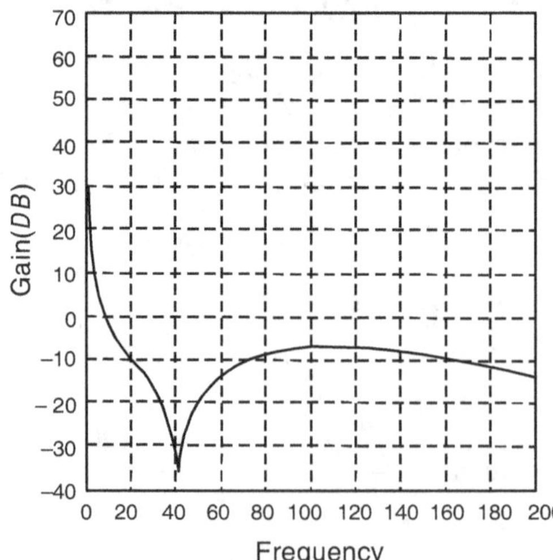

Fig. 6.5 Gain plot for roll control system

The attenuation of 8.5db after TWD frequency was obviously not adequate to attenuate the higher frequency component due to hydraulic pressure oscillations. The oscillation problem was solved using second order filter in the control loop which gave a gain peak of −25.5db after TWD frequency. With this filter, the flight performance showed a very smooth roll control.

6.5.3   Flexible Mode Instability due to Control-structure Interaction

The control structure interaction generally enters the control loop through sensor and actuator mountings. As a general guideline, the rate gyros and angle sensors are placed near the antinode of the first mode and accelerometers are put at nodes, so that the sensor senses a minimum signal due to flexible modes. Many times these locations are practically not possible due to non-availability of space. In such a case, the detailed analysis is carried out incorporating the structural mode dynamics and adequate stability margins are provided. As a general rule, the stability studies are also carried out by perturbing the mode shape data within the expected bounds. The same thing is achieved by shifting the sensor location by a specified amount forward and rearward and carrying out stability studies.

Inspite of all the above studies showing satisfactory stability margins, one of the missile flight showed unstable bending mode leading to diverging oscillations and structural failure. (Fig. 6.6)

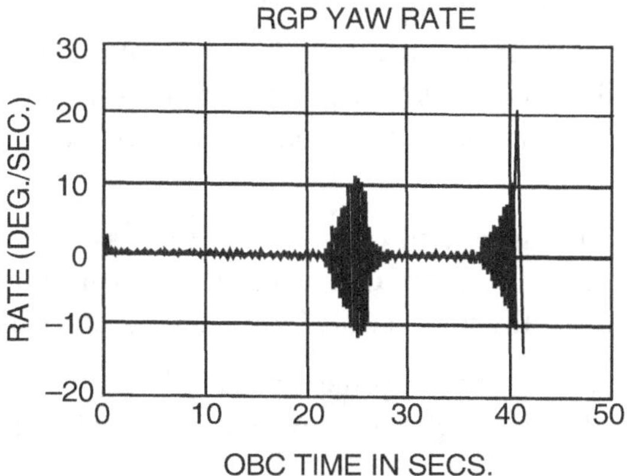

Fig. 6.6 Bending mode deflection vs. time

For the missile under study, two sensor packages were used (Fig. 6.7). One sensor package was put near the antinode to be used for the control system during the first stage flight. The other sensor package was placed inside the equipment bay nearer to the nose and was to be used after the first stage was separated. The sensed angular rate by the rate gyros will be as follows:

$$Q_1 = Q_R + \frac{d\phi_i(l_1)}{dl} q_i(t) \qquad \qquad ... (6.2)$$

$$Q_2 = Q_R + \frac{d\phi_i(l_2)}{dl} q_i(t) \qquad \qquad ... (6.3)$$

where Q_R is rigid body rate

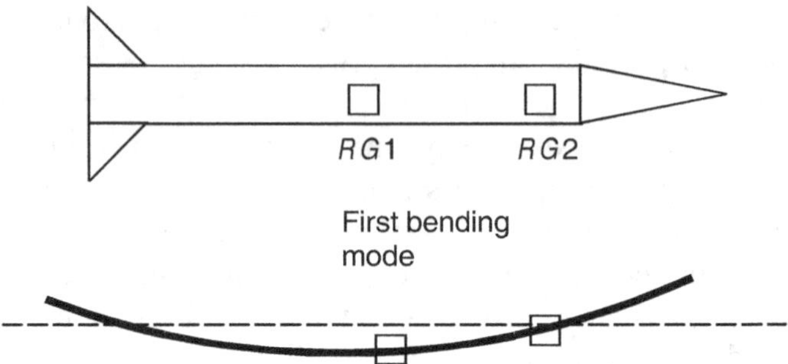

Fig. 6.7 Vehicle sketch showing rate gyros along X-axis

Assuming dominant contribution only from the first mode, the amplitude of the rates sensed by the two rate gyros will be in the ratio of mode slopes (*i.e.*, $\frac{d\phi_1(l_1)}{dl} / \frac{d\phi_1(l_2)}{dl}$) at two sensor locations. For the preflight predicted mode shape data, this ratio was expected to be around (1/8). The flight data showed that the amplitude ratio is about (–4/8) indicating that the rate gyro placed near the antinode was not only reading higher slope but the slope had an opposite sign to the expected one.

The ground resonance test was carried out to verify this result. The test showed that the sensor was in fact sensing as per flight due to local deformation in the mode shape. This highlights the importance of fully instrumented ground resonance test (GRT).

The ground resonance test also showed another anomaly. The rate gyro package consisting of two rate gyros (*Rp* and *Ry*) for pitch and yaw control system was mounted as shown in Fig. 6.8(a).

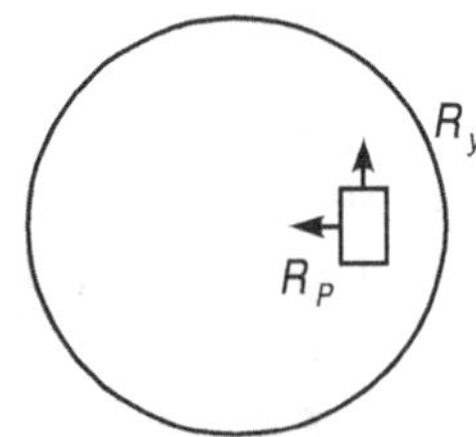

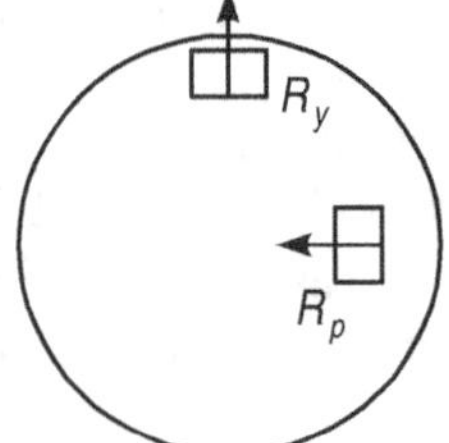

Fig. 6.8 (a) Rate gyro
Initial mounting

Fig. 6.8 (b) Revised
rate gyro mounting

The repeated test results showed that the rate gyro with sensitive axis mounted along the radius showed the correct polarity but the rate gyro whose sensitive axis was mounted parallel to tangent was sensing a phase reversal. The problem was solved by putting two separate rate gyros 90° apart with both having sensitive axis along the radius as shown in Fig. 6.8(b).

6.5.4 Roll Oscillations at High Angle of Attack

A roll control system design is relatively simple. The missile under consideration is a subsonic missile having imaging infrared seeker. It was observed that when the missile was executing a high lateral acceleration (latax) manoeuvre it was experiencing roll oscillations with high roll rate (Fig. 6.9)

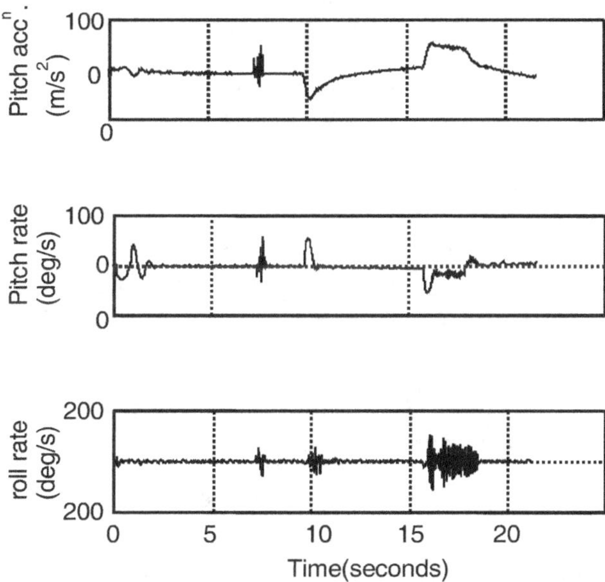

Fig. 6.9 Roll oscillation at latax manoeuvre

The high roll rates were not acceptable for the seeker. These roll oscillations could not be reproduced in trajectory simulation. Working backwards, the required disturbance torque to produce the observed motion was computed. It was observed that the disturbance torque is changing directions. To understand this characteristics, the wind tunnel testing was carried out and the test was repeated number of times for each setting.

The test results are shown in Fig. 6.10 which shows that the repeatability for side force and rolling moment coefficient was satisfactory only upto an angle of attack $\alpha = -16°$ and beyond that the values were widely different for the same setting and sometimes even changing the sign. It was thought that non-repeatable data was due to vortices being set at high angle of

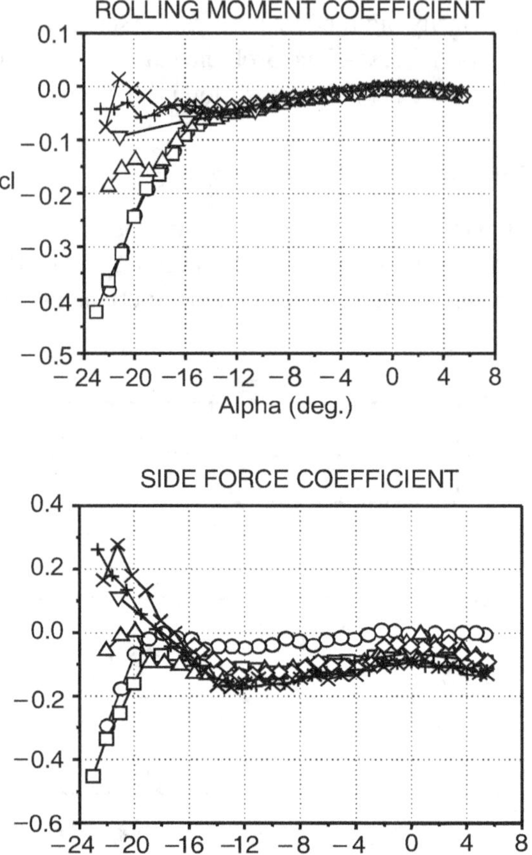

Fig. 6.10 Repeatability of side force and roll moment coeff before providing a band

attack giving unpredictable performance. Aerodynamic scientists provided a ring on the missile (Fig. 6.11) to fix the vortex separation.

The wind tunnel tests were repeated and the results are shown in Fig. 6.11 which show that the results are quite repeatable. Flight test with ring fitted on the missile is yet to be carried out.

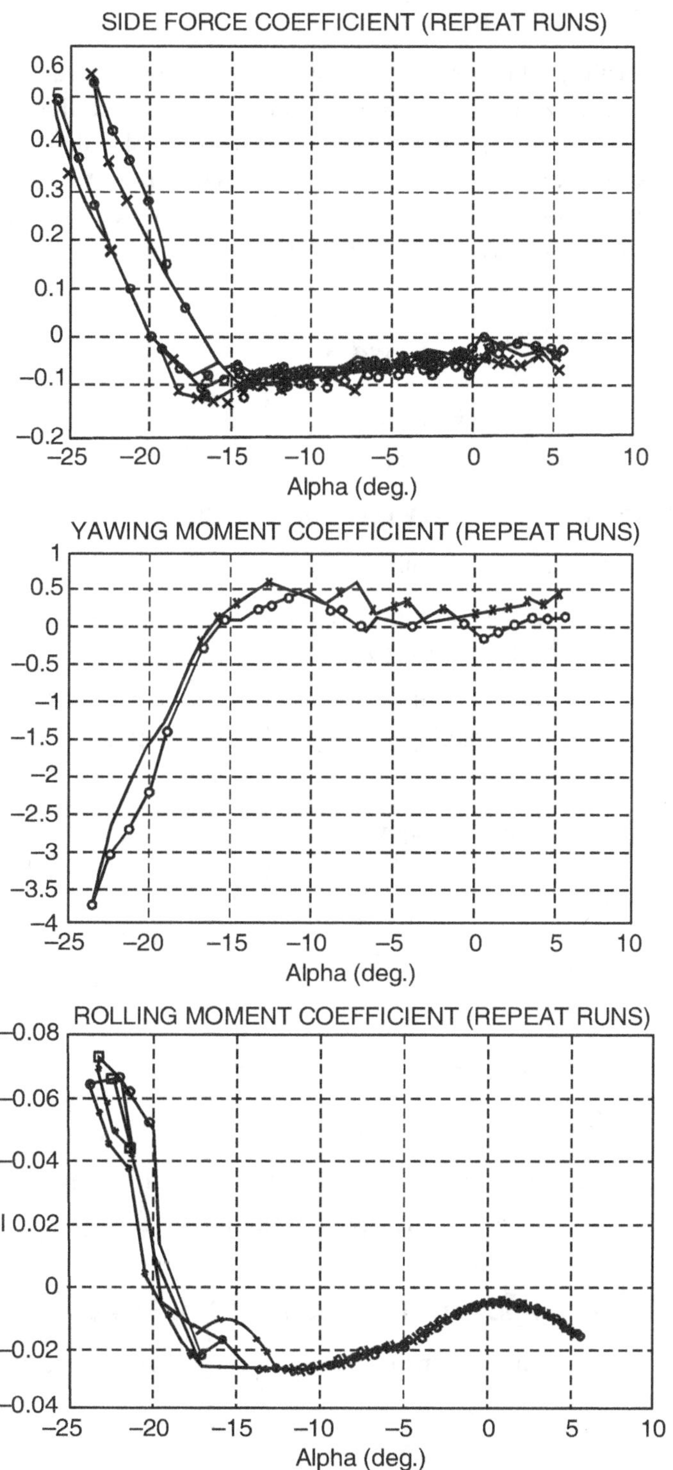

Fig. 6.11 Repeatability of side force and roll moment coeff. after providing a band

6.5.5 Flexible Structure Aerodynamics

Large L/D ratio missiles, while executing manoeuvres undergo structural bending due to aerodynamic load distribution and vehicle inertial parameters. The flexed shape depends on the EI distribution along the length and gives rise to changes in local angle of attack and thus change in aerodynamic load. This gave rise to mismatch in performance of control system during flight and simulation. The effect of vehicle flexing on aerodynamic parameters was subsequently estimated and indicated by the ζ factor where

$$\zeta_{CN\alpha} = \frac{C_{N\alpha}\,flex}{C_{N\alpha}\,rigid} \qquad\qquad \text{... (6.4)}$$

$$\zeta_{CN\delta} = \frac{C_{N\delta}\,flex}{C_{N\delta}\,rigid} \qquad\qquad \text{... (6.5)}$$

The missile under consideration had two control surfaces:

(*i*) Wings for lateral acceleration control and

(*ii*) Tail control panels used for providing damping torque for lateral acceleration control system and also control torque for roll.

The aerodynamic loads acting on Tail Control Panels (TCP) created offset loading on the main fin resulting in its twisting leading to a significant reduction in net control torque due to TCP deflections.

The detailed evaluation of ζ factors showed that

$$\zeta_{CN\alpha} = 0.933$$

$$\zeta_{CN\delta_{wing}} = 1.136$$

$$\zeta_{CN\delta TCP} = 0.671$$

$$\Delta_{Cp} = -0.0976$$

$$\zeta_{CM\delta TCP} = 0.686 \qquad\qquad \text{... (6.6)}$$

This indicates that there is a significant reduction in control effectiveness of TCP and the control gains need to be designed taking into account this ζ factor.

6.5.6 Spin Tolerant Logic for a Reaction Control System

Reaction control systems are generally used for stages operating outside the sensible atmosphere.

Considering a roll control system, the usual control logic is as follows:

$$e = \phi + K_R \dot\phi \text{ assuming } \phi_c = 0 \qquad \ldots (6.7)$$

Then
$$M = -M_x \text{ if } e > d_z$$
$$= M_x \text{ if } e < -d_z$$
$$= 0 \text{ if } |e| < d_z \qquad \ldots (6.8)$$

The block diagram of the control system is shown in Fig. 6.12.

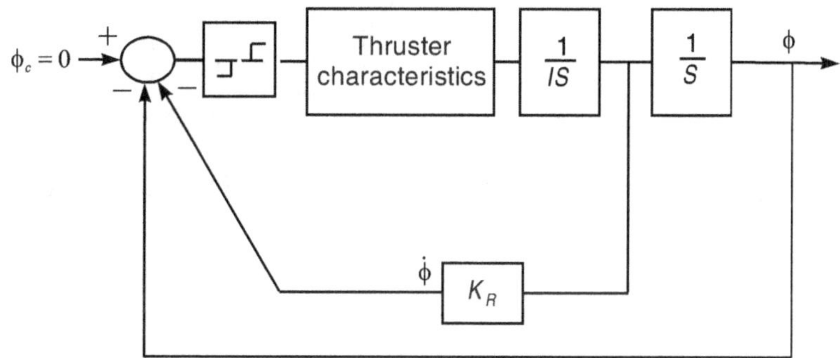

Fig. 6.12 Block diagram of roll control systems

The necessity of using spin tolerant control logic for this system is discussed for the following design parameters:

$$K_R = 0.5$$
$$d_z = 1°$$

Rate gyro saturation limit $= 30°/\text{sec}$

Angular sensor range $= \pm 180°$

For the vehicle under consideration, the above control system was used for a second stage roll control system. The first stage was using aerodynamic control and electro hydraulic actuation system. After the stage burn out, one of the actuator developed a problem leading to maximum deflection of the control surface. The vehicle developed a spin rate and rate gyro was saturated. After the stage separation, the reaction control system (RCS) came into operation, however, the performance was not satisfactory due to following reasons:

(1) Rate gyro saturation results into a change in switching boundaries (Fig. 6.13).

(2) The sign of angle changes as soon as the angle passes through $\pm 180°$. This results in change of sign of the error function and leads to firing of wrong RCS motor. Thus, the RCS will give torque to increase the spin rate instead of reducing it.

$$e = \phi + K_R \dot\phi_{sat} \qquad \ldots (6.9)$$

$$= \phi + 15° \qquad (\text{using } K_R = 0.5 \quad \dot\phi_{sat} = 30°/s)$$

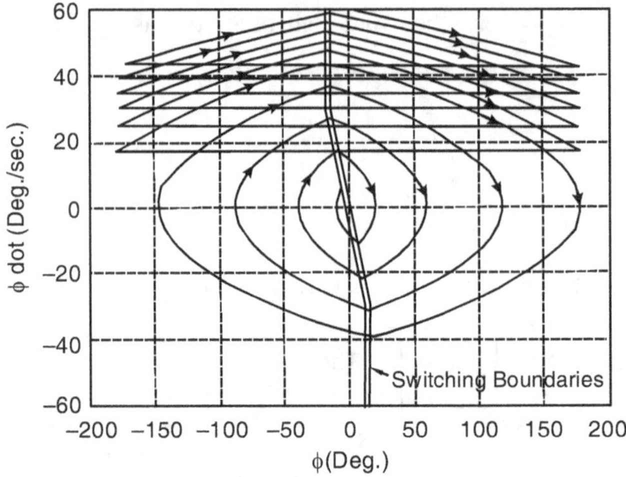

Fig 6.13 Systems response without spin tolerant control logic

For a dead zone $d_z = 1°$

$$e > 1.0° \text{ for } -14° < \phi < +180° \qquad\qquad ... (6.10)$$

$$< -1.0° \text{ for } -180 < \phi < -16°$$

Thus, in a total 360°, RCS will be attempting to reduce the spin rate over 194° but acting to increase the rate over 164° and dead zone of 2°. Fig. 6.13 shows the sketch of the missile state in phase plane for an initial condition $\phi = 0$, $\dot{\phi} = 60°/\text{sec}$. It will take a long time to kill the spin rate and come to normal operation scenario.

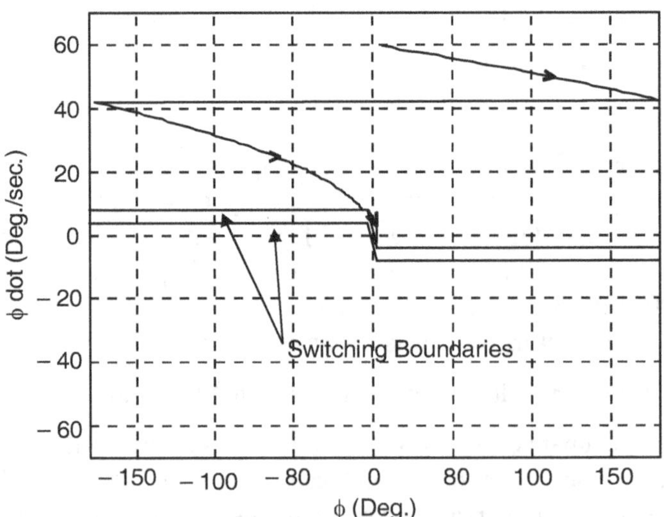

Fig 6.14 Systems response with spin tolerant control logic

This scenario can be corrected by providing a simple spin tolerant logic as follows:

If $\qquad\qquad\qquad\qquad |\dot{\phi}| > \dot{\phi}_{sat}$

$$\phi_s = \phi_{sat}\text{SGN}(\phi) \qquad \qquad \text{... (6.11)}$$

Then use this ϕ_s in the error function calculation. The phase plane sketch of switching boundary will appear as in Fig. (6.14) for $\phi_{sat} = 3°$ and the missile state will appear as shown. It may be seen that the RCS is now attempting to bring down the spin rate at all times. As soon as the state enters within the switching boundaries, the RCS stops and the missile state is brought down on its own without expenditure of fuel. The value of ϕ_{sat} can be set slightly above the dead zone value to have a satisfactory operation of the whole system.

It may be seen that the logic appears very simple and its significance may not be understood by the newer designers. However, it has great significance from the point of view of making the system very robust and tolerant of system failures. The mission may lead to a complete success even in case of some short duration failures. This logic is also useful for quick settling down of the system state after stage separation disturbances to normal operating zone.

6.5.7 Software Protections to avoid Abnormal Functioning of Flight Control System

In Section 6.5.6, it is demonstrated how a control system can get into difficulty due to change in sign of Euler angles when it passes through $\pm 180°$. It is also demonstrated how the problem can be handled by limiting the angle variable to be used in error function even though the sensor is sensing the large angles. Many times flight software uses quaternions instead of Euler angles. Due to relatively less frequent use of quaternions compared to Euler angles, the protective features in respect of quaternions are not familiar to general Scientific community. It can be easily proved that the comparative protective features for quaternions and Euler angles are as given below: (Also see Section 3.22 of Chapter 3).

(i) Limiting the Angular Range between 0 to $\pm 180°$

If $\theta > 180°$, $\theta = \theta - 360°$

If $\theta < -180°$, $\theta = \theta + 360°$ $\qquad \qquad \text{... (6.12)}$

This is equivalent to:

Limiting q_0 to $q_0 \geq 0$, where the quaternion is given by $Q = [q_0, q_1, q_2, q_3]$ and q_0 is a scalar part of the quaternion. To ensure this condition, the software protection is given by:

If $q_0 < 0$, multiply all four components of quaternions by (-1).

(ii) Protection of Angular Errors to the Range 0 to $\pm 180°$

Inspite of limiting all the angles to a range of 0 to $\pm 180°$, the angular errors can violate this range in specific cases. For example, let the desired $\theta_d = 175°$ and the actual vehicle angle is $\theta = -175°$. The vehicle needs to be rotated through $10°$ only to align with the desired orientation.

However, the computed error will give $\theta_d - \theta = 175 - (-175) = 350°$.

The control system, if not given appropriate protection, will attempt to align the vehicle with the desired orientation by rotating through 350° and will develop large angular rates and in the process leading to failure.

Hence, it is essential to provide protection to angular errors as well, even though the angles themselves are already given the protection.

In case of quaternions, the same thing is achieved by providing protection to error quaternion q_e where,

$$q_e = Q_M^{-1} Q_D \qquad \qquad \text{... (6.13)}$$

where Q_M is actual quaternion and Q_D is desired quaternion.

If $q_{e0} < 0$, multiply all components of q_e by (–1). $\qquad \qquad$ ... (6.14)

6.6 CLOSING REMARKS

We have discussed in this Chapter, the essential steps to be followed for validating the autopilot design. We have also described some of the problems faced during flight trials and subsequent solutions which were adopted for overcoming the design problems. Some of the solutions adopted for overcoming the problem appear trivial in nature and their significance will not be understood. However, these should be adopted as a matter of standard practice to avoid potential problems. It is worthwhile consolidating such practices and documenting them along with the description of problems one is likely to face if we do not adopt the same.

I believe that this Chapter will generate an awareness among the designers to this effect.

6.7 ACKNOWLEDGEMENTS

The conclusions arrived at on the problems of TWD oscillations and flexible mode oscillations are a result of collective experimental studies by large number of scientists.

Author thanks Shri Patrick D'Silva for supplying data on ç-factor and Dr. Paneerselvam, Shri DN Thakur and Shri RN Bhattacharjee for studies on roll oscillations at high angle of attack.

REFERENCES

1. **P.S. Subramaniam, N.V. Kadam:** Hardware-in-loop simulation for Prithvi Missile DRDL. 3300.1007.513 1986.

2. **S.K. Choudhuri, N.V. Kadam:** The plan for hardware-in-loop simulation for Prithvi Missile DRDL.3203.1004.513 Dec. 1987.

3. **N.V. Kadam:** Practical design of flight control systems-some problems and their solutions. Special issue of Defence Science Journal on Guidance and Control of Missiles. Defence Science Journal. Vol. 55, No.3, July 2005, pp. 1-11.